STUDIES IN PHYSIOLOGY

Presented to John C. Eccles

Edited by

D. R. Curtis and A. K. McIntyre

With 80 Figures

Springer-Verlag Berlin · Heidelberg · New York 1965

Softcover reprint of the hardcover 1st edition 1965
Library of Congress Catalog Card Number 65—17 381

ISBN 978-3-642-48614-2 ISBN 978-3-642-48612-8 (eBook)
DOI 10.1007/978-3-642-48612-8

Title No. 1279

PREFACE

This collection of papers is presented to Sir JOHN ECCLES by his former and present collaborators to commemorate the award of the 1963 Nobel Prize in Medicine, which was shared with A. L. HODGKIN and A. F. HUXLEY. Sir JOHN's interest in, and influence on, the study of physiology, particularly that of the nervous system, is reflected by the range of topics discussed, and by the distribution of the various authors in laboratories throughout the world. Those who have been privileged to work with him in Oxford, Sydney, Dunedin or Canberra have enjoyed a good discipline in scientific thought, as well as in the use of neurophysiological techniques. Basic knowledge is always transferable, and the inspiration which comes from association with a great scientist is not confined to any one field of physiology.

The contributors to this volume were requested to review briefly that aspect of physiology of current interest to them, and it is hoped that the resulting papers will serve as an up to date review of many physiological problems.

The editors are greatly indebted to the publishers, Springer-Verlag, particularly to Dr. H. GÖTZE for his advice and interest. It is also a pleasure to thank Mrs. H. WALSH for her unfailing assistance during the preparation of this book.

Canberra
Melbourne
1965.

CONTENTS

LIST OF CONTRIBUTORS

Dr. Per Andersen	Laboratory of Neurophysiology, University of Oslo, Karl Johansgt. 47. Oslo. Norway.
Dr. I. A. Boyd	Institute of Physiology, University of Glasgow. Glasgow. Scotland.
Prof. V. B. Brooks	Department of Physiology, New York Medical College, Fifth Avenue at 106th Street. New York, 10029. U.S.A.
Prof. C. Mc. C. Brooks	Department of Physiology, State University of New York, Downstate Medical Center, 450 Clarkson Avenue, Brooklyn 11203. New York. U.S.A.
Dr. A. J. Buller	Department of Physiology, King's College. London, W.C.2. England.
Mr. J. S. Coombs	Department of Physiology, Australian National University, G.P.O. Box 4. Canberra. Australia.
Dr. D. R. Curtis	Department of Physiology, Australian National University, G.P.O. Box 4. Canberra. Australia.
Prof. D. Denny-Brown	Harvard Neurological Unit, Boston City Hospital. Boston 18. Massachusetts. U.S.A.
Prof. C. B. B. Downman	Royal Free Hospital School of Medicine, 8 Hunter Street. London. England.
Prof. D. M. Easton	Department of Biological Sciences, The Florida State University. Tallahassee. Florida. U.S.A.
Dr. R. M. Eccles	Department of Physiology, Australian National University, G.P.O. Box 4. Canberra. Australia.
Dr. P. Fatt	Department of Biophysics, University College, Gower Street. London, W.C.1. England.
Prof. R. Granit	The Nobel Institute for Neurophysiology, Karolinska Institutet. Stockholm 60. Sweden.
Prof. H. E. Hoff	Department of Physiology, Baylor University College of Medicine. Houston. Texas. U.S.A.
Dr. J. I. Hubbard	Department of Physiology, Australian National University, G.P.O. Box 4. Canberra. Australia.
Prof. A. Iggo	Department of Veterinary Physiology, Royal (Dick) School of Veterinary Studies, Summer Hall. Edinburgh 9. Scotland.
Dr. M. Ito	Department of Physiology, Faculty of Medicine, University of Tokyo, Bunkyo-ku. Tokyo. Japan.
Prof. J. C. Jaeger	Department of Geophysics, Australian National University, G.P.O. Box 4. Canberra. Australia.

Prof. B. Katz	Department of Biophysics, University College, Gower Street. London, W.C.1. England.
Dr. K. Koketsu	Research Laboratories, Department of Psychiatry, Illinois Neuropsychiatric Institute, 912 South Wood Street. Chicago 12. Illinois. U.S.A.
Prof. P. G. Kostyuk	Institute of Physiology, Academy of Sciences of the Ukrainian SSR. Kiev 24. USSR.
Dr. W. Kozak	Department of Neurophysiology, Nencki Institute of Experimental Biology, 3, Pasteur St. Warsaw 22. Poland.
Dr. K. Krnjević	A.R.C. Institute of Animal Physiology. Babraham. England.
Prof. Stephen W. Kuffler	Neurophysiology Laboratory, Dept. of Pharmacology, Harvard Medical School, 25 Shattuck Street, Boston. Mass. U.S.A.
Dr. S. Landgren	Department of Physiology, Medicinaregatan 11. Göteborg SV. Sweden.
Prof. B. Libet	Department of Physiology, University of California, San Francisco Medical Centre. San Francisco 22. California. U.S.A.
Dr. A. W. Liley	Postgraduate School of Obstetrics and Gynaecology, National Womens Hospital, Green Lane West. Auckland, S.E.4. New Zealand.
Prof. D. P. C. Lloyd	The Rockefeller Institute, 66th Street and York Avenue. New York 10021. U.S.A.
Dr. Y. Løyning	Department of Physiology, Australian National University, G.P.O. Box 4. Canberra. Australia.
Prof. A. Lundberg	Department of Physiology, University of Göteborg, Medicinaregatan 11. Göteborg SV. Sweden.
Prof. W. V. Macfarlane	Waite Institute, University of Adelaide, Private Bag 1. Glen Osmond. S.A. Australia.
Prof. A. K. McIntyre	Physiology Department, Monash University, P.O. Box 92. Clayton. Victoria. Australia.
Dr. F. Magni	Istituto di Fisiologia, Della Universita de Pisa, Via S. Zeno N. 11—13. Pisa. Italy.
Dr. R. Miledi	Department of Biophysics, University College, Gower Street. London, W.C.1. England.
Dr. S. Obrador	Eduardo Dato 23. Madrid. Spain.
Dr. W. J. O'Connor	Department of Physiology, Medical School, University of Leeds. Leeds 2. England.
Dr. O. Oscarsson	Institute of Physiology, University of Lund. Lund. Sweden.
Dr. T. Oshima	Section of Neurophysiology, Institute of Brain Research, School of Medicine, University of Tokyo, Bunkyo-ku. Tokyo. Japan.
Prof. J. J. Pritchard	Department of Anatomy, Queen's University. Belfast. Northern Ireland.
Dr. W. Rall	Bldg. 31, Room 9A 17, National Institutes of Health. Bethesda. Maryland. U.S.A.
Dr. R. F. Schmidt	Institut für Allgemeine Physiologie, Universität Heidelberg, Akademiestraße 5, Postfach 1347. Heidelberg. Germany.

Dr. T. A. Sears — The Institute of Neurology, The National Hospital, Queen Square. London, W.C.1. England.

Dr. C. N. Shealy — Western Reserve University, Division of Neurosurgery, University Hospital of Cleveland, 2065 Adelbert Road. Cleveland. Ohio 44106. U.S.A.

Dr. P. E. Voorhoeve — Department of Physiology, University of Leiden, Wassenaarseweg 62. Leiden. Netherlands.

Dr. R. A. Westerman — Department of Physiology, Monash University, P.O. Box 92. Clayton. Victoria. Australia.

Prof. Wm. D. Willis — Department of Anatomy, Southwestern Medical School, The University of Texas, 5323 Harry Hines Boulevard. Dallas 35. Texas. U.S.A.

Prof. J. Z. Young — Department of Anatomy, University College, Gower Street. London, W.C.1. England.

RHYTHMIC THALAMIC ACTIVITY

by Per Andersen

The ability to produce rhythmic activity is a fundamental property of the nervous system, especially of the higher levels of the central nervous system. Best known is the rhythmic activity of about 10/sec that can be recorded from the thalamus and the cortex. The rhythmic activity in the thalamus appears in two separate, but related, forms, both having a frequency around 10/sec. First, there is the rhythmic discharge in response to a synchronous afferent volley. This activity will be called the rhythmic burst discharges. Second, there is the spontaneously occuring rhythmic activity that appears in short periods. This activity will be called the thalamic spindle activity.

I. The evoked rhythmic burst discharges

In his pioneering study, Adrian (1941) reported that a single afferent volley in a cutaneous sensory nerve produced a series of rhythmical waves in the sensory cortex and that a related activity was found simultaneously in the white matter, signalling the discharges of thalamic neurones. Bremer found essentially the same types of responses in the auditory cortex and in the medial geniculate nucleus, respectively (Bremer 1937; Bremer and Bonnet 1950). With newer neurophysiological techniques, it has been possible to throw further light upon the mechanism underlying the rhythmic thalamic activity, and to locate the neural substrate necessary for its production (Andersen, Brooks and Eccles 1963; Andersen, Brooks, Eccles and Sears 1964; Andersen and Sears 1964).

Most of the information has been gathered from the posterolateral part of the ventral nucleus (VPL) which relays somaesthetic impulses to the sensory cortex. The possibility of orthodromic as well as antidromic activation of the VPL cells provides a necessary measure of control over the experimental situation.

A single shock to a foreleg cutaneous nerve elicits in the VPL a response (Fig. 1 A, B, D, E, F), consisting of a series of discharges of about 10/sec, confirming the earlier reports (Adrian 1941; Bremer 1953). The activity consists of a group of cell discharges followed by a large positive wave (the P-wave), subsequently a new group of discharges followed by another P-wave, and so on, repeating itself from 4—20 times (Fig. 1 A, B, D). The P-waves last from 80—150 msec, being longer the deeper the anaesthesia (ANDERSEN et al. 1963; ANDERSEN et al. 1964). Concomitant with the burst discharges in the thalamus, an evoked slow wave appears in the postcruciate corticogram (Fig. 1 A, lower line). When the cortex was removed, and recording was made from the killed ends of the thalamocortical fibres in the white matter, a sharp positive wave

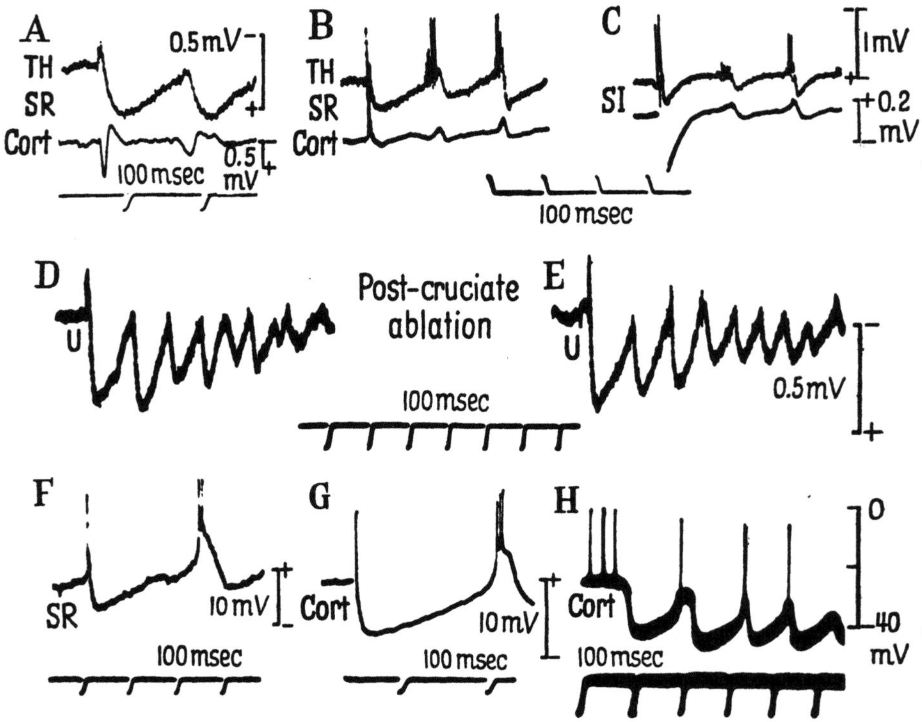

Fig. 1. Evoked rhythmic burst discharges in the VPL of the thalamus. *A—E*. Extracellular records. *A*. Records obtained from the VPL nucleus (TH) and pericruciate cortex (CORT) in response to a single shock to the superficial radial nerve (SR). Negativity upwards. *B*. As *A*, but lower line is recording from white matter, and has positivity upwards. *C*. Similar to *B*, but following stimulation of white matter underlying area SI. *D*. Rhythmic VPL response to a single ulnar (U) volley before, and after (E) removal of all pericruciate cortex. *F—H*. Intracellular records. *F*. Response of a VPL cell to a SR volley. *G*. Response of VPL cell to antidromic activation in a chronically decorticate preparation. *H*. Similar to *G*, but another preparation and slower sweep speed. Voltage and time calibrations as indicated

was obtained simultaneously with the thalamic burst discharges, signalling the rhythmic volleys along these fibres. Simultaneous recording from the dorsal column nuclei showed no repetitive activity of this kind. In conclusion, a single afferent volley entering the VPL is transformed to a rhythmic train of thalamocortical volleys at a frequency of about 10/sec.

Verifying the observations of ADRIAN (1941, 1951) and BREMER (1953), extensive ablation of the cortex does not change the rhythmic thalamic activity (Fig. 1 C, E) (ANDERSEN et al. 1963). Therefore, in the absence of any rhythmical input to the thalamus, the mechanism for the rhythmic activity probably resides within the thalamus itself.

Intracellular records show that the initial discharge is due to an excitatory postsynaptic potential (EPSP) generated by the afferent volley. Subsequently, there appears a large and long-lasting hyperpolarization (Fig. 1 F), fulfilling all criteria for an inhibitory postsynaptic potential (IPSP) (ECCLES 1963). Following the IPSP, a depolarizing wave is often present, giving rise to a series of discharges of the cell (Fig. 1 F). Subsequently, a new IPSP develops — and the cycle repeats itself several times before the activity vanishes.

A possible explanation of the depolarization with a burst of spikes occurring after an IPSP is the phenomenon of post-anodal exaltation (PAE) (ECCLES 1963; ANDERSEN and ECCLES 1963; ANDERSEN et al. 1964). An additional possibility is a synaptic influence derived from excitatory interneurones.

As seen in Fig. 1 C, G and H, thalamic rhythmic activity, similar to that evoked by an orthodromic volley, can be produced by an antidromic volley (ANDERSEN et al. 1964). Since typical responses also can be obtained in chronically decorticate animals (ANDERSEN et al. 1964), the effect is evoked by an antidromic volley along the axons of the VPL neurones, and the inhibition is, therefore, due to a pathway employing recurrent collaterals of these axons.

There was invariably a latency difference of about 1.5 msec between the antidromic invasion of a VPL neurone and the onset of the IPSP (ANDERSEN et al. 1964). Thus, it is supposed that one inhibitory interneurone is intercalated in the recurrent inhibitory pathway (ANDERSEN and ECCLES 1963), its extensive axonal ramifications explaining the observed greater distribution of IPSPs than EPSPs in orthodromically activated cells.

Similar EPSP-IPSP sequences are seen in the anterior thalamic nucleus in response to midline thalamic stimulation at 7/sec (PURPURA and COHEN 1962; PURPURA and SHOFER 1963).

Mechanism of the rhythmic burst discharges

It is postulated that an orthodromic volley brings a restricted number of neurones to discharge. Through their axon collaterals, a set of inhibitory interneurones is activated, producing IPSPs in many neurones. In these cells, the IPSPs are followed by a post-anodal exaltation (PAE) that brings a proportion of the cells to discharge. The firing of these neurones will subsequently be followed by new IPSPs leading to a PAE with discharges, and so on. The activity will fade away because of slight asynchrony in the discharge of the different cells due to a varying duration of the IPSPs.

Histological evidence of interneurones in the thalamus is given by the survival of a large number of thalamic cells, mostly small, following decortication with subsequent retrograde degeneration of cells sending their axons to the cortex (SHEPS 1945; MCLARDY 1950; CLARK and POWELL 1953). Small neurones within VPL with extensive axon arborizations were described by CAJAL (1911).

II. Spontaneous rhythmic thalamic activity

In a lightly anaesthetized animal there appears periods of rhythmic activity whose basic components are the same as for the evoked rhythmic burst discharges (Andersen and Sears 1964). The spontaneous activity appears as a series of burst discharges each separated from the next by a P-wave (Fig. 2A). Because of its shape, and its simultaneity with the barbiturate cortical spindles, each period of spontaneous rhythmic activity is called a thalamic spindle (Andersen

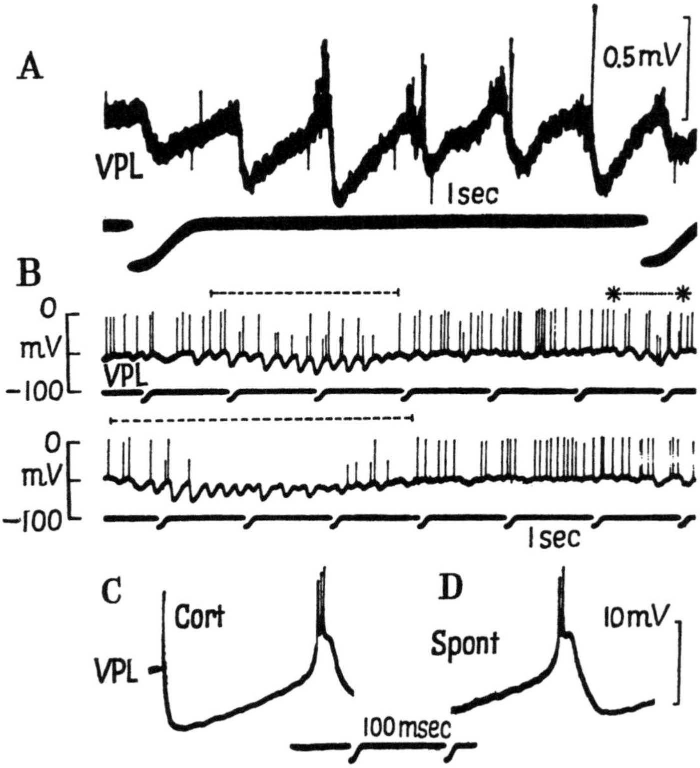

Fig. 2. Spontaneous rhythmical discharges in VPL. *A*. Extracellular record of the first half of a thalamic spindle. Negativity upwards. *B*. Intracellular DC-recording of a VPL cell, continuous recording. Two periods of spindle activity are marked with dotted lines above the record. An abortive spindle is marked with asterisks. *C*. Response of a VPL neurone to antidromic activation. *D*. Excerpt of a spontaneous spindle from the same cell as in *C*, showing the similar sequence of an IPSP followed by the post-anodal depolarization with spike discharges

and Sears 1964). The P-waves of these spindles are most likely identical to the spontaneous series of slow waves recorded from the thalamus by Morison, Finley and Lothrop (1943).

Intracellular recording from neurones in various thalamic nuclei showed the spindle activity as a series of augmenting and summating IPSPs (Fig. 2B). Since the cell only fired on the crest between two successive IPSPs, the interspindle irregular pattern of firing was changed to rhythmic discharges. Comparison between an IPSP-PAE sequence evoked by an antidromic volley (Fig. 2C) and an excerpt from a spindle (Fig. 2 D) shows that the same sequence was found in the two types of rhythmic activity. On the basis of these observations, an inhibitory phasing theory was advanced (ANDERSEN and SEARS 1964). It explains the mechanism of the thalamic spindle activity as due to the operation of a system of neurones, each having the possibility of influencing its neighbours through a recurrent inhibitory pathway. According to this theory, a spindle starts with the discharge of one or a few neurones. Through their axon collaterals, these cells activate some inhibitory interneurones that, in turn, produce IPSPs in a greater number of neighbouring neurones. The PAE that subsequently develops in these cells brings a certain proportion of them to discharge, which through their recurrent inhibitory pathway produces IPSPs in an even larger number of neurones. In this manner the rhythmic activity spreads until a majority of cells beat in unison. The duration of the IPSPs will act as the *timing device* for this rhythm. Through its elicitation of the PAE, the IPSP also acts as the *triggering device* for the neuronal discharges.

III. Simulation of the rhythmic activity in a model

A test of the whole assembly of thalamic neurones with their rhythmic interactions is made impossible by the requirement of simultaneous recording from a large number of relatively closely spaced cells. Therefore, a tentative test has been made by the simulation of a neuronal network in which the individual cells have been given characteristics according to the theory described above (ANDERSEN and RUDJORD 1964). Eighty cells were defined on an electronic computer, IBM 1620. The initial part of the thick line in Fig. 3 A indicates that the cells in the network discharged randomly, at a mean probability of firing (PF) of 20 per cent. Since the scanning time was equivalent to 10 msec in an animal preparation, this frequency corresponds to 20/sec. Following the discharge of any neurone, up to 10 other cells (Fig. 3 B) underwent a change in the PF indicated by the thick line: an immediate drop to zero (simulating an IPSP) for 10 units of time (equalling 100 msec) followed by a period of increased PF (simulating the post-anodal exaltation) before the cell again resumed its random firing at 20 per cent probability. The number of discharging cells in each unit of time was printed out and subsequently plotted (Fig. 3 C). Since for technical reasons the starting of the computer initiated the discharge of about half of the neurones (falling just to the left of the Y-axis) the first part of the diagram is comparable to an afferent volley exciting a certain proportion of thalamic cells. Each peak in the diagram indicates the near simultaneous firing of many neurones. Thus, the model behaves similarly to the evoked burst discharges.

Leaving the simulated neuronal network to operate by itself, there appeared short periods of rhythmic activity (arrows) composed of peaks in the diagram, as in the initial part or the graph. Although varying in length and frequency of occurrence, the periods are all characterized by a gradual onset and offset, thus resembling the thalamic spindles. By varying the parameters of this model, two factors were found to be important for the occurrence of periodic rhythmic activity. First, more pronounced rhythmic activity was observed the greater the difference was between the random PF and the post-inhibitory period of enhanced PF. Second, the larger the number of neurones being inhibited following the discharges of one cell, the more pronounced was the rhythmic activity.

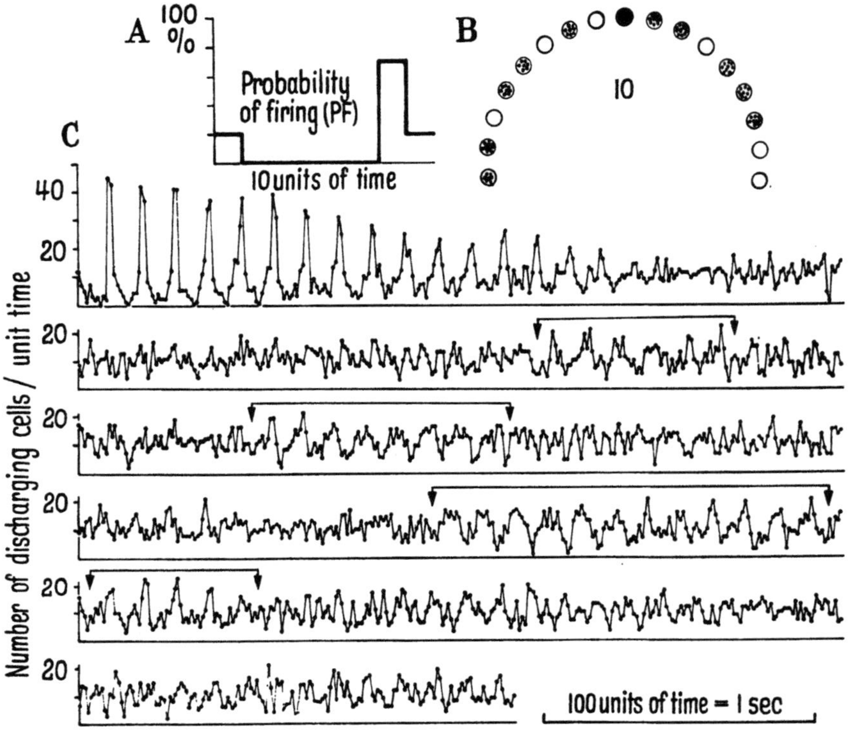

Fig. 3. Model simulation of a rhythmically active network. *A*. Change in probability of firing (PF) of a cell subsequent to the discharge of a neighbouring cell. Zero PF is maintained for 10 units of time, equalling 100 msec in animal experiments. *B*. The circles denote 17 of the 80 cells, the dotted circles marking the maximum number (10) of inhibited cells subsequent to the discharge of one neurone (black circle). C. Graph showing the total number of cells discharging within each unit of time. The initial period of rhythmic activity simulates the evoked burst discharges, whereas the periods marked with arrows are similar to the thalamic spindles

In conclusion, the computer model with a network having relatively simple characteristics gives rhythmical discharges, both evoked and "spontaneous". Furthermore, it has underlined the role that the post-inhibitory excitation and the distribution of the inhibition play in such a network. Although the simulation uses a highly simplified network, the results still suggest that future animal experiments on thalamic rhythms should be focussed upon the mechanism underlying the post-anodal exaltation and the connections and properties of the inhibitory interneurones.

Acknowledgement

This investigation was supported in part by Public Health Service Research Grant NB 04764-01 from the United States National Institutes of Health.

References

ADRIAN, E. D. (1941). *J. Physiol.* (Lond.) **100**, 159.
ADRIAN, E. D. (1951). *J. Physiol.* (Lond.) **113**, 9 P.
ANDERSEN, P., BROOKS, C. McC. & ECCLES, J. C. (1963). In: *Progress in brain research*, ed. by Schadé, J. P. Amsterdam: Elsevier Publ. Co.
ANDERSEN, P., BROOKS, C. McC., ECCLES, J. C. & SEARS, T. A. (1964). *J. Physiol.* (Lond.) **174**, 348.
ANDERSEN, P. & ECCLES, J. C. (1962). *Nature* (Lond.) **196**, 645.
ANDERSEN, P. & RUDJORD, T. (1964). *Nature* (Lond.) **204**, 289.
ANDERSEN, P. & SEARS, T. A. (1964). *J. Physiol.* (Lond.) **173**, 459.
BREMER, F. (1937). *Bull. Acad. Méd. Belg.* **2**, 68.
BREMER, F. (1953). *Some problems in Neurophysiology*, p. 79. London: London University Press.
BREMER, F. & BONNET, V. (1950). *Electrenceph. clin. Neurophysiol.* **2**, 389.
CAJAL, S. R. (1911). *Histologie du Système Nerveux de l'Homme et des Vertébrés*, vol. 2. p. 993. Paris: A. Maloine.
CLARK, W. E. LE GROS & POWELL, T. P. S. (1953). *Proc. Roy. Soc.* B **141**, 467.
ECCLES, J. C. (1963). *The physiology of synapses*, p. 316. Berlin: Springer-Verlag.
MCLARDY, T. (1950). *J. Neurol.* **13**, 198.
MORISON, R. S., FINLEY, K. H. & LOTHROP, G. N. (1943). *J. Neurophysiol.* **6**, 243.
PURPURA, D. P. & COHEN, B. (1962). *J. Neurophysiol.* **25**, 621.
PURPURA, D. P. & SHOFER, R. J. (1963). *J. Neurophysiol.* **26**, 494.
SHEPS, J. G. (1945). *J. comp. Neurol.* **83**, 1.

DIFFERENCES IN THE DIAMETER AND CONDUCTION VELOCITY OF MOTOR AND FUSIMOTOR FIBRES IN NERVES TO DIFFERENT MUSCLES IN THE HIND LIMB OF THE CAT

by I. A. BOYD

Two distinct types of motor nerve fibre supplying the muscle spindles of the cat were described by BOYD (1962). It was concluded from indirect evidence that the parent fibres of both types were contained within the gamma efferent group in the muscle nerves. A simultaneous study of the composition of the de-afferented nerves from the same cats showed that there were two distinct types of gamma efferent fibre in most nerves (BOYD and DAVEY 1962), fibres with myelin sheaths about 1 μ in thickness, and fibres of approximately the same axon diameter but with myelin sheaths about 0.1 μ in thickness. Compound action potentials were recorded from a number of normal nerves and in some it proved possible to differentiate two groups of gamma efferent fibres in terms of conduction velocity and of threshold to stimulation (Boyd and Eccles 1962).

In further experiments compound motor action potentials were recorded from the de-afferented nerves to fourteen different muscles in two cats anaesthetised with pentobarbitone sodium, in response to stimulation of the ventral spinal roots. The nerve electrodes were placed proximal to the bifurcation of individual motor fibres. The nerves and spinal roots were immersed in paraffin pools at 37° C. To obtain monophasic recording each nerve was severed from its muscle and crushed between the electrodes, and 0.1 % procaine hydrochloride was applied over the distal electrode if necessary. Values of conduction velocity obtained orthodromically and antidromically were similar.

The nerves were fixed *in situ* by perfusing 12 % formalin through the abdominal aorta and the lumbar plexus and muscle nerves were removed intact. Each nerve was dissected back about 2 cm into its parent trunk and a portion which had not previously been in paraffin or handled in any way was stained with osmium tetroxide. Transverse sections of each nerve were prepared and fibre-size histograms constructed for all the myelinated motor fibres. All very small fibres were traced through serial transverse sections.

The histological results are summarised in Fig. 1. There was a range of 1 to 3 μ between the alpha and gamma groups containing no fibres in all the nerves to the tibialis posterior and flexor digitorum brevis muscles and also in some of the nerves to other muscles. The histogram of the alpha fibres in one extensor digitorum longus nerve had two peaks. The range and mean diameter of the alpha fibres in a particular nerve in different cats were fairly consistent. There was clearly a difference, however, in both the maximum and the mean diameter of the alpha fibres in different nerves in the lower leg measured at about the same level. The mean diameter in the soleus nerve was 75 % of that in the gastrocnemius nerve, comparable with the figure of 78 % given by Eccles and Sherrington (1930). The nerves to the small muscles of the foot contained small alpha fibres, probably because of their greater distance from the spinal cord.

The groups of thickly and thinly myelinated gamma fibres overlapped considerably in most nerves. Two distinct gamma peaks were present, however, in the histograms of nerves containing a large number of thinly myelinated gamma fibres, and in which the mean diameters of the two gamma groups were well separated, e. g. popliteus, tibialis posterior, flexor and extensor digitorum brevis.

The maximum and mean total diameters of the thickly myelinated gamma fibres differed from nerve to nerve in a manner similar to the alpha fibres. The nerves to the peroneal group of muscles, and to the small muscles of the foot, however, had thickly myelinated gamma fibres which were large relative to their alpha fibres. The mean diameter of the thinly myelinated gamma fibres varied from nerve to nerve, also. It was difficult, however, to obtain reliable mean values in nerves such as those to the plantaris, flexor digitorum longus and the peroneal group of muscles, which contained relatively few thinly myelinated gamma fibres.

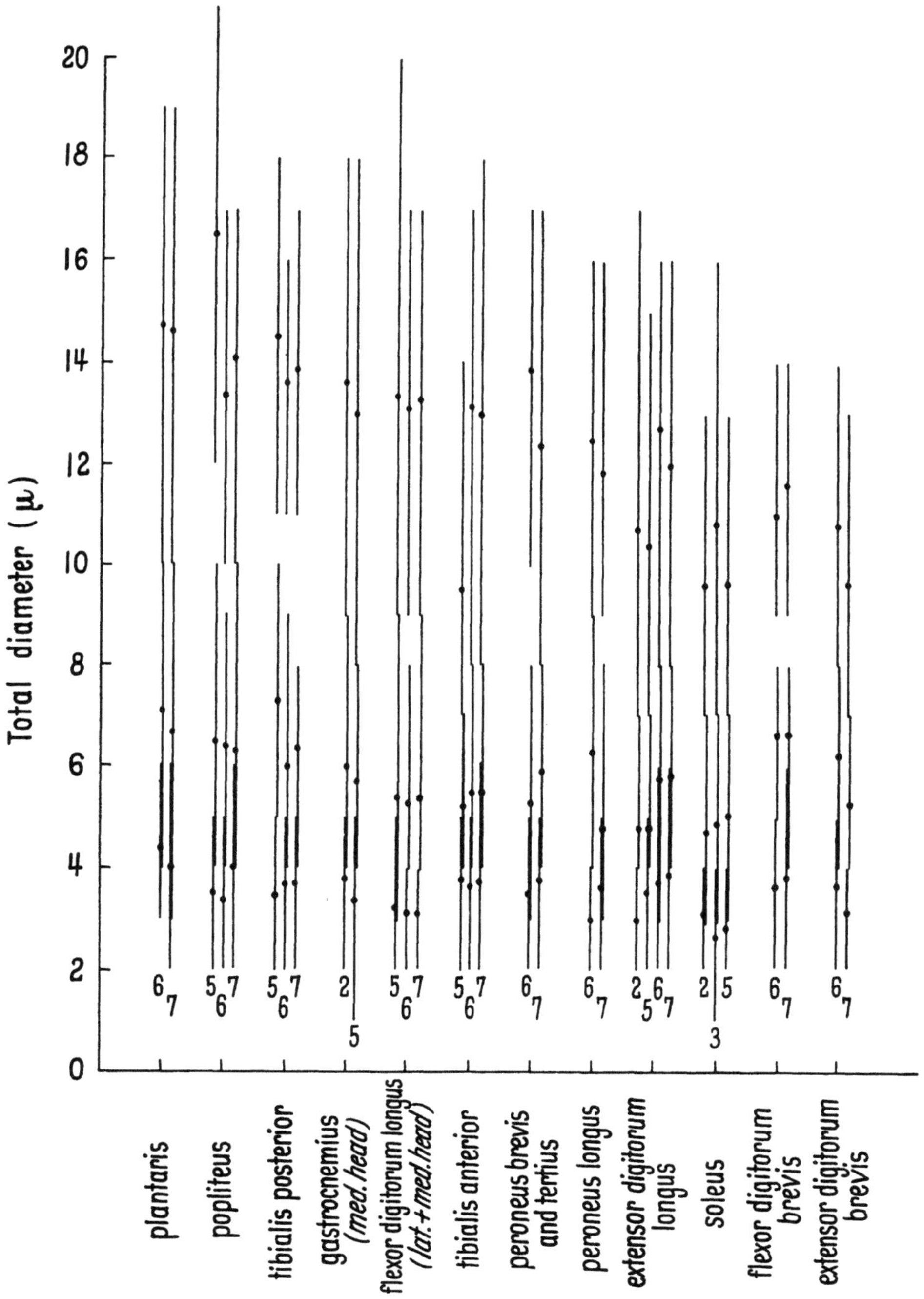

Fig. 1. The total diameter of the motor fibres in fourteen anatomically different nerves from de-afferented hind limbs of five cats. Each row of three vertical lines indicates range of diameter of α fibres, thickly myelinated γ fibres, thinly myelinated γ fibres, from above downwards, in one nerve. Where ranges overlapped line has double thickness. Points indicate mean diameter of each group. Cat from which each nerve was taken indicated by number at base of each row of lines. Degeneration times for cats 2, 3, 5, 6, 7 were 45, 56, 58, 35, 35 days. Regenerating fibres present in cat 1; de-afferentation incomplete in cat 4. Nerves from lower leg arranged in order of decreasing mean diameter. Only one branch of the nerve to tibialis anterior (possibly to red head) was removed in cat 5

Measurements of conduction velocity are summarised in Fig. 2. In every compound potential the alpha and fast gamma waves were quite separate. Hence,

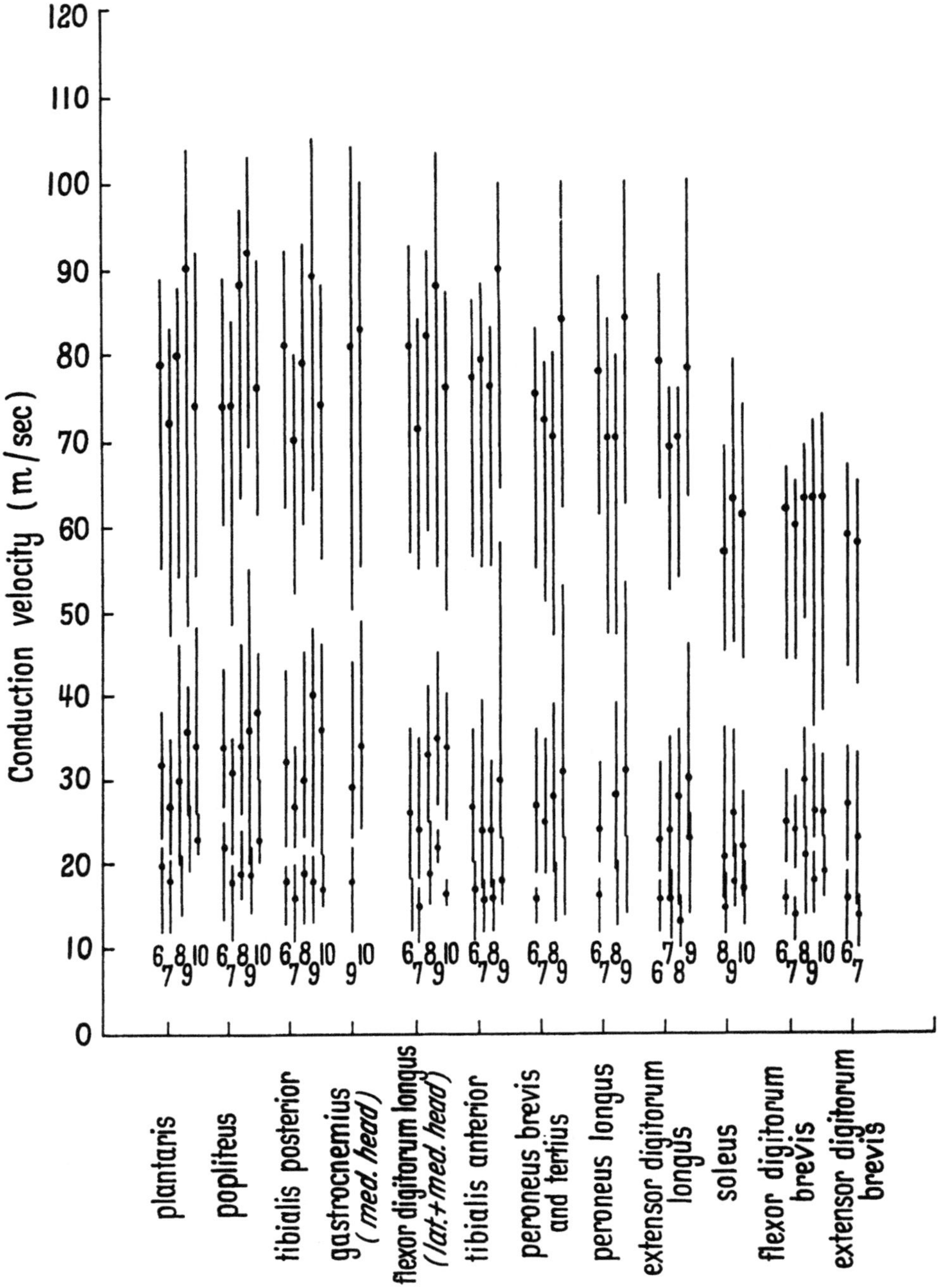

Fig. 2. The conduction velocity of the motor fibres in fourteen anatomically different nerves from five cats. Cats 6 and 7, de-afferented; cats 8, 9, 10, normal. Each row of three vertical lines indicates the range of conduction velocity of α fibres, fast γ fibres and slow γ fibres, from above downwards, in one nerve. Points indicate mean conduction velocity for each group. Aggregation of values obtained when stimulating different ventral roots partially obscures division between fast γ and slow γ groups in some nerves. Values for γ fibres not obtainable in a few nerves due to poor recording conditions. No allowance made for excitation time since recording distance was large (150 mm to 300 mm)

values for the maximum conduction velocity in alpha and fast gamma groups, and for mean velocity in the alpha group, are accurate. In one nerve to the extensor digitorum longus muscle the alpha wave had two peaks. The slow gamma wave in all nerves, and the fast gamma wave in nerves containing a small number of fibres, had a series of peaks and the mean values are approximate only. The separation of the gamma action potentials into two groups was most obvious in nerves, such as those to tibialis posterior and flexor digitorum brevis, whose fibre-size histograms contained two gamma peaks. Only a very small slow gamma wave, or sometimes none at all, could be recorded from nerves which contained few thinly myelinated gamma fibres. It is possible that the degree of overlap between the two gamma groups in several nerves has been underestimated.

The range of velocities in any group was less than the corresponding range of diameters, because some distortion of fibres was inevitable during preparation of histological sections. The scatter in the maximum and mean conduction velocities of the alpha fibres in the same nerve in different cats was considerable, however. There was no obvious difference in the conduction velocities in deafferented, as compared with normal, nerves. The overall mean conduction velocity of both alpha and gamma fibres in different nerves varied approximately in proportion to the differences in fibre diameter. The mean conduction velocity of the alpha fibres in the soleus nerve was 74% of that in the gastrocnemius nerve, comparable with the figure of 72% given by ECCLES, ECCLES and LUNDBERG (1958).

The relation between mean conduction velocity and mean total diameter is shown in Fig. 3. The mean ratio of conduction velocity in m/sec to total diameter in μ, measured in the same nerves, was 5.7 for the alpha fibres and 4.5 for both groups of gamma fibres (cf. 6.0, HURSH 1939). If values of mean conduction velocity and mean diameter in all the cats studied were aggregated the ratio was 5.9 for alpha fibres, 5.0 for fast gamma fibres and 5.2 for slow gamma fibres.

The nerves to all limb muscles contain a mixture of axons belonging to both phasic and tonic alpha motoneurones (GRANIT, HENATSCH and STEG 1956) which have high and relatively low conduction velocities, respectively (ECCLES et al. 1958). The nerve to the slow soleus muscle contains predominantly small tonic fibres, whereas the nerves to fast muscles contain a high proportion of phasic fibres. It is possible that in the nerve to extensor digitorum longus the phasic and tonic fibres are so distributed as to produce the two alpha peaks sometimes observed in its fibre-size histogram, or in its compound potential.

The differences in the diameter and conduction velocity of the two groups of gamma fibres in different nerves are certainly related to variations in structure and complexity of innervation of muscle spindles in different muscles. The precise relationship of the fast (thickly myelinated) gamma fibres and the slow (thinly myelinated) gamma fibres to the "dynamic" and "static" fusimotor fibres of Matthews (1962), or to the gamma$_1$ and gamma$_2$ motor fibres actually entering the spindles (Boyd 1962), has yet to be elucidated.

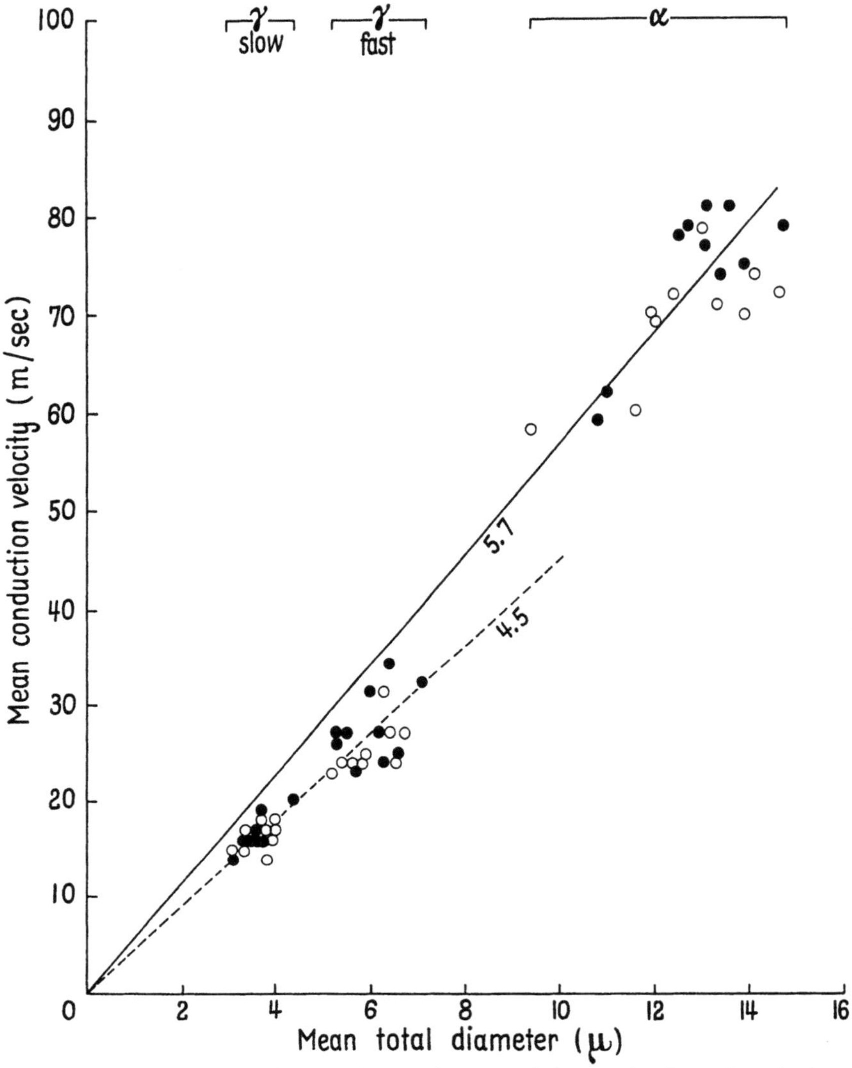

Fig. 3. The relation between mean conduction velocity and mean total diameter for the myelinated efferent fibres in de-afferented nerves to ten different muscles in each of two cats. ●, Cat 6; ○, Cat 7. Measurements of conduction velocity and diameter were made on the same nerve; each point shows the mean values for one motor group in one particular nerve

References

Boyd, I. A. (1962). *Phil. Trans.* B. **245**, 81.

Boyd, I. A. & Davey, M. R. (1962). *Proceedings of Symposium on Muscle Receptors, Hong Kong*, 1961. p. 189. Hong Kong University Press.

Boyd, I. A. & Eccles, J. C. (1962). *J. Physiol.* **165**, 29 P.

Eccles, J. C., Eccles, R. M. & Lundberg, A. (1958). *J. Physiol.* **142**, 275.

Eccles, J. C. & Sherrington, C. S. (1930). *Proc. Roy. Soc.* 106, 326.

Granit, R., Henatsch, H. D. & Steg, G. (1956). *Acta physiol. scand.* **37**, 114.

Hursh, J. B. (1939). *Amer. J. Physiol.* **127**, 131.

Matthews, P. B. C. (1962). *Quart. J. Physiol.* **47**, 324.

SOME FACTORS GOVERNING SENSORY CONVERGENCE IN THE CAT'S MOTOR CORTEX

by VERNON B. BROOKS

The cat's motor cortex contains neurones whose axons descend in the pyramidal and extrapyramidal corticofugal paths. We have been studying the sensory activation of the pyramidal system as an example of an integrative pathway. Our method has been that of recording extracellular responses from cells in the anterior sigmoid gyrus and in the postcruciate strip, rostral to the postcruciate dimple. The rest of the posterior sigmoid gyrus constitutes the primary somatic receiving area (SI). Neurones in both areas respond to stimulation of skin and joints (MOUNTCASTLE 1961; BROOKS 1963), but the motor cortex also receives proprioceptive (OSCARSSON and ROSEN 1963), as well as auditory and visual inputs (BUSER and IMBERT 1961; PATTON, TOWE and KENNEDY 1962; BROOKS 1963; JUNG, KORNHUBER and DA FONSECA 1963). Another difference between motor and sensory cortex involves the spatial origins of adequate peripheral somesthetic and kinesthetic stimuli. The sensory cortex overwhelmingly receives somatotopic representation of the contralateral body, i. e., impulses from receptors in circumscribed, continuous parts of the body are relayed to a definite part of the contralateral sensory cortex. The sizes of receptive fields of neurones in sensory cortex reflect density of peripheral sensory endings. Fields in densely innervated skin, such as the paws, are smaller than those from proximal, relatively sparsely innervated parts of a limb (MOUNTCASTLE 1961). For all or at least most (BUSER and IMBERT 1961; BROOKS and LEVITT 1964) neurones in the sensory cortex, one type of peripheral stimulus is exclusively adequate (e. g. hair-bending): the cells are modality-specific. The motor cortex, in contrast, contains many cells whose receptive fields are multimodal and/or large, discontinuous, and ipsi- or bilateral (BUSER and IMBERT 1961; PATTON et al. 1962; BROOKS 1963; BROOKS and LEVITT 1964).

A particular neurone in the motor cortex may thus be activated by stimuli of the same modality from two or more different parts of the body, or by stimuli of different modalities from the same or from different parts of the body. Such non-somatotopic arrangement becomes increasingly probable in the more rostral motor cortex (BUSER and IMBERT 1961; BROOKS 1963). A greater percentage of cells with these properties are, moreover, found deep rather than superficial in the grey matter, which is suggestive of the neurones giving rise to the pyramidal tract (PT cells). Cellular identification by pyramidal backfiring has confirmed that PT-cells are more often, though not exclusively, targets for sensory convergence than are other cells in the motor cortex (PATTON et al. 1962; BROOKS 1963). Many PT- and non-PT cells respond to peripheral stimuli with brief, early and longer, late discharges that reflect arrival of two distinct input volleys. In the precruciate cortex deep-lying PT-cells are more likely than superficially located cells to receive early as well as late inputs from the periphery, whereas in the postcruciate cortex only the deep-lying PT-cells receive both inputs (Towe, Patton and Kennedy 1963). Thus there exist within radial columns of the motor cortex subgroups of cells receiving different kinds of input. Somatotopic peripheral receptive fields of cortical neurones in the same and in adjacent columns may overlap extensively. A minimal relationship exists however, between these restricted (local) peripheral receptive fields of cells within a column and the non-somatotopic (wide) fields of cells in that column: the somatotopic fields are usually within the boundaries of the wider ones (Brooks, unpublished).

Several pathways serve the convergent messages. The contralateral somatotopic component which arrives at the cortex with the shortest latency appears to ascend through the rostral extension of the medial lemniscal system (Buser and Imbert 1961; Patton et al. 1962; Brooks 1963), but does so independently of similar messages destined for the sensory cortex (Buser, Ascher, Bruner, Jassik-Gerschenfeld and Sindberg 1963). This early component is relayed to the ipsilateral cortex through the corpus callosum (Patton et al. 1962). Subsequent responses from wide fields, that may include other parts of the body, arrive later in the motor cortex through relay paths from subcortical centres, of which antero-lateral system (Mountcastle 1961) is probably one important component (Brooks 1963). Cortico-cortical connections from other areas with convergent input (e. g., S II) are not involved (Patton et al. 1962). Somatic, visual and auditory sensory cortex are linked to the motor cortex through subcortical loops likely involving midline thalamic structures. Separation of input controls is apparent, however: pyramidal activation by sound or light can be influenced from specific or visual sensory cortex, but the late somesthetic response cannot (Buser et al. 1963).

Overlap of receptive fields implies that a large number of cells are excited by a peripheral stimulus. This multiplicity is reduced, however, by operation of surround, or afferent, inhibition. Since sensory convergence is more common in motor than in sensory cortex, it is not surprising that peripherally produced surround inhibition is more common for cells in motor (Brooks 1963) than in sensory (Mountcastle 1961) cortex. This inhibitory influence from the edges of excitatory fields on to cortical cells could arise in "forward" fashion as arrival of afferent inhibitory messages from subcortical centres (Mountcastle 1961), or it could be generated through "recurrent" loops. These could run either from active cortical cells to subcortical afferent relay stations (Andersen, Eccles and Sears 1964) or through short intracortical loops by way of recurrent axon collaterals (Brooks 1963). Recent work in our laboratory has revealed that activation of the recurrent axon collaterals of PT-cells (brought about by antidromic repetitive stimulation of the medullary pyramids) does in fact constrict peripheral receptive fields of cortical cells. The foci and edges of the receptive fields (BROOKS and ASANUMA 1965 a) shrink with a time course comparable to the effect produced by peripheral conditioning and testing stimulation (MOUNTCASTLE 1961). Fig. 1 illustrates an example of constriction of a "wide fixed" receptive field of a cortical neurone before and during maximal recurrent inhibition. The maximal discharges, whose rates are identified in the key, were elicited by electrical stimulation of skin in 17 positions, from which data the figurines

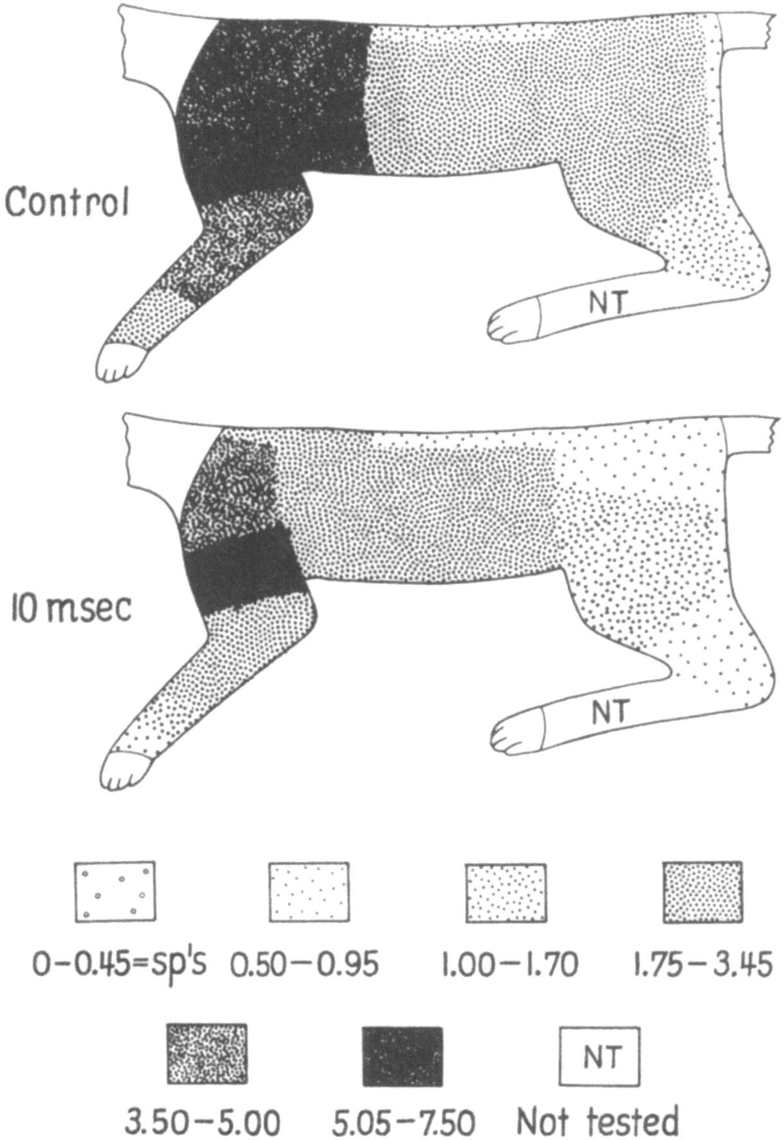

Fig. 1. Recurrent inhibition of a PT-cell with a wide fixed peripheral receptive field. Upper figurine: Intensity of maximal discharges (see key) in response to single electrical stimuli applied to skin (latencies up to 20 msec.). Lower figurine: Equivalent responses plotted during maximal recurrent inhibition (10 msec. after last of 11 conditioning pyramidal stimuli recurring at 5 msec. intervals). Diagrams based on 20 replicate determinations. (PT-cell A 50-2, lateral ant. sigmoid gyrus. Taken in part from BROOKS and ASANUMA, 1965 a

were constructed. Weak excitatory signals from near the field edges are suppressed more effectively than strong signals from the field focus (Brooks and Asanuma 1965 b).

The degree of inhibition produced by a given intensity of recurrent conditioning depends on the state of activity of the test-cell rather than on fixed anatomical connections involving the receptive field. This statement is based on experiments with neurones whose receptive fields enlarge after repeated peripheral stimulation. One such trial with a "labile" field (Brooks 1963) is illustrated in Fig. 2. Control test responses to stimulation of the edge of the initial

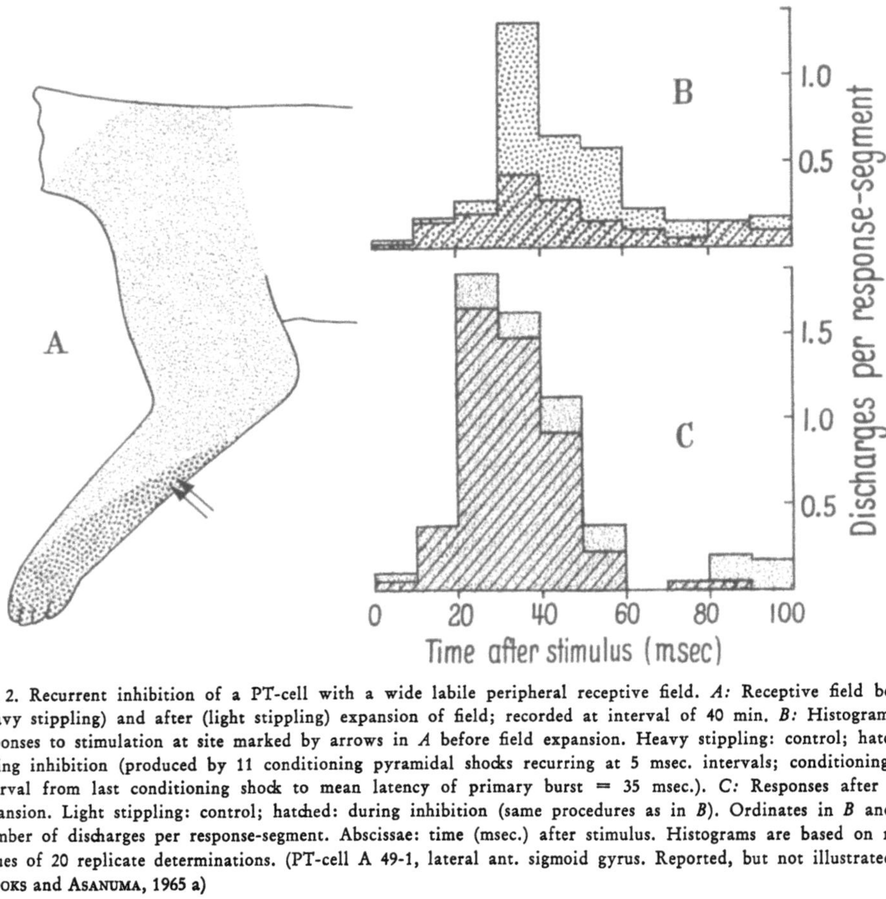

Fig. 2. Recurrent inhibition of a PT-cell with a wide labile peripheral receptive field. *A:* Receptive field before (heavy stippling) and after (light stippling) expansion of field; recorded at interval of 40 min. *B:* Histograms of responses to stimulation at site marked by arrows in *A* before field expansion. Heavy stippling: control; hatched: during inhibition (produced by 11 conditioning pyramidal shocks recurring at 5 msec. intervals; conditioning-test interval from last conditioning shock to mean latency of primary burst = 35 msec.). *C:* Responses after field expansion. Light stippling: control; hatched: during inhibition (same procedures as in *B*). Ordinates in *B* and *C:* Number of discharges per response-segment. Abscissae: time (msec.) after stimulus. Histograms are based on mean values of 20 replicate determinations. (PT-cell A 49-1, lateral ant. sigmoid gyrus. Reported, but not illustrated, in Brooks and Asanuma, 1965 a)

receptive field (arrows in *A*) are shown stippled in *B*. After expansion, the fixed stimulus site was now near the center of the field, and the more vigorous response appeared with a briefer latency (stippled in *C*). Comparison of the hatched histograms in *B* and *C* shows that the same conditioning pyramidal stimulation produced less inhibition of the stronger response to stimulation of the enlarged field.

We concluded that the inhibitory pathways runs through short intra-cortical recurrent loops, and not through afferent paths or long loops, because antidromic stimulation of the rostral end of chronically sectioned internal capsules yields very similar inhibitory curves of spontaneous or driven activity as does antidromic stimulation of medullary pyramids (ASANUMA and BROOKS 1965). Pharmacological evidence suggests that this inhibitory system operates with a transmitter similar to that in spinal postsynaptic inhibition because intracortical administration of tetanus toxin reduces or abolishes the inhibition. It differs in that i. v. injection of strychnine has little effect (BROOKS and ASANUMA 1965 b).

The point of this brief survey is to show that an efferent system, in this case the pyramidal tract, receives greater sensory convergence than does the primary sensory cortex, but that the grouping of inputs to cells within a column of motor cortex cells is not random. Sensory confluence is accompanied by frequent occurrence of surround inhibition, mediated partly through recurrent axon collaterals. These findings on the static properties of the motor cortex are clarifying our understanding of input channelling. Study of unit activity has demonstrated the interplay between initial arrival in motor cortex of information from focal field points and later influx of messages from other parts of the body. The resulting integration of supraspinal motor control will be revealed perhaps best through knowledge of the dynamic, time-dependent (WERNER and MOUNTCASTLE 1963) properties of many cells assembled in functionally related cortical columns and of their effects on the relevant spinal interneurones.

Acknowledgement

This research was supported in part by U. S. Public Health Research Grants NB-02618 and NB-05508 from the National Institute for Neurological Diseases and Blindness.

References

ANDERSEN, P., ECCLES, J. C. & SEARS, T. A. (1964). *J. Neurophysiol.* **27**, 63.
ASANUMA, H. & BROOKS, V. B. (1965). *Arch. ital. Biol.* **103**, 220—246.
BROOKS, V. B. (1963). *Electroenceph. clin. Neurophysiol.* (Suppl. 24) 13.
BROOKS, V. B. & LEVITT, M. (1964). *The Physiologist.* **7**, 95.
BROOKS, V. B. & ASANUMA, H. (1965 a). *Arch. ital. Biol.* **103**, 247—278.
BROOKS, V. B. & ASANUMA, H. (1965 b) *Amer. J. Physiol.* (in Press).
BUSER, P. & IMBERT, M. (1961). In: *Sensory Communication* (ed. by ROSENBLITH, W. A.) p. 607 Cambridge, Mass.: M.I.T. Press.
BUSER, P., ASCHER, P., BRUNER, J., JASSIK-GERSCHENFELD, D. & SINDBERG, R. (1963). In: *Progress in Brain Research,* Vol. 1 (ed by MORUZZI, G., FESSARD, A. & JASPER, H. H.) p. 207. Amsterdam: Elsevier.
JUNG, R., KORNHUBER, H. H. & DA FONSECA, J. S. (1963). In: *Progress in Brain Research,* Vol. 1 (ed. by MORUZZI, G., FESSARD, A. & JASPER, H. H.) p. 294. Amsterdam: Elsevier.
MOUNTCASTLE, V. B. (1961): In: *Sensory Communication* (ed. by ROSENBLITH, W. A.) p. 403. Cambridge, Mass.: M.I.T. Press.
OSCARSSON, O. & ROSEN, I. (1963). *J. Physiol.* (Lond.) **169**, 924.
PATTON, H. D., TOWE, A. L. & KENNEDY, T. T. (1962). *J. Neurophysiol.* **25**, 501.
POGGIO, G. F. & MOUNTCASTLE, V. B. (1963). *J. Neurophysiol.* **26**, 775.
TOWE, A. L., PATTON, H. D. & KENNEDY, T. T. (1963). *Exper. Neurol.* **8**, 220.
WERNER, G. & MOUNTCASTLE, V. B. (1963). *J. Neurophysiol.* **26**, 958.

CONTROL OF THE NEURONES OF THE SUPRAOPTIC AND PARAVENTRICULAR NUCLEI

by CHANDLER McC. BROOKS and KIYOMI KOIZUMI

There has been much interest of recent years in hypothalamico-hypophyseal functional relationships and in the production and action of neurosecretions. Numerous reviews have appeared (HARRIS 1955; SAWYER 1961; BROOKS, GILBERT, LEVEY and CURTIS 1962; SCHARRER and SCHARRER 1963; ROTHBALLER 1963). It has been demonstrated beyond question that the supraoptic nucleus participates in the production of antidiuretic hormone (ADH) and in the control of its output. Cells of the nucleus appear to be osmosensitive (JEWELL and VERNEY 1957; CROSS and GREEN 1959) and the output of ADH is cholinergically mediated (PICKFORD and WATT 1951).

The paraventricular nucleus, though considered to be somewhat concerned with osmotic control, is primarily involved in oxytocin production and release (OLIVECRONA 1957; CALDEYRO-BARCIA and HELLER 1961; BROOKS et al. 1962). It has been shown that in the rabbit, goat, cow and sheep, suckling elicits reflex release of oxytocin (HARRIS 1955; FOLLEY 1956; CROSS 1961). There is abundant evidence of a reflex control of the milk ejection response and an oxytocin output in labor (CALDEYRO-BARCIA et al. 1963).

There have been attempts to record activity in the pituitary stalk and from the hypophysis. Abstracts have been published describing electrical activity in the unmyelinated axons of the pituitary stalk of the goosefish (LOPHIUS) (POTTER and LOWENSTEIN 1955) and evoked and spontaneous potentials in the pituitary gland of the cat (SIGG and SHAPIRO 1961). KANDEL (1964) has published an extensive article on the electrical properties of hypothalamic neuroendocrine cells of the goldfish. Activity of single neurones of the supraoptic nucleus has been described by the present authors (BROOKS, USHIYAMA and LANGE 1962; SUDA, KOIZUMI and BROOKS 1963; KOIZUMI, ISHIKAWA and BROOKS 1964) and others (CROSS and GREEN 1959; VON EULER 1953).

Extensive studies have not been made of the control of unit activity in the supraoptic and paraventricular nuclei nor of their relationship to discharges along axons of the pituitary stalk which are supposed to release stored hormone (ADH and oxytocin) from the pars neuralis. The present work reports such studies. The origins of possible controlling influences have been identified. Activity in the supraoptic and paraventricular nuclei was related to discharge along the stalk (ISHIKAWA, KOIZUMI and BROOKS 1965) and to the reflexly induced release of posterior hypophyseal hormones.

Method

Cats anesthetized with chloralose (35 mg/kg) or a chloralose-urethane mixture (35 mg chloralose and 0.5 g urethane/kg) given intraperitoneally were used in this study. When assaying oxytocin output, postpartum cats (2 to 3-1/2 weeks) were used. In most studies of the supraoptic nucleus and in recording from the pituitary stalk the brain was exposed in the usual manner and the left hemisphere then removed to permit visual control of electrode placement (Koizumi et al. 1964). For stimulations and recordings from the paraventricular nuclei, stereotaxic electrode placement was done without brain tissue ablation. Flaxedil (4 mg/kg) and artificial respiration were used in many instances to minimize body movement.

Peripheral nerves (gastrocnemius, posterior tibialis) were prepared for stimulation in the usual way. Occasionally the skin was stimulated directly through a needle electrode. A balloon was inserted in the uterus to permit uterine distention. In stimulation of nipples gentle intermittent suction was applied or electrical pulses through bipolar electrodes penetrating the skin of the mammary gland close to the nipples were used. The midbrain reticular formation, cingulate gyrus, paramedian lobe of the cerebellum, motor and prefrontal areas of the cerebral cortex and the thalamus were stimulated in the usual manner (Koizumi et al. 1964).

Blood pressure from a catheter in the femoral artery and pressures from the mammary gland, one duct of which was cannulated with a capillary drawn from nylon tubing, were recorded continuously by means of transducers and a Gilson recorder. Responses of the mammary glands were calibrated by injections of 5 or 10 milliunits of oxytocin (Pitocin — Parke-Davis Co.). These injections and other materials were administered through the left radial vein. For close intra-arterial injection a small polyethylene tubing was introduced through the wall of the left carotid artery and advanced as far forward as possible. Circulation was not otherwise disturbed and materials were liberated very close to the hypothalamic nuclei.

Recording electrodes were steel needles ground to a tip diameter of 1 to 2 microns and insulated except at the tip (Suda et al. 1963). Electrical activities of neurones were monitored through condensor-coupled amplifiers and were displayed on two oscilloscope screens; one showed the shapes of action potentials, the other the continuous firing pattern. Particular care was taken in identifying locations of the recording electrode tip. Small amounts of iron were deposited by passage of a 20 microampere current for 45 seconds. The brain was perfused by way of the carotid arteries with a mixture of 3 per cent potassium ferrocyanide-ferricyanide solution and physiological saline. The tissue was then fixed in 10 per cent neutral formalin. Blocks of tissue containing the area to be examined were frozen, sectioned at 40 μ, and stained with thionine. The point of iron deposit was identifiable as a very small deep blue stained spot.

Results

Previous work on the osmosensitive cells of the supraoptic nucleus region showed them to be affected by many peripheral and centrally originating influences (Brooks et al. 1962; Suda et al. 1963; Koizumi et al. 1964). Stimulation of the frontal and prefrontal regions of the cerebral cortex, the cingulate gyrus, somatic afferents of leg nerves and certain regions of the midbrain reticular formation caused marked augmentation of neurone firing. These excitations were frequently succeeded by brief periods of inhibition, but in more than half of the osmosensitive cells studied, approximately 100 to date, excitation alone occurred. Stimulation of the posterior lobe of the cerebellum had predominantly a pure inhibitory action and only a few osmosensitive cells were excited by cerebellar stimuli. Inhibitory effects from the reticular formation were obtained but these require further study.

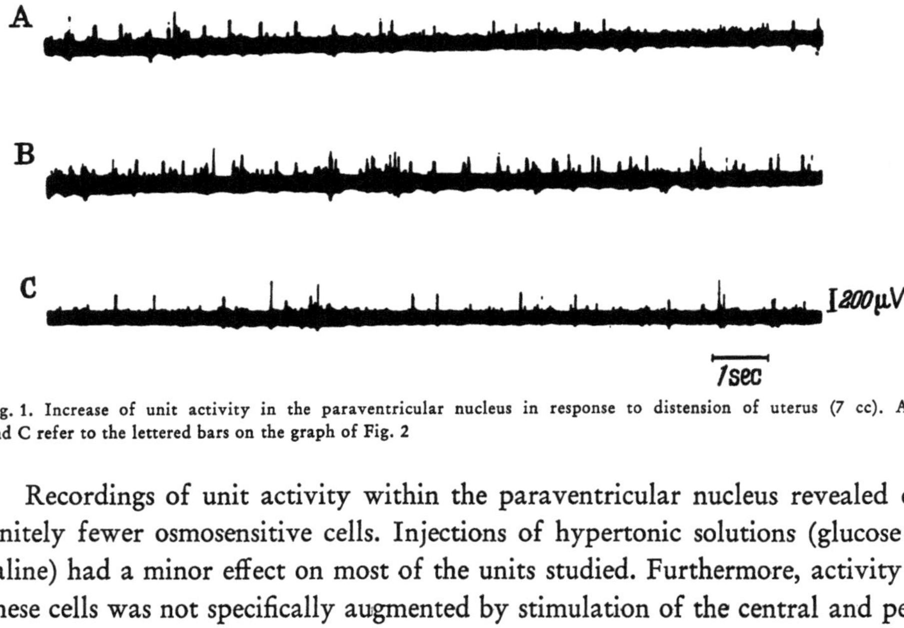

Fig. 1. Increase of unit activity in the paraventricular nucleus in response to distension of uterus (7 cc). A, B and C refer to the lettered bars on the graph of Fig. 2

Recordings of unit activity within the paraventricular nucleus revealed definitely fewer osmosensitive cells. Injections of hypertonic solutions (glucose or saline) had a minor effect on most of the units studied. Furthermore, activity of these cells was not specifically augmented by stimulation of the central and peripheral structures which excited units of the supraoptic nucleus. Such stimulation also did not produce a "milk ejection" response. Some paraventricular units were definitely inhibited by cortical and thalamic stimulations and by stimulation of the skin but, in general, paraventricular units were unresponsive.

Gentle intermittent suction applied to the nipples, however, had a strong excitatory action on many paraventricular nucleus units. Quick distention of the uterus (5 to 15 cc in 10 to 30 seconds) also produced a marked augmentation of unit discharge (Fig. 1). These reflexly induced activities were accompanied by a milk ejection response and an increase in intramammary pressure equivalent to that produced by 3 to 7 milliunits of oxytocin (Fig. 2).

Close intra-arterial injection of acetylcholine (40—80 micrograms of acetylcholine chloride) augmented activity in both the osmosensitive cells of the supraoptic nuclei and the cells of the paraventricular nucleus reflexly activated by mammary gland and uterine stimulation. A small milk ejection response also occurred. As reported by ABRAHAMS and PICKFORD (1956) and KOIZUMI et al. (1964) adrenaline had a slight inhibitory action on supraoptic nucleus cells but we found it to have no clear effect on activity of paraventricular nucleus units. After adrenaline administration milk ejection sometimes occurred unassociated with a change in unit activity and it was therefore thought to be of possible peripheral origin.

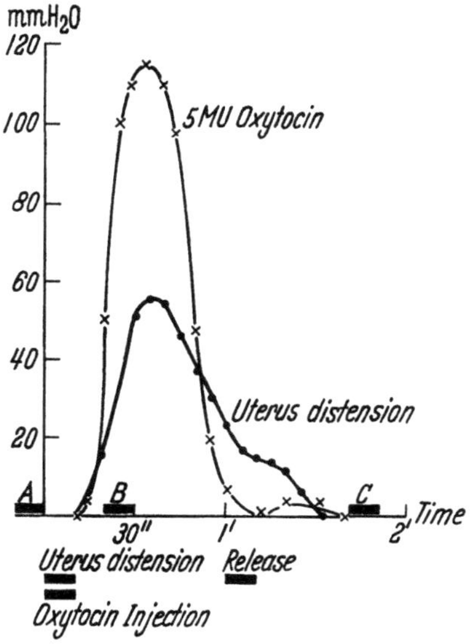

Fig. 2. Milk ejection response produced by distension of uterus (7 cc) — closed circles, and by 5 M. U. oxytocin — crosses. The time of uterus distension and release, and the oxytocin injection, are indicated by the lower bars of the graph. The bars A, B, C refer to the records of Fig. 1

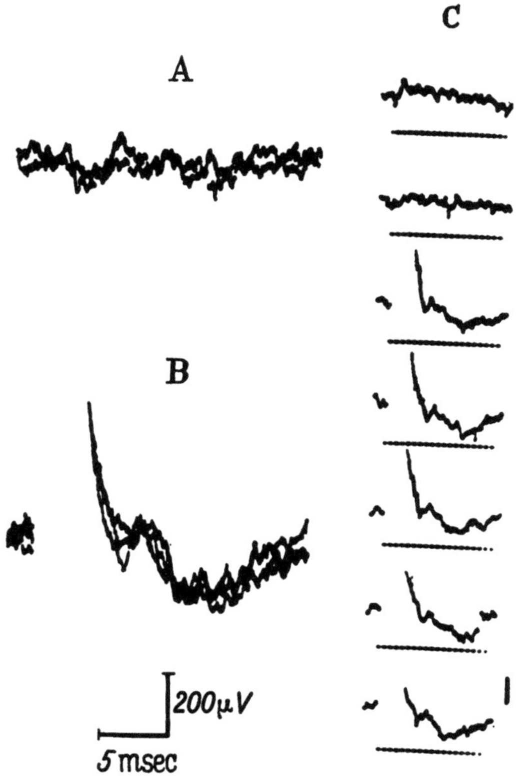

Fig. 3. Potentials evoked in fibres of the pituitary stalk by single pulse stimulation of supraoptic nucleus. Superimposed repetitive tracings taken before (A), and with supraoptic nucleus stimulation (B). Examples of a series of individual tracings of evoked responses in stalk (C). Distance between recording and stimulating electrodes was 8 mm (ISHIKAWA et al. 1965)

Difficulties were encountered in attempting to record from axons or units of the pituitary stalk. It was, however, possible to do so by using hemispherectomized preparations in which the stalk could be seen. Very fine steel monopolar microelectrodes were placed on or in the stalk by using a micromanipulator and a dissecting microscope. Tonic discharge were thus recorded and these were definitely augmented by repetitive (1 or more per second) stimulations of both the supraoptic and paraventricular nucleus. Distention of the uterus also caused discharge in the stalk as a milk ejection reaction was evoked. Single evoked potentials in stalk fibers were induced by single pulse stimulations of the supraoptic nucleus. Estimated conduction time was approximately 1.4 to 0.6 meters per second (Fig. 3).

Discussion

It has been shown that the osmosensitive units of the supraoptic nuclei are affected by stimulation of numerous afferent neurones and central structures. Inhibitory and excitatory control apparently exists. Difficulties in assaying antidiuretic hormone or urine output in cats prevented clear demonstration of an association between degree of neurone activity and hormone release. Much evidence, however, has shown that ADH output is affected by exercise, emotional and other disturbances (Rydin and Verney 1938; O'Connor and Verney 1945).

The fact that discharge in units of the stalk has now been clearly demonstrated as associated with activity in supraoptic und paraventricular nuclei and reflex discharge of oxytocin provides the long lacking direct evidence that the fibers of the stalk function as do nerve fibers. Conduction of activity along the stalk of the mammalian hypophysis has now been recorded, and its rate calculated.

It is generally accepted that some ADH is produced in and liberated from the paraventricular nucleus. This nucleus, however, is considered to be primarily concerned with oxytocin production and release. Our results support these conclusions in that we found few osmoreceptors in the paraventricular nuclei. Units there also failed to respond to stimulation of peripheral nerves and centers as did the osmosensitive cells of the supraoptic nucleus. Paraventricular cells were activated by stimuli which evoked a milk ejection response and an associated discharge was also recordable from the fibers of the pituitary stalk. Thus this nucleus is under specific peripheral control and in this work very clear evidence was obtained of a relationship between peculiar types of peripheral stimulation, augmented activity in units of the paraventricular nuclei, discharge along the stalk and an oxytocin release which caused milk ejection.

Summary

The previously unreported observations described can be summarized as follows:

Many fewer osmosensitive units were found in the paraventricular nuclei than in the supraoptic nuclei.

The activity of units of the paraventricular nuclei was not much altered by stimulation of the same central structures and peripheral nerves which markedly increased and decreased activity in supraoptic nucleus cells.

Paraventricular nucleus cell discharge rate was much augmented by application of gentle intermittent suction to the nipples and by distention of the uterus. This was accompanied by a milk ejection response, in the postpartum cat, equivalent to that produced by injection of 3 to 7 milliunits of oxytocin.

A discharge in pituitary stalk fibers was also associated with the increased paraventricular cell activity and the milk ejection response.

It was also found that evoked responses in these stalk fibers resulted from single pulse stimulation of the supraoptic nucleus.

Units of the paraventricular nucleus were acetylcholine sensitive, as were those of the supraoptic nuclei, but they were less affected by adrenaline injection than were the supraoptic nucleus cells.

Acknowledgements

The authors are much indebted to Drs. T. Ishikawa and H. H. Lu who participated in these investigations.

This work was supported by a grant from USPHS-NB-00847 (08—09).

References

Abrahams, V. C. & Pickford, M. (1956). *J. Physiol.* (Lond.) **131**, 712.

Brooks, C. McC., Gilbert, J. L., Levey, H. A. & Curtis, D. R. (1962). *Humors, Hormones and Neurosecretions.* New York: State University of New York.

Brooks, C. McC., Ushiyama, J. & Lange, G. (1962). *Amer. J. Physiol.* **202**, 487.

Caldeyro-Barcia, R. & Heller, H. (1961). *Oxytocin.* Oxford: Pergamon Press.

Caldeyro-Barcia, R., Sica-Blanco, Y., Alvarez, H., Pose, S. V., Poseiro, J. J., Méndez-Bauer, C., Fielitz, C., Gonazález-Panizza, V. H., Cabot, H. M., Branda, L., Escarcena, L., Coch, J. A. & Brovetto, J. (1963). In: *Perspectives in Biology* (ed. by Cori, C. F., Foglia, V. G., Leloir, L. F. & Ochoa, S.) p. 62. Amsterdam: Elsevier Publ. Co.

Cross, B. A. (1961). In: *Oxytocin* (ed. by Caldeyro-Barcia, R. & Heller, H.) p. 24. Oxford: Pergamon Press.

Cross, B. A. & Green, J. D. (1959). *J. Physiol.* (Lond.) **148**, 554.

Folley, S. J. (1956). *The Physiology and Biochemistry of Lactation.* Edinburgh: Oliver and Boyd.

Harris, G. W. (1955). *Neural Control of the Pituitary Gland.* London: Eward Arnold Ltd.

Ishikawa, T., Koizumi, K. & Brooks, C. McC. (1965). *J. Neurophysiol.*-submitted.

Jewell, P. A. & Verney, E. B. (1957). *Phil. Trans.* B. **240**, 197.

Kandel, E. R. (1964). *J. gen. Physiol.* **47**, 691.

Koizumi, K., Ishikawa, T. & Brooks, C. McC. (1964). *J. Neurophysiol.* **27**, 878.

O'Connor, W. J. & Verney, E. B. (1945). *Quart. J. exp. Physiol.* **33**, 77.

Olivecrona, H. (1957). *Acta physiol. scand.* **40**, Suppl. 136, 1 .

Pickford, M. & Watt, J. A. (1951). *J. Physiol.* (Lond.) **114**, 333.

Potter, D. D. & Lowenstein, W. R. (1955). *Amer. J. Physiol.* **183**, 652.

Rothballer, A. B. (1963). *Res. Publ. Ass. nerv. ment. Dis.* (in the Press).

Rydin, H. & Verney, E. B. (1938). *Quart. J. exp. Physiol.* **27**, 343.

Sawyer, W. H. (1961). *Pharmacol. Rev.* **13**, 225.

Scharrer, E. & Scharrer, B. (1963). *Neuroendocrinology.* New York: Columbia Univ. Press.

Sigg, E. B. & Shapiro, T. D. (1961). *Physiologist.* **4**, 110.

Suda, I., Koizumi, K. & Brooks, C. McC. (1963). *Jap. J. Physiol.* **13**, 374.

Von Euler, C. (1953). *Acta physiol. scand.* **29**, 133.

MAMMALIAN SLOW AND FAST SKELETAL MUSCLE

by A. J. Buller

In gladly accepting the opportunity to contribute to Professor John Eccles' presentation volume it is appropriate that I should write on the subject to which he introduced me, mammalian skeletal muscle. Although many would not primarily couple the name of Eccles with investigations in the field of muscle physiology his contributions to this field have extended over many years, passing from his early work in collaboration with Sybil Cooper on the true form of the isometric responses of mammalian muscle (Cooper and Eccles 1930), through his wartime observations on muscle atrophies arising from disuse and tenotomy (Eccles 1944) to the most recent work in which I had the priviledge of joining him, concerning the influence of motor innervation on the speed of a muscle's contractile response (Buller, Eccles and Eccles 1960). All of these observations were undertaken using mammalian muscle, and have been concerned more with the principles of organisation and functioning of muscles as they exist in the intact organism than with the biophysical questions of contractile mechanisms. To this end Eccles in his experiments has insisted upon keeping the muscles as near to "normality" as possible by maintaining an adequate circulation, a physiological temperature and generally using stimulation via the motor nerves. Such experimental conditions contrast sharply with those used by such eminent muscle physiologists as Professor A. V. Hill and his school. Their observations have been largely concerned with the responses of excised amphibian muscle at low temperatures to direct (all over) stimulation. However it must be remembered that the intention of such workers was to investigate the intricacies of the intracellular mechanical response and to this end they selected suitable experimental preparations, while the fascination for Eccles remained the total functional interrelationship between the nervous and muscular systems of the mammal, and ultimately of man. Such contrasting approaches to the study of muscle function are blatantly not mutually exclusive, but such radically different experimental conditions have inevitably slowed the exchange of ideas between workers concerned with what may be termed the biophysics of muscle and those concerned with the organisation of muscles within the body. Worse, it has occasionally led to complete dismissal of certain data from discussion because it did not conform to the experimental pattern required by one or other group.

Eccles' most recent excursion into the field of muscle physiology (Buller, Eccles and Eccles 1960) has demonstrated a neural influence on the speed of striated muscle contraction and has raised questions which, for their solution, will require the attention of workers with both approaches. It is my intention to attempt to formulate some of these problems in this brief review.

The limb muscles of the cat are made up of motor units differing appreciably in the number of muscle fibres they contain and consequently in their tension development. Polyneuronal innervation is slight or absent in such muscles (Brown and Matthews 1960). Recently Bessou, Emonet-Denand and Laporte (1963) have demonstrated that the speed of the twitch contraction of individual motor units within a single muscle (1st lumbrical of the foot) may vary over a three fold range, with the time to peak tension ranging from 15 to 45 msec. The speed of contraction of a single motor unit was found to be linearly related to the conduction velocity of its motor axon, the range of conduction velocities observed extending from 42 metres/sec. for the slowest contracting motor units to 76 metres/sec. for the fastest. While the range and distribution of motor unit contraction speeds is not yet known for larger limb muscles such as the soleus and gastrocnemius, it is known that the slower total twitch contraction of the former muscle is associated with a slower mean conduction velocity of its motor axons (Eccles, Eccles and Lundberg 1958). It is therefore very likely that the anatomical soleus muscle is composed of a collection of motor units having a spectrum of contraction times and whose individual motor axons conduct with a velocity proportional to the contraction time of the muscle fibres they innervate. Likewise, so called fast muscles such as gastrocnemius may well contain a spectrum of motor units but with both the mean contraction time of its motor units and the mean conduction velocity of their motor axons faster than in soleus. The concept of a wide spectrum of motor unit contraction speeds must undoubtedly replace the earlier concept of two populations of identically "fast" and identically "slow" motor units. It is important to stress that the terms "fast" and "slow" used here both refer to motor units composed of twitch fibres. The recent demonstration by Hess and Pilar (1963) of muscle fibres in the extrinsic eye muscles of the cat which are analogous with the slow muscle fibres of the frog has produced a further difficulty of nomenclature, but ambiguity is avoided in the present text since so far no such fibres have been observed in mammalian limb muscles.

The apparent correlation between the conduction velocity of motor nerve fibres and the twitch speed of the skeletal muscle fibres they innervate prompts the question is the relationship cause and effect? Certainly when the twitch response of soleus is speeded up by cross innervation with a rapidly conducting motor nerve the motor nerve retains its high conduction velocity (Buller et al., 1960). It so happens that there is one period in a cat's life when the twitch speeds of potentially fast and slow muscles are very similar, namely the first few days after birth. The conduction velocities of the motor nerves to soleus, gastrocnemius and flexor hallucis longus in day old kittens has been examined by Ridge (1965), who found that while the absolute conduction velocities are very much lower than in the adult a clear distinction exists between the values found for the motor nerves supplying muscles which are destined to become fast and those supplying soleus. With the growth of the kitten the absolute conduction velocities increase towards the adult values but the motor nerve fibres to soleus always conduct proportionately slower. However it has also been observed that the skeletal muscles mature to their adult speeds some weeks before the motor nerves have reached their final velocity. There is apparently no simple cause and effect relationship, and further observations are required. For example it would be interesting to know the conduction velocity of the motor nerves fibres innervating a chronically tenotomised soleus muscle, since this procedure has also been shown to speed the twitch contraction (VRBOVÁ 1963 a). It would also be fascinating to know whether all the muscle fibres innervated by a single motor nerve fibre behave mechanically in an identical fashion.

At this stage it is pertinent to enquire whether a total transformation from fast to slow muscle and vice versa may be brought about by cross innervation experiments. In other words can a normal fast muscle be caused to behave *completely* as a normal slow muscle simply by implanting the slow muscle's motor nerve into it, and vice versa? Following the experiments of BULLER and LEWIS (1964) the answer appears to be no. While with the measurements at present available the mechanical behaviour of a fast muscle appears to be completely transformed to that of a slow muscle by cross innervation, we have been unable to convert a normally slow muscle to a fast muscle by such means. While the time to peak of the twitch response is invariably shortened, the rate of rise of tension during an isometric tetanus is never as rapid as in a normal fast muscle, and the tetanus twitch ratio of such cross innervated muscles is abnormally high. Some intermediate or "bastard" form of muscle has been produced by innervating normally slow muscle fibres with a rapidly conducting motor nerve. That these changes are not due to quirks of reinnervation may be shown by allowing self reinnervation of a slow muscle by a slowly conducting nerve, when the normal characteristics of slow muscle are re-established. These findings may be summarised by saying that the fast muscle fibres are more labile than the slow, apparently having their mechanical responses completely determined by the motor nerve fibres which innervate them. These observations also stress the point that in calling a muscle fast or slow one should consider its total mechanical behaviour and not only the time to peak of the twitch response. As a result of her study of the effects of chronic tenotomy on the rabbit soleus muscle VRBOVÁ (1963 b) was led to state that tenotomy converts the mechanical response of soleus to that of a fast muscle. This statement was based only on the time to peak of the twitch response and the apparent fusion frequency during tetanic stimulation (which parallels the twitch speed). Further observations (BULLER and LEWIS, unpublished observations) make it apparent that the tenotomised soleus muscle is far from being a *normal* fast muscle, and it is therefore perhaps unfortunate that a conversion to fast muscle should have been inferred.

While it appears clear that the mechanical responses of skeletal muscle fibres are at least partially determined by their motor innervation the modus operandi by which this is brought about remains obscure.

Two possible mechanisms which are not mutually exclusive immediately present themselves, namely the pattern of motor nerve impulses and the influence of some hypothetical trophic substance. Neither possibility has so far been satisfactorily investigated. VRBOVÁ (1963 a) reported continuous electromyographic activity in the soleus of the conscious rabbit which disappeared after tenotomy. As reported above tenotomy also produces a speeding of the soleus twitch response, and VRBOVÁ was inclined to couple the decrease in myographic activity and the speeding as cause and effect. However, as has been stated above, the chronically tenotomised soleus muscle is far from being a *normal* fast muscle, and by the time its twitch is speeded (some 3—4 weeks after tenotomy) it is a degenerate residuum capable of producing only some 5—10% of its former twitch tension. The evidence is obviously insufficient to establish the unique dependance of a muscle's mechanical response upon the total number of nerve impulses transmitted by its motoneurones. It would be instructive to have some quantitative data on the E. M. G. activity of normal fast and slow skeletal muscles in the conscious animal, and compare this with the patterns seen in muscles following cross innervation. However, even with such information, it would be difficult to exclude the possibility of some trophic mechanism. The evidence in favour of a "non impulse" function for the motor nerve in regard to acetylcholine sensitivity has been stated by MILEDI (1963). The case appears good, and if such a mechanism can be established for one function it is possible that a comparable case may in the future be made for a "non impulse" influence of the motor nerve on muscle speed. The intra-axonal canals described by WEISS (1963) might conceivably offer a transport system for such chemical substances.

So far consideration has been made only of what may be termed the organisational side of the problem, the means by which the motor nerve fibres bring their influence to bear on the muscle. Of equal fascination is the nature of the influence they exert within the muscle fibres.

It is at this juncture that the studies of A. V. HILL and many others on the intimate mechanism of muscular contraction become pertinent to this particular story. Formerly attempts were made to analyse the mechanical performance of muscle in terms of certain defined and non-interacting characteristics such as the tension length curve, the force velocity relation and the series elastic elements. To these must be added the concept of active state introduced by HILL (1938). More recently JEWELL and WILKIE (1958) were unable to derive the form of the isometric myogram of frog's sartorius muscle from the measured series compliance and force velocity curve, and this led them to speculate on the influence of muscle's past history as well as the force in determining the velocity of shortening at any instant. PRINGLE (1960) has also found the classical analytical concepts of muscle listed above inadequate to explain certain findings in cicada muscle, and has introduced the concept of activation as a possible alternative to Hill's active state. It is apparent that our understanding of the contractile mechanism is far from complete. In addition the application of much of the detailed knowledge already derived from amphibian muscle to the problem under review is rendered difficult by the complex arrangement and form of the mammalian muscles. While it appears desirable that measurements should be made of the force velocity relationships of mammalian fast and slow muscle, (CLOSE, R., 1964) together with measurements of series and parallel elastic components and active state it would seem that there is small likelihood of changes in any one of these characteristics providing the unique distinction between the various types of mammalian motor units. Equally if not more rewarding at this stage might be the consideration of formalised models of striated muscle such as that proposed by A. F. HUXLEY (1957). Based on the sliding filament theory of contraction this models equates the speed of contraction and relaxation to the formation and breaking of actomyosin bridges. It might well be that manipulation of the rate constants for these two reactions could produce a model in which the contractile responses of cross innervated muscle are mimicked. Such a model might then suggest further useful experiments.

It is apparent that many problems remain to be answered and many techniques will be required for the solution, but this is the thrill of biological research. As Sir JOHN ECCLES once succinctly put it "It's by doing experiments that we set ourselves new problems".

References

BESSOU, P., EMONET-DENAND, FRANÇOISE & LAPORTE, Y. (1963). *C. R. Acad. Sc.* **256**, 5625.

BROWN, M. C. & MATTHEWS, P. B. C. (1960). *J. Physiol.* (Lond.) **151**, 436.

BULLER, A. J., ECCLES, J. C. & ECCLES, R. M. (1960). *J. Physiol.* (Lond.) **150**, 417.

BULLER, A. J. & LEWIS, D. M. (1964). *J. Physiol.* (Lond.) **170**, 67 P.

COOPER, SYBIL & ECCLES, J. C. (1930). *J. Physiol.* (Lond.) **69**, 377.

CLOSE, R. (1964). *J. Physiol.* (Lond.) **173**, 74.

ECCLES, J. C. (1944). *J. Physiol.* (Lond.) **103**, 253.

ECCLES, J. C., ECCLES, R. M. & LUNDBERG, A. (1958). *J. Physiol.* (Lond.) **142**, 275.

HESS, A. & PILAR, G. (1963). *J. Physiol.* (Lond.) **169**, 780.

HILL, A. V. (1938). *Proc. Roy. Soc.* B **126**, 136.

HUXLEY, A. F. (1957). *Progress in Biophysics* **7**, 255.

JEWELL, B. R. & WILKIE, D. R. (1958). *J. Physiol.* (Lond.) **143**, 515.

MILEDI, R. (1963). In: *The effect of use and disuse on neuromuscular functions* (ed. by GUTMANN, E. & HNÍK, P.). Prague: Czechoslovak Academy of Sciences.

PRINGLE, J. W. S. (1960). In: *Models and Analogues in Biology.* Symposium of the Society for Experimental Biology. Cambridge: Cambridge University Press.

RIDGE, R. M. A. P. (1965). *J. Physiol.* (Lond.) **176**, 8 P.

VRBOVÁ, GERTA (1963 a). *J. Physiol.* (Lond.) **166**, 241.

VRBOVÁ, GERTA (1963 b). *J. Physiol.* (Lond.) **169**, 513.

WEISS, P. (1963). In: *The effect of use and disuse on neuromuscular functions* (ed. by GUTMANN E. & HNÍK, P.). Prague: Czechoslovak Academy of Sciences.

SOME METHODS OF REDUCING INTERFERENCE CAUSED BY STIMULUS ARTIFACTS

by J. S. Coombs

If two platinum wires are used as a pair of stimulating electrodes and a recording electrode is placed nearby, there is a considerable possibility of a large artifact in the output of the recording amplifier at the time of electrical stimulation and during the period immediately following. A disturbance at the time of stimulation may not be troublesome, since signals indicating physiological activity are generally looked for after stimulation has ceased. Hence, the stimulus artifact itself is not the disturbing factor, but any stimulus-produced after-effects which are not of physiological origin can cause serious distortions, and if possible should be reduced to negligible dimensions compared with the desired physiological signals. In this article, two causes of disturbance are considered, namely, the inability of some amplifiers to recover promptly from overload and the effect of polarization at electrodes.

I. Amplifier overload

If an amplifier experiences an overloading input signal, then, on cessation of this signal, the output may not return immediately to its normal baseline state. If the recovery is essentially completed within a fraction of a millisecond, it is satisfactory for most physiological applications. This would be the desired performance of a neurophysiological amplifier, since overload conditions would be automatically catered for, causing minimal disturbance of the electrical record corresponding to physiological activity. However, some amplifiers do not behave in this way, and produce a spurious output for a considerable period following the removal of an overloading input signal. In such cases, it is necessary for the operator to protect the amplifier from overload. With strong stimulation, this may require attention each time changes are made in the relative positions of the stimulating and recording electrodes.

There are two common procedures for reducing the extent of the stimulus artifact within the amplifier: (1) the use of an isolated stimulating unit, possibly with a bridge balancing arrangement to control the distribution of potentials in the stimulating circuit with respect to earth, and (2) the use of a difference (or differential) preamplifier with one electrode recording physiological signals together with stimulus artifact, and the other, if possible, recording an equal stimulus artifact alone.

In the case of monopolar recording using, say, a single "ball" electrode against an indifferent earthing electrode for the preparation, it should be possible using the bridge balancing technique in the stimulating circuit to arrange that the signals received at the recording site from the two stimulating electrodes are of approximately equal magnitude and opposite polarity, thereby providing a null balance. In practice, an appropriate adjustment of the earthed moving arm of the potentiometer connected across the stimulating leads cannot always be found, because of the presence of other disturbing effects such as inadequate isolation of the stimulating unit or polarization discussed below. If an electrode is moved, rebalancing of the bridge may be required. Even if a null balance is not achieved, the stimulus signal at the recording electrode may be reduced sufficiently to prevent amplifier overload.

Bipolar recording with the two electrodes connected to the inputs of a difference preamplifier is often used. Common-mode signals originating as mains interference or from some other source are then reduced significantly in the output, provided the signals are of the same magnitude at each recording electrode. In general, it is unlikely that stimulus artifact signals from a pair of near-by stimulating electrodes will be of equal magnitude at each recording electrode. Bridge balancing of the stimulator circuit with respect to earth can make the stimulus signal small at one or other of the recording electrodes, but not at both at the same time, nor can it make the signals equal. To get approximate equality of these undesired signals, it is necessary for the second recording electrode to be of the "wandering" variety, which can be adjusted in position so that it records an equal stimulus artifact without the addition of the physiological signal being recorded by the main electrode. This is not always possible, and it is certainly not convenient, especially if experimental requirements make it necessary to shift an electrode site, for the "wandering" procedure would need to be repeated to obtain minimal interference again. In view of these difficulties, it is desirable to use an amplifier with a fast recovery from overload, since it is then not so necessary to take the evasive actions described.

The undesirable after-effects in an amplifier following the removal of an overloading input signal are produced while capacitors in the system are being recharged to their normal operating states after changes caused during the overload period. For recovery to be completed in a fraction of a millisecond, the individual time constants of all capacitor circuits, which can be influenced by over-load, should be less than 0.1 millisecond. If a capacitor-coupled amplifier is used for physiological purposes, such an upper limit for the coupling time constant would be much too small. Hence, capacitor couplings must be protected from the influence of overload. If the signal at a capacitor coupling was positive going at the time of overload, the passage of grid current from the following valve would change the charge on the capacitor, resulting in an output different from the baseline upon removal of the overloading signal. The rate of recovery depends on the time constant of the coupling circuit. If the signal at the capacitor coupling was negative going, no extra current equivalent to the grid current in the previous case would flow at the time of overload. Hence, in an amplifier with one capacitor coupling, it is desirable that the overloading signal should produce a negative going change at the capacitor coupling. This can be achieved with a difference preamplifier by appropriate connexions of the input leads. In an amplifier with multi-stage capacitor couplings, it is the last stage which will overload first. The overload signal should be negative going at this stage, and earlier stages must not be overloaded.

These considerations suggest that direct-coupled amplifiers may be more immune to the effects of overload than capacitor-coupled amplifiers. However, capacitors are sometimes used in direct-coupled amplifiers as shunts across high value coupling resistors to neutralize the effect of stray capacitance, so it is important to check that the time constants of these circuits are small compared with 0.1 millisecond. Sometimes, a parallel combination of, say, 100 K and 0.1 μF is used directly in series with the input grid of an oscilloscope to provide protection for the amplifier from a gross positive-going overload. This circuit, with a time constant of 0.01 second, could cause trouble in a physiological application immediately following an overload signal. However, 100 K and 0.001 μF in parallel might be acceptable.

During an overload period, some stages are driven to saturation and others to cut-off. Hence, the normal current drains of the individual stages are changed considerably. If, due to the presence of decoupling circuits or for some other reason, the effective supply voltages after removal of the overload state are different from their normal values, undesired after-effects in the output of the amplifier will be observed.

II. Polarization at electrodes

If two platinum wires in contact with tissue are used as the stimulating electrodes, polarization effects will take place at the metal-liquid interfaces during the passage of the stimulating current tending to reduce its value. Following stimulation, the effective polarization batteries will drive a reverse current round the circuit composed of preparation and stimulator (Fig. 1). This current will die away gradually. A recording electrode placed near the stimulating electrodes will be in this changing current field, and will record a changing potential at a time when physiological signals alone are the desired input. Hence, polarization effects at the stimulating electrodes can cause an interference during the period following stimulation. Non-polarizable stimulating electrodes would cure the effect, but, in practice, platinum wire electrodes are often more convenient. The passage of the reverse current after stimulation can be prevented by including a silicon diode in the stimulating circuit in series with one of the electrodes (Fig. 2). During stimulation, the low forward resistance of the diode has only a very small influence on the passage of the stimulating current, but following stimulation the high reverse resistance of the diode reduces the magnitude of the

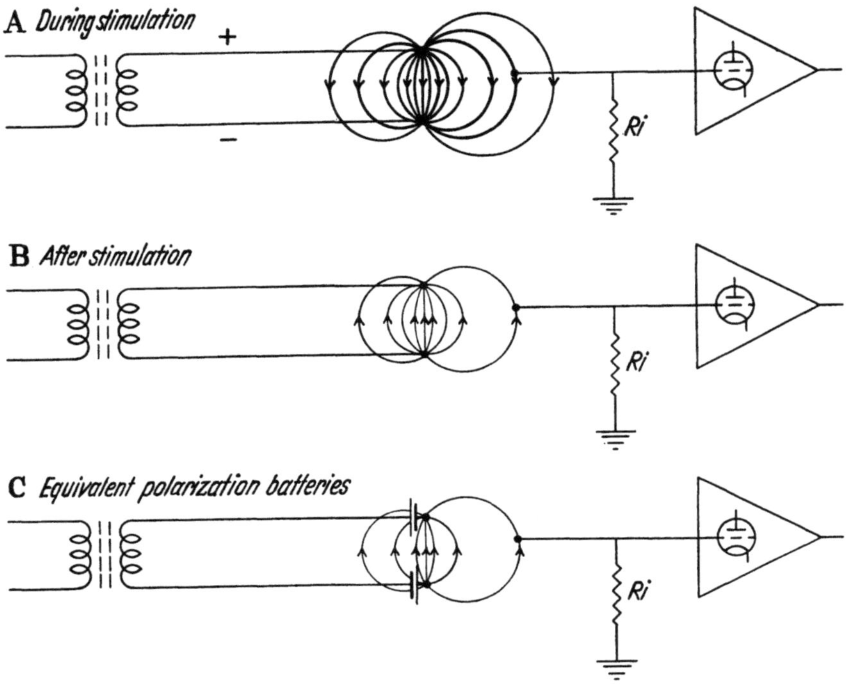

Fig. 1. Current distributions in the region surrounding stimulating and recording electrodes during (A) and after (B) stimulation. (C) The equivalent polarization batteries which produce the same 'reverse' current as that present after stimulation

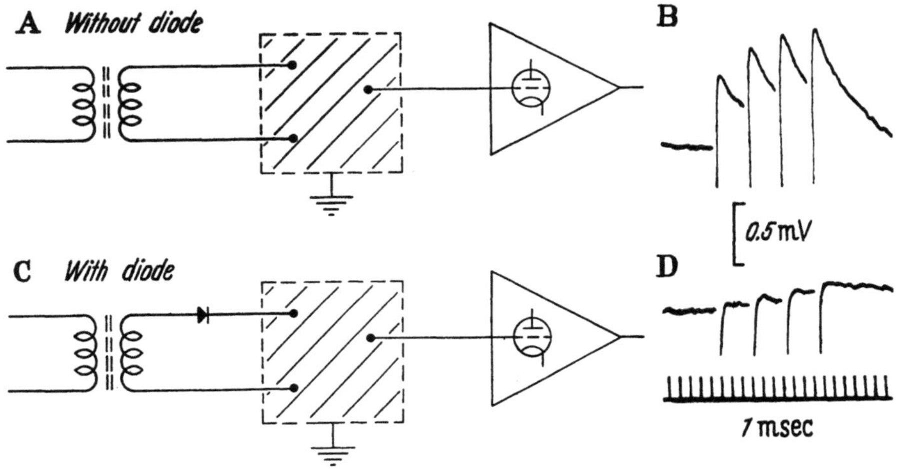

Fig. 2. The effect of a series diode which prevents the reverse current, upon the recorded response. (A) — Circuit without diode. (B) Corresponding distorted record. (C) Circuit with diode. (D) Improved response. The recordings B, D were obtained in the feline cuneate nucleus in response to stimulation of the somatosensory cortex. The build-up of potential in D could be of physiological significance, but was completely masked in B

reverse current to a low value, and thereby reduces the interference during this period. The effect of the diode can be dramatic, but only if other conditions are optimal. Likewise, the advantages of an amplifier with a rapid recovery from overload or the effect of bridge balancing in the stimulator circuit can be seen clearly only if there is no masking influence due to polarization currents. If the stimulator circuit is isolated from earth, one diode in series with either stimulating electrode is sufficient. However, if a bridge balancing circuit across the stimulator with one point earthed is used, it is necessary to have a diode in series with each stimulating electrode.

The after-effects of polarization can also be detected at a recording electrode if too much current flows through it during the period of electrical stimulation (Fig. 3). The current flow can be reduced by increasing the input resistance of

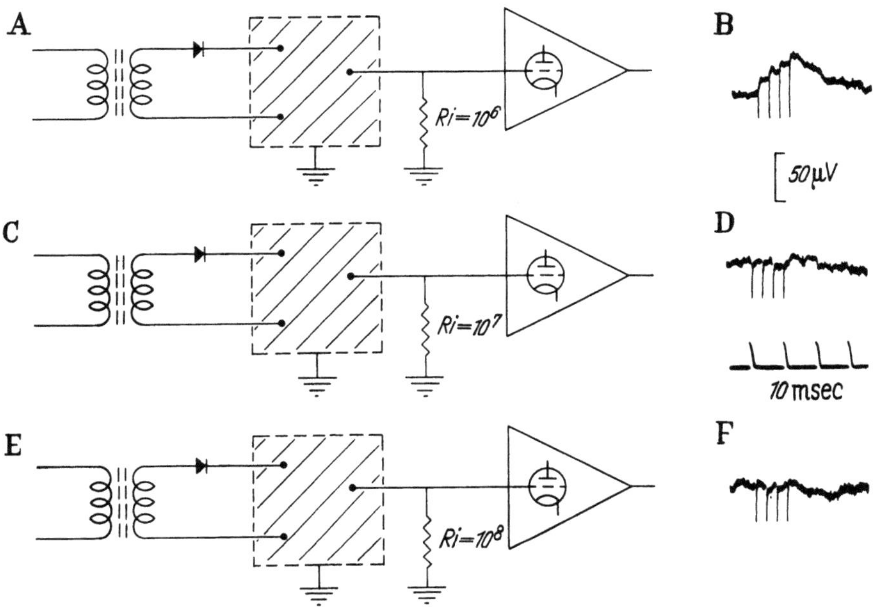

Fig. 3. Stimulating and recording circuits (A, C, E) and corresponding potentials (B, D, F) recorded in the cuneate nucleus in response to stimulating the somatosensory cortex. The current flow through the combination of recording electrode and R_i produced an appreciable polarization potential only when $R_i = 10^6$ ohms. When $R_i = 10^7$ ohms and 10^8 ohms, the current flow was sufficiently small to produce a negligible polarization artifact

the amplifier (Ri), or by using the bridge balancing technique in the stimulator circuit to decrease the magnitude of the voltage appearing across the recording electrode.

In practice, it is desirable to use a high input impedance for the recording amplifier to prevent polarization effects at a recording electrode, a diode in the stimulating circuit to prevent the flow of a reverse current in the stimulator following polarization of the stimulating electrodes, and an amplifier with a fast recovery from overload.

THE ACTIONS OF AMINO ACIDS UPON MAMMALIAN NEURONES

by DAVID R. CURTIS

The presence of certain acidic and neutral amino acids in mammalian central nervous tissue, and the marked pharmacological action of some of these substances upon nerve cells, has led to considerable speculation concerning the possible function of amino acids as central transmitter substances. It is intended in this brief review to outline the action of amino acids upon neurones within the feline central nervous system, and to discuss the question of a transmitter function for these substances.

Although many investigations in which amino acids were administered systematically, by direct injection into brain tissue, or by topical administration to the surface of cortical or spinal tissue suggested that some amino acids were depressants and other were excitants of neurones (HAYASHI 1952, 1954, 1959a, b; PURPURA, GIRADO, SMITH, CALLAN and GRUNDFEST 1959; VAN HARREVELD 1959; HONOUR and MCLENNAN 1960; KUNO and MUNEOKA 1962; KITA, KAMIYA and KIYOTA 1963; ECCLES, SCHMIDT and WILLIS 1963; CRAWFORD 1963; see also HILLMAN and MCILWAIN, 1961), it was not until these substances were administered close to individual cells that the mode and site of action could be determined with any precision. By means of the technique of microelectrophoresis (CURTIS 1964), amino acids have been ejected into the vicinity of spinal interneurones, Renshaw cells and motoneurones (CURTIS, PHILLIS and WATKINS 1959, 1960; CURTIS and WATKINS 1960a, 1963), mid-brain neurones (CURTIS and KOIZUMI 1961), neurones of the lateral geniculate nucleus (CURTIS and DAVIS 1962; CURTIS and WATKINS 1963) and of the ventrobasal complex of the thalamus (ANDERSEN and CURTIS 1964), neurones of the hippocampus (ANDERSEN, CRAWFORD and CURTIS, unpublished; STEFANIS 1964) and caudate nucleus (CURTIS, VOORHOEVE and WILSON, unpublished; SALMOIRAGHI, BLOOM and COSTA 1964), of the cerebral and cerebellar cortices (KRNJEVIĆ and PHILLIS 1963; CRAWFORD, CURTIS, VOORHOEVE and WILSON 1963; CRAWFORD and CURTIS 1964), the olfactory bulb (VON BAUMGARTEN, BLOOM and SALMOIRAGHI 1963) and retina (NOELL 1960). In most cases extracellular drug ejection has been combined with recording of extracellular spike responses by means of five barrel micropipettes, but with spinal motoneurones and Betz cells it has proved possible to combine extracellular amino acid administration with the recording of intracellular response from single neurones. Although there are differences in opinion regarding the central action of some amino acids (see CRAWFORD and CURTIS 1964), it is remarkable that the pharmacological action of any one of these compounds is the same upon cells throughout the feline central nervous system. Furthermore, the relative potency of a particular amino acid remains fairly constant and, in addition, there is a similarity between the excitant potencies of some acidic amino acids and the depressant potencies of the corresponding neutral amino acids (CURTIS and WATKINS 1960a).

Neutral amino acids related to γ-amino-n-butyric acid (GABA) reversibly depress the activity of many types of central neurone. An amino group and an acidic group separated by two or three carbon atoms are essential requirements for this action, and the most active depressants, in approximate order of potency, include 3-amino-1-propanesulphonic acid, GABA, β-alanine, taurine, γ-amino-β-hydroxybutyric acid and δ-amino-n-valeric acid (CURTIS and WATKINS 1960, 1961; KRNJEVIĆ and PHILLIS 1963; CRAWFORD and CURTIS 1964). These substances increase the membrane conductance of neurones and thus lower the electrical excitability of the membrane. Antidromic invasion of the cell body is blocked, the amplitudes of excitatory and inhibitory postsynaptic potentials are reduced, and the action of excitant amino acids depressed. Furthermore there is a reduction in the effectiveness of acetylcholine upon cholinoceptive neurones (CURTIS et al. 1959). The depressant actions upon motoneurones are not associated with an increase in membrane potential. Similar results have been reported for Betz cells (KRNJEVIĆ 1963), and, since it has proved technically impossible to restrict amino acid administration to synapses upon the dendrites of these latter cells, no proof has yet been obtained of the postulate that GABA and closely related ω-amino acids specifically block excitatory synaptic action at the axodendritic synapses of cortical neurones (PURPURA, GIRADO and GRUNDFEST 1957; PURPURA et al. 1959; see also IWAMA and JASPER 1957; BINDMAN, LIPPOLD and REDFEARN 1962). Earlier studies, in which GABA and β-alanine failed to modify the spike potentials recorded from presynaptic fibres within the spinal cord (CURTIS et al. 1959), were interpreted as indicating that depressant amino acids had no action upon these fibres. However, a more recent investigation (CURTIS and RYALL, unpublished) has shown that GABA does depress the electrical excitability of presynaptic fibres (or terminals) and it is therefore possible that these depressant amino acids have a similar action upon both pre- and postsynaptic structures.

Acidic amino acids related to glutamic acid depolarize central neurones, and in many cases this depolarization is sufficient to initiate action potentials. Some of these amino acids also depolarize spinal afferent terminals (CURTIS and RYALL, unpublished). An amino group and two acidic groups are required for excitant activity: the acidic groups may be carboxyl, sulphonic or sulphinic, and the optimal separation between the α-amino and ω-acidic group is two or three carbon atoms (CURTIS and WATKINS 1960a, 1963). One remarkable feature is that N-alkylation of D-aspartic acid produces very potent excitants. N-methyl-D-aspartic acid (WATKINS 1963) is the most powerful excitant amino acid investigated; in addition N-iminomethyl-, N-ethyl-, and N-n-propyl-D-aspartic acids are also more active than the parent compound. Another noteworthy feature concerns the relative activities of optical isomers. Whereas the L forms of aspartic, glutamic and cysteic acids are slightly more potent than the respective D forms (CURTIS and WATKINS 1960a, 1963; CRAWFORD and CURTIS 1964), the D-enantiomorphs of the N-alkyl derivatives of aspartic acid, and also of homocysteic acid, are much more active than the L forms.

It was a natural consequence of these observations to consider GABA (or β-alanine, taurine, γ-amino-β-hydroxybutyric acid) and L-glutamic acid (or L-aspartic acid), all of which have been detected in mammalian brain tissue (WAELSCH 1957; HAYASHI 1959b; but see MITOMA 1960), as important substances in the regulation of impulse transmission (see particularly ELLIOTT and JASPER 1959; ROBERTS 1960). One argument against the function of these substances as synaptically released transmitters is the comparatively high concentrations (10^{-4}—10^{-3} M) which are required to demonstrate pharmacological effects (CURTIS et al. 1960; KRNJEVIĆ and PHILLIS 1963). Although N-methyl-D-aspartic acid, D-homocysteic acid, 3-amino-1-propanesulphonic acid and other amino acids are effective in lower concentrations, there is no evidence that these very active compounds occur within nervous tissue. Furthermore, it is essential that a proposed transmitter has the same postsynaptic action as the naturally occuring transmitter with which it is to be identified, and in the case of amino acids this condition has not been satisfied.

Postsynaptic inhibition of mammalian central neurones is associated with a membrane hyperpolarization (see ECCLES 1964), and the failure of GABA and of β-alanine to hyperpolarize spinal motoneurones (CURTIS et al. 1959) indicates that these amino acids, and closely related substances, are not inhibitory transmitters within the spinal cord. Supportive evidence is provided by the observation that strychnine, which effectively abolishes spinal postsynaptic inhibition, does not modify the depression of spinal neurones by GABA. (This latter argument presupposes a postsynaptic site of action of strychnine which prevents the access of inhibitory transmitter to receptors.) The finding that GABA decreases the excitability of spinal afferent terminals does not support the proposal that this substance could be the depolarizing transmitter responsible for spinal "presynaptic" inhibition (ECCLES et al. 1963). The evidence with regard to the function of GABA as an inhibitory transmitter of cortical neurones is less definite, and it has been proposed recently that the failure of a large series of pharmacological agents, including tetanus toxin and strychnine, to influence some types of cortical inhibition and the depressant effect of GABA upon cortical neurones, indicates a possible role of GABA as an inhibitory transmitter (KRNJEVIĆ, RANDIC and STRAUGHAN 1964). Such negative evidence lends little support to this proposal, and it has yet to be demonstrated that GABA hyperpolarizes cortical neurones.

The observation that the time courses of action of D- and L-glutamic acid (and aspartic acid) were identical, thus indicating that enzymes were unlikely to be important in the removal of these substances after electrophoretic ejection, was initially considered as evidence against a transmitter function for the L enantiomorphs (CURTIS et al. 1960). However, it is uncertain whether enzymic removal is necessarily always associated with central transmitter action. In the case of spinal motoneurones further evidence against a transmitter role for L-glutamic acid was provided by the demonstration that the reversal (equilibrium) potential for the amino acid-induced depolarization differed from that

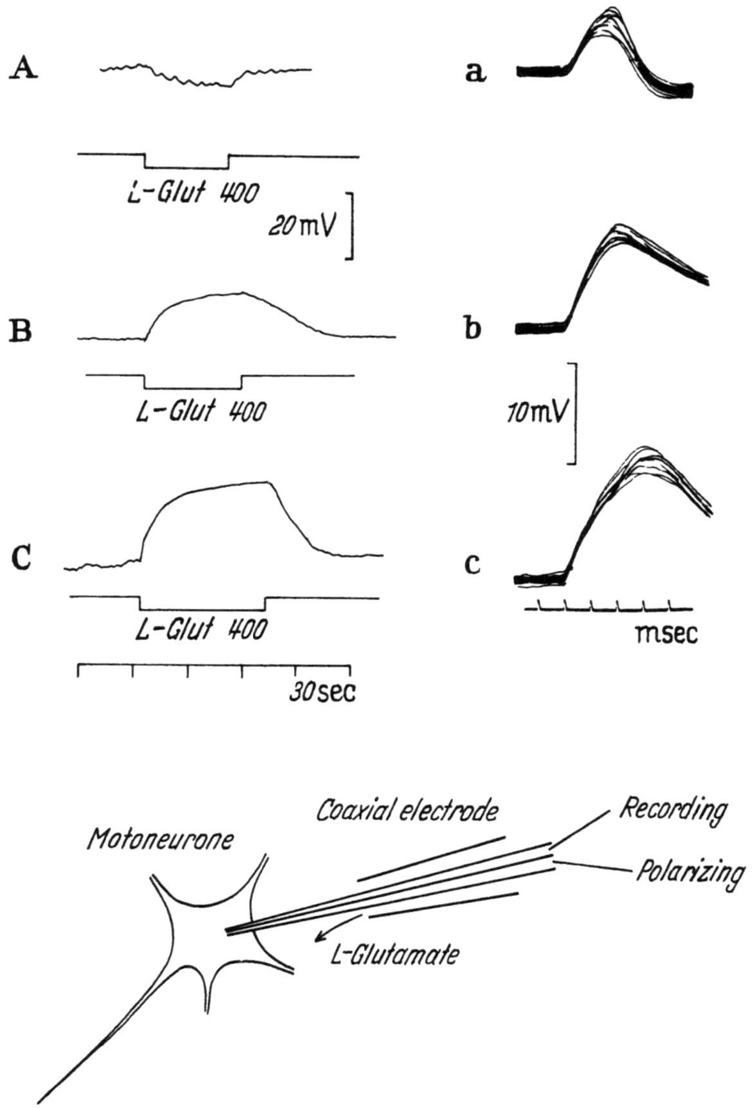

Fig. 1. Potentials recorded from a Flexor digitorum longus (FDL) motoneurone by means of one barrel of a double intracellular electrode in response to the extracellular administration of L-glutamic acid (A—C) and the electrical stimulation of the lowest threshold FDL afferent fibres (a—c). An electrophoretic current of 400 nA was used to eject the amino acid from the outer barrel of a co-axial electrode; the time course of ejection is indicated by the lower record of A, B and C and the amino acid induced potentials have been corrected for the artifact introduced by the coupling resistance between inner and outer barrels. The responses B, b were recorded at a resting potential of 58 mV; those of A, a while depolarizing current of 30 nA was passed through the other barrel of the intracellular electrode, those of C, c whilst a hyperpolarizing current of 30 nA was flowing. Calibrations: 20 mV for A—C beneath A, 10 mV for a—c beneath b; 30 sec for A—C beneath C, msec for a—c beneath c. In all records depolarization is indicated as an upward deflection

of the depolarization induced by excitatory synaptic action. Figure 1 illustrates results from an experiment in which a motoneurone was impaled by a double barrel micropipette. This electrode was the inner barrel of a co-axial (see inset diagram), the outer barrel of which contained a solution of sodium L-glutamate. The electrophoretic ejection of glutamate ion (current of 400 nA) resulted in a membrane depolarization of approximately 14 mV, at a resting potential of 58 mV (B). At this resting potential the monosynaptic excitatory postsynaptic potential (EPSP) had a magnitude of 8.5 mV (b). When a hyperpolarizing current of 30 nA was passed through one of the barrels of the double electrode, thus increasing the potential across the cell membrane, both the glutamate-induced (C-23 mV) and the excitatory postsynaptic (c) potentials were increased in magnitude. In contrast, the passage of a depolarizing current across the cell membrane (30 nA) produced a reduction in the EPSP (a) but a reversal of the potential induced by glutamic acid (A-5 mV). Such experiments have been carried out successfully on only a few cells, but in all cases an amino acid-induced hyperpolarization was demonstrated at a membrane potential at or near the reversal potential of the EPSP (COOMBS, ECCLES and FATT 1955 b). Furthermore, when large electrophoretic currents were used to eject N-methyl-D-aspartic acid near motoneurones the maximum attainable depolarization was lower than the reversal potential of the EPSP, and EPSP's superimposed upon this amino acid-depolarization were still depolarizing in nature. Although the arrangement of the cell body and dendrites of a neurone, and the distribution thereon of excitatory synapses, introduces errors into the estimation of reversal potentials by passing current through a microelectrode which is presumably located in the soma (see also BENNETT 1964), it is unlikely that the amino action in the above experiments was confined to the soma of the impaled neurone. Thus the results suggest that the reversal potential for amino acid depolarization is less positive than that for synaptic excitation, it is therefore unlikely that these two types of depolarization are produced by an identical alteration in membrane permeability. It follows that amino acid depolarization is not a consequence of the interaction between amino acid molecules and the specific transmitter receptors of excitatory synapses. Similar experiments have not been carried out upon other central neurones and, indeed, the experimental difficulties are so formidable that it will be extremely difficult to establish whether or not the action of L-glutamic acid upon Betz cells is identical with that of the as yet unidentified synaptic excitants.

In summary, the available evidence is not greatly in favour of a transmitter function for neutral and acidic amino acids within the mammalian central nervous system (CURTIS and WATKINS 1960 b). Recent findings (RYALL 1962, 1964; WEINSTEIN, ROBERTS and KAKEFUDA 1963; WHITTAKER 1964) demonstrated that unlike acetylcholine, the subcellular distributions of GABA and L-glutamic acid in mammalian brain do not indicate any special storage of these amino acids in synaptic endings. There is no evidence that the techniques of homogenization of brain in sucrose solutions, differential centrifugation and density layer separation are too drastic to preserve synaptic endings which contain these amino acids as stored transmitters, and the results suggest a distribution of these substances throughout the cellular cytoplasm, and presumably a function for amino acids which is predominantly metabolic in nature (see STRECKER 1957; BAXTER and ROBERTS 1960; HIRSCH and ROBINS 1962). The comparatively high brain content of L-glutamic acid and GABA (approximately 10^{-2} M and 2×10^{-3} M respectively if uniformly distributed; WAELSCH 1957), and the pharmacological actions of these amino acids in concentrations of the order of 10^{-3}—10^{-4} M, suggest that these and other active amino acids are stored within cells, and that the normal extracellular amino acid level is low (see also VAN HARREVELD 1959). The distribution of these substances between neurones and glial cells remains unknown, but if intracellular amino acids are concerned with the maintenance of electrolyte distributions across cell membranes, the finding that glial cells have resting potentials similar to those of neurones (HILD and TASAKI 1962) might indicate that the amino acid content of these cells is similar. If this be so, and the extracellular space within the brains is of the order of 20%, it may be calculated that a neurone of diameter 30 μ contains at least 10^{-13} mole of L-glutamic acid and 2×10^{-14} mole of GABA. If released into the surrounding extraneuronal space these amounts would be sufficient to excite or depress nearby neurones.

It may now be asked how amino acids modify neuronal activity and whether amino acid receptors have functional significance. The high intracellular amino acid content, the failure of intracellularly administered L-glutamic acid to excite motoneurones (COOMBS, ECCLES and FATT 1955 a) and the very rapid action of amino acids such as D- and L-glutamic, D- and L-aspartic, D- and L-cysteic, GABA and β-alanine indicate that these substances interact with receptors located upon the external surface of nerve cell membrane. The relationship between the structure and activity of a large series of amino acid has led to the formulation of the possible structure of such receptors (CURTIS and WATKINS 1960 a), and it has been proposed that the amino-acid receptor interaction leads to changes in the permeability of certain components of the membrane. In the case of GABA, and related depressants, this increase in permeability is presumably limited to ions which are in electrochemical equilibrium at or near the resting potential, possibly chloride ions. On the other hand the depolarization produced by excitant amino acids must be associated with an increased membrane permeability towards sodium ions, and the finding that the reversal potential for this depolarization is lower than that for synaptic excitation suggest that there is also an increased permeability to potassium and/or chloride ions. The antagonism observed between depressant and excitant amino acids (see CURTIS and WATKINS 1960 a; CRAWFORD and CURTIS 1964) is probably in part due to competition for the same receptor sites, and in part due to the shunting effect of the membrane conductance increase produced by the depressants upon the depolarization induced by the excitants.

The apparent lack of action of amino acids upon mammalian nerve fibres (MCLENNAN 1957; CURTIS et al. 1959), and upon axons in the vicinity of spinal motoneurones (CURTIS, unpublished observations) was originally taken to indicate that amino acid receptors were located in the region of synapses (see CURTIS and WATKINS 1960a). If this be the case, the effet of amino acids upon presynaptic fibres is presumably related to the presence of axo-axonic synapses upon the terminals (see ECCLES 1964). However recent findings indicate that amino acids can influence the properties of nerve fibres (SASAKI 1958; KERKUT and PRICE 1963) and it is possible that amino acid receptors may be components of the electrically excitable component of nerve membrane. Whatever the site of these receptors, however, it is unlikely that they are identical with the specific subsynaptic receptor sites with which transmitters interact.

Although it is probable that amino acids do not function as synaptically released transmitter substances (mediators) it is conceivable that these compounds could operate as modulators (see TOMAN 1963). The difficulties inherent in such a concept have been discussed previously (CURTIS 1961) and it has yet to explained how the simultaneous excitation (or depression) of a group of anatomically related, but possibly functionally dissimilar, neurones could be of significance in the operation of the nervous system. Futhermore, at the present time there is no more evidence for a function of amino acids as central modulators then there is for sodium, potassium, calcium and chloride ions. The unique occurrence of GABA in nervous tissue, the known metabolic pathways involving GABA, L-glutamic acid and glutamine, and the convulsant action of certain hydrazides which inhibit pyridoxal dependent enzymes have given undue emphasis to the possible function of GABA as a regulator of neural activity (see papers and discussions in ROBERTS 1960). A direct relationship between GABA levels and the convulsive actions of these inhibitors cannot always be demonstrated (KAMRIN and KAMRIN 1961; MAYNERT and KAJI 1962; ROBERTS 1962; MEDINA 1963) and indeed, as has been pointed out (CURTIS et al. 1960), alterations in the brain level of any one amino acid under such conditions may not be significant. Estimations of the extracellular concentration of various amino acids would be required; furthermore it cannot be determined whether measured alterations in amino acid levels cause or merely arise from disturbances of nerve cell function. Similar considerations hold for the possible involvement of amino acids in the initiation of epileptic discharges (TOWER 1960; WOLFE and ELLIOTT 1962). There is evidence that extracellular release of excitant amino acids could initiate spreading depression (VAN HARREVELD 1959; CURTIS and WATKINS 1961) but the significance of this phenomenon in the normal functioning of the mammalian cerebral cortex remains unknown.

References

ANDERSEN, P. & CURTIS, D. R. (1964). *Acta physiol. scand.* **61**, 85.
BAXTER, C. F. & ROBERTS, E. (1960). In *The Neurochemistry of Nucleotides and Amino Acids*, (ed. by BRADY, R. O. & TOWER, D. B.) New York: John Wiley.
BENNETT, M. V. L. (1964). *Ann. Rev. Physiol.* **26**, 289.
BINDMAN, L. J., LIPPOLD, O. C. J. & REDFEARN, J. W. T. (1962). *J. Physiol. (Lond.)* **162**, 105.
COOMBS, J. S., ECCLES, J. C. & FATT, P. (1955 a). *J. Physiol. (Lond.)* **130**, 291.
COOMBS, J. S., ECCLES, J. C. & FATT, P. (1955 b). *J. Physiol. (Lond.)* **130**, 374.
CRAWFORD, J. M. (1963). *Biochem. Pharmacol.*, **12**, 1443.
CRAWFORD, J. M. & CURTIS, D. R. (1964). *Brit. J. Pharmacol.* **23**, 313.
CRAWFORD, J. M., CURTIS, D. R., VOORHOEVE, P. E. & WILSON, V. J. (1963). *Nature (Lond.)* **200**, 579.
CURTIS, D. R. (1961). In: *Regional Neurochemistry* (ed. by KETY, S. S. & ELKES, J.). Oxford: Pergamon Press.
CURTIS, D. R. (1964). In *Physical Techniques in Biological Research*, vol. 5, (ed. by NASTUK, W. L.). New York: Academic Press.
CURTIS, D. R. & DAVIS, R. (1962). *Brit. J. Pharmacol.* **18**, 217.
CURTIS, D. R. & KOIZUMI, K. (1961). *J. Neurophysiol.* **24**, 80.
CURTIS, D. R. & WATKINS, J. C. (1960 a). *J. Neurochem.* **6**, 117.
CURTIS, D. R. & WATKINS, J. C. (1960 b). In *Inhibition in the Nervous System and γ-Aminobutyric Acid*, (ed. by ROBERTS, E.) p. 424. New York: Pergamon Press.
CURTIS, D. R. & WATKINS, J. C. (1961). *Nature (Lond.)* **191**, 1010.
CURTIS, D. R. & WATKINS, J. C. (1963). *J. Physiol. (Lond.)* **166**, 1.
CURTIS, D. R., PHILLIS, J. W. & WATKINS, J. C. (1959). *J. Physiol. (Lond.)* **146**, 185.
CURTIS, D. R., PHILLIS, J. W. & WATKINS, J. C. (1960). *J. Physiol. (Lond.)* **150**, 656.
ECCLES, J. C. (1964). *Physiology of Synapses.* Berlin: Springer-Verlag.
ECCLES, J. C., SCHMIDT, R. F. & WILLIS, W. D. (1963). *J. Physiol. (Lond.)* **168**, 500.
ELLIOTT, K. A. C. & JASPER, H. H. (1959). *Physiol. Rev.* **30**, 383.
VAN HARREVELD, A. (1959). *J. Neurochem.* **3**, 300.
HAYASHI, T. (1952). *Jap. J. Physiol.* **3**, 46.
HAYASHI, T. (1954). *Keio J. Med.* **3**, 183.
HAYASHI, T. (1959 a). *Neurophysiology and neurochemistry of convulsion.* Tokyo: Dainohon-Tosho Co.
HAYASHI, T. (1959 b). *J. Physiol. (Lond.)* **145**, 570.
HILD, W. & TASAKI, I. (1962). *J. Neurophysiol.* **25**, 277.
HILLMAN, H. H. & MCILWAIN, H. (1961). *J. Physiol. (Lond.)* **157**, 263.
HIRSCH, H. E. & ROBINS, E. (1962). *J. Neurochem.* **9**, 63.
HONOUR, A. J. & MCLENNAN, H. (1960). *J. Physiol. (Lond.)* **150**, 306.
IWAMA, K. & JASPER, H. H. (1957). *J. Physiol. (Lond.)* **138**, 365.
KAMRIN, R. P. & KAMRIN, A. A. (1961). *J. Neurochem.* **6**, 219.
KERKUT, G. A. & PRICE, M. A. (1963). *Life Sciences* **10**, 722.
KITA, T., KAMIYA, H. & KIYOTA, C. (1963). *Biochem. Pharmacol.* **12**, 213.
KRNJEVIĆ, K. (1963). In *Neuropsychopharmacology*, vol. 3 (ed. by BRADLEY, P. B., FLÜGEL, F. & HOCK, P.), p. 260. Amsterdam: Elsevier.
KRNJEVIĆ, K. & PHILLIS, J. W. (1963). *J. Physiol. (Lond.)* **165**, 274.
KRNJEVIĆ, K., RANDIC, M. & STRAUGHAN, D. W. *Nature (Lond.)* **201**, 1294.
KUNO, M. & MUNEOKA, A. (1962). *Jap. J. Physiol.* **12**, 397.
MCLENNAN, H. (1957). *J. Physiol. (Lond.)* **139**, 79.
MAYNERT, E. W. & KAJI, H. K. (1962). *J. Pharmacol. exp. Ther.* **137**, 114.
MEDINA, M. A. (1963). *J. Pharmacol. exp. Ther.* **140**, 133.
MITOMA, C. (1960). In *Inhibition in the Nervous System and γ-Aminobutyric acid*, (ed. by ROBERTS, E.) p. 236. New York: Pergamon.

Noell, W. K. (1960). Quoted by Rosen, F. In *Inhibition in the Nervous System and γ-amino-butyric acid,* (ed. by. Roberts, E.) p. 482. New York: Pergamon Press.
Purpura, D. P., Girado, M. & Grundfest, H. (1957). *Science* **125**, 1200.
Purpura, D. P., Girado, M., Smith T. G., Callan, D. A. & Grundfest, H. (1959). *J. Neurochem.* **3**, 238.
Roberts, E. (1960). *Inhibition in the nervous system and γ-aminobutyric acid.* New York: Pergamon Press.
Roberts, E. (1962). In *Neurochemistry,* (ed. by Elliott, K. A. C., Page, I. H. & Quastel, J. H.). Springfield, Illinois: C. C. Thomas.
Ryall, R. W. (1962). *Nature, (Lond.)* **196**, 680.
Ryall, R. W. (1964). *J. Neurochem.* **11**, 131.
Salmoiraghi, G. C., Bloom, F. E. & Costa, E. (1964). *Science* — in press.
Sasaki, S. (1958). *Medical Science (Japan)* **9**, 28.
Stefanis, C. (1964). *Fred. Proc.* — in press.
Strecker, H. J. (1957). In *Metabolism of the Nervous System,* (ed. by Richter, D.), p. 459. London: Pergamon.
Toman, J. E. P. (1963). *Ann. Rev. Pharmacol.* **3**, 153.
Tower, D. B. (1960). *Neurochemistry of Epilepsy.* Springfield, Illinois: C. C. Thomas.
Von Baumgarten, R., Bloom, F. E., Oliver, A. P. & Salmoiraghi, G. C. (1963). *Pflügers Archiv ges. Physiol.* **277**, 125.
Waelsch, H. (1957). In *Metabolism of the Nervous System,* (ed. by Richter, D.), p. 431. London: Pergamon.
Watkins, J. C. (1963). *J. med. pharm. Chem.* 5, 1187.
Weinstein, H., Roberts, E. & Kakefuda, T. (1963). *Biochem. Pharmacol.* **12**, 503.
Whittaker, V. P. (1964). In *Progress in Brain Research,* vol. 8. *Biogenic Amines* (ed. by Himwich, H. E. & Himwich, W. A.) p. 90. Amsterdam: Elsevier.
Wolfe, L. S. & Elliott, K. A. C. (1962). In *Neurochemistry,* (ed. by Elliott, K. A. C., Page, I. H. & Quastel, J. H.). Springfield, Illinois: C. C. Thomas.

AN EXAMPLE OF SENSORY INTEGRATION OF THE STRETCH REFLEX

by D. Denny-Brown

The enormous increase in knowledge of the processes of synaptic excitation and inhibition in the last ten years, attributable in so large a part to the stimulus and contributions of Sir John Eccles, provides ever widening insight into the extraordinary complexity of reflex organization. In order to study the part played by the different types of inhibitory effect in any particular part of motor behavior some relatively simple reflex motor performance that is automatically performed and readily reproducible is necessary. The flexion reflex and the reflex of crossed extension serve well enough for study of the interaction of excitation and inhibition, but are not sufficiently specific to enable more than very general conclusions regarding the organization of the special stimulus situations in other than the most simple motor acts. The scratch reflex, studied so intensively by SHERRINGTON, is difficult to restrict to an isolated group of neurones. Reflex stepping, with an ubiquitous inborn rhythm, has a large element of the unknown, and its most specific precipitation, the extensor thrust, requires an optimal recovery of reflex status seldom attained except in the chronic spinal dog.

We therefore invite attention to a very specific response in the cat with hemisection of the spinal cord, first observed by us in nine experiments in collaboration with Dr. LUIS SAENZ-ARROYO (DENNY-BROWN, TWITCHELL and SAENZ-ARROYO 1949). We have observed it since in two experiments assisted by Dr. G. RUSHWORTH, and have discussed it briefly in two recent communications (DENNY-BROWN 1962b, 1964).

After hemisection of the spinal cord at high cervical level (C3, C4) in the cat the limbs on the side of hemisection make a remarkable degree of recovery in the subsequent 3 to 6 weeks. Early spastic paralysis is soon modified by the appearance of positive supporting reactions and scratch reflex. The natural tactile avoiding reaction is at first abolished, and all reflex function is biased toward extensor effects, that we have described in some detail elsewhere (DENNY-BROWN 1964). Between the 9th and 24th day there appears an interesting reaction in the paws on the side of hemisection. A firm stroke by the finger of the examiner passing from the posterior foot-pad onto the pads of the toes results in a brief dorsiflexion of the toes with protrusion of the claws, followed by a powerful plantar flexion of the toes onto the stimulating finger. We call this the "clawing response". It appears in either fore- or hind limb, and is facilitated by holding or supporting the animal so that the limb being tested is uppermost. The lowermost limb (foot of the animal if held vertically head uppermost, forepaws if held head down) tends to give a positive proprioceptive supporting reaction instead of a clawing response. Once it has appeared the clawing response continues to be elicitible for the remaining period of survival of the animal. It can be used in attack or defense by the animal, for the hemisected limbs can ultimately be used in projected types of movement. By this time some withdrawal of the limb, associated with suppression of the clawing response, resulted from any heavy contact stimulation of the dorsum of the paw. This return of the avoiding response never became as facile as on the normal side.

With the appearance of this clawing response and the positive supporting reactions the initial spasticity of the hemisected limbs lessens, though it does not entirely disappear. Nevertheless decerebration of the animal at intercollicular level, at any time after the first week of spinal hemisection, produces the usual decerebrate rigidity of approximately equal intensity in the hemisected and opposite limbs.

Of special importance to our present discussion, the clawing response persists after decerebration, together with the scratch reflex. Neither reaction is then so facile as before decerebration, but nevertheless their response is unmistakable. This is the more remarkable because the paws of the normal decerebrate preparation are most unreactive. The decerebrate clawing response then provides, not only evidence of a coordinated proprioceptive subthalamic reaction to a very specific cutaneous stimulus combination, but in a form available for many types of recording that are not yet possible in the intact animal.

We would like to cite one type of simple experiment that we have found to be readily repeatable. A black female cat was subjected to hemisection of the spinal cord on the left side between 3 and 4 cervical segments on July 14, 1948, by Dr. Saenz-Arroyo and myself, under pentobarbital anesthesia. We always use a needle, blunted at the end and sharpened on one side, to make the hemisection without cutting the pia, so as to avoid disturbing the blood supply to the cord. The completeness of the lesion was subsequently verified histologically. The day after the hemisection the left limbs were flaccid, with large slow tendon reflexes. On the second day after operation the left limbs were spastic with brisk tendon reflexes and ample flexion response to pin prick. By the 5th day positive supporting had appeared in the left hind limb, with righting from lying on the left side. By the 8th day the hind limb showed spreading and additional extension when the animal was held suspended and dropped. The clawing response first appeared in the left hind foot on the 9th day, and in the left fore foot on the 10th day. On the 12th day the animal was decerebrated by a spatula, after marking a trephine opening under ether anesthesia. After 34 minutes there was

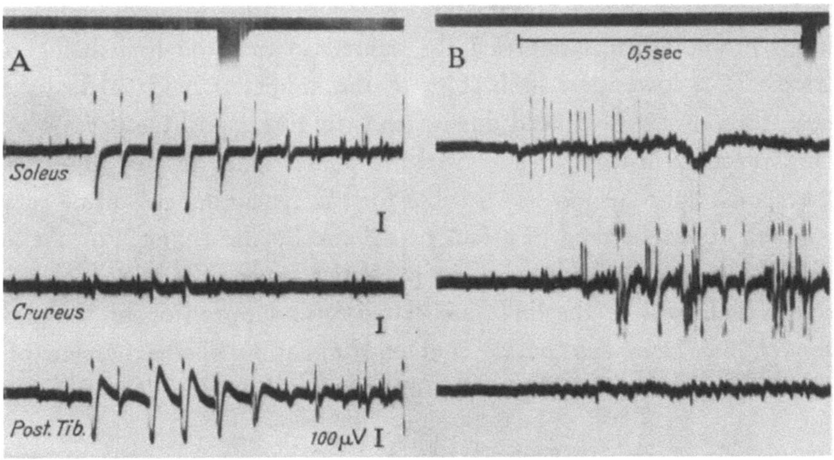

Fig. 1. Electromyograms of soleus, crureus, and posterior tibial muscles in the left hind limb to show, in A a clonic ankle jerk from tap on tendo achillis, and B the initiation of a positive proprioceptive supporting reaction in the left crureus by dorsiflexing the toes of the left foot. Cat with left hemisection of spinal cord at C 3—4 level, decerebrated 12 days later

strong decerebrate rigidity, comparable in degree in all limbs. Many EMG records were made of stretch reflexes with concentric needle electrodes. The reflexes in the left limbs were more clonic than those in the right, but were otherwise very similar. The clonic tendon reflex at the ankle on the left side is shown in

Fig. 1 A, compared with the onset of a positive supporting reaction in the same muscles in Fig. 1 B. The clawing response was recorded in the long flexors of the left fore paw (Fig. 2 A). The burst of contraction as the claws flexed was not as intense as in the intact animal, but the positive flexion contraction was unmistakable to the stimulating fingers of the examiner.

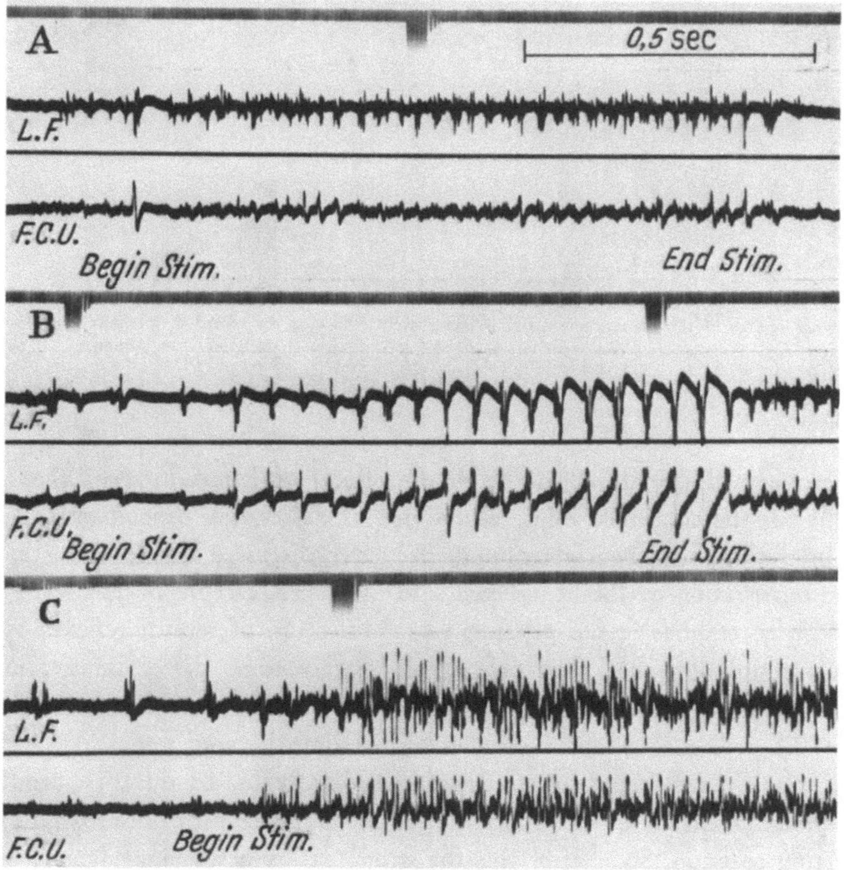

Fig. 2. Cat, decerebrate, 12 days after hemisection as in figure 1. Three sets of electromyograms of long flexors of the digits (LF), and flexor carpi ulnaris (FCU), in the left forelimb. In A a pressure stimulus on the posterior foot pad moves onto the pads of the toes, eliciting a "clawing response". In B is shown the effect of the same stimulus after section of the nerves to the dorsum of the foot. In C all nerves to the foot have been sectioned above the wrist

The dorsal branches of the ulnar and radial nerve supplying the skin of the dorsum of the paw were then divided. The clawing response became immediately very brisk and clonic (Fig. 2 B). Tendon reflexes in the long flexors of the digits, which had been brisk before this nerve section (Fig. 3 A) then become also very clonic (Fig. 3 B).

The next step was to section both ulnar and median nerves just above the wrist. The effect was dramatic. The clawing response disappeared, and was replaced by a simple stretch reflex in the long flexors (Fig. 2 C) without protrusion of the claws. The tendon reflex of the long flexors became small, but associated with a large reduplicated response in triceps muscle, not previously present (Fig. 3 C), representative of the spread of the positive supporting reaction.

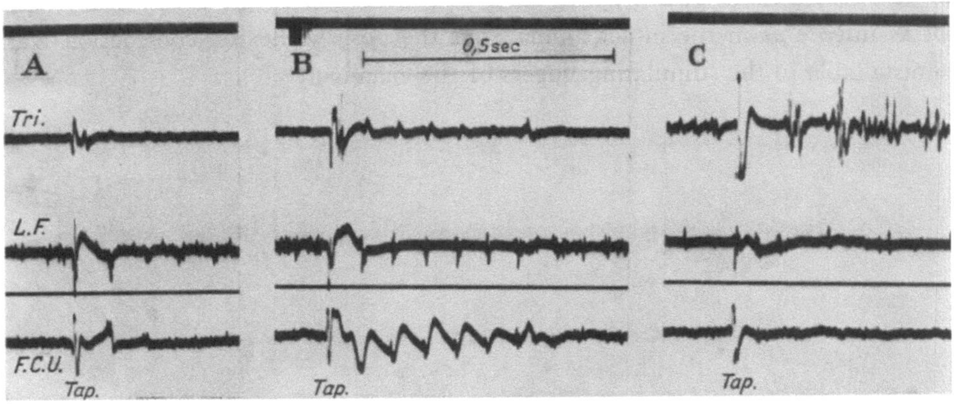

Fig. 3. Companion figures to figure 2, including EMG of triceps brachii, to show A, a tendon reflex elicited by a tap on under surface of the digits ("finger jerk") immediately following the record A in figure 2. B shows the effect of the same stimulus after section of nerves to the dorsum, and C the same after section of all nerves to the fore foot

Discussion

The recovery of spinal reflexes from the effects of hemisection of the spinal cord reveals the continued bilateral activity of descending reticulo-spinal pathways in the opposite ventral column. We have discussed the resulting bias of gamma innervation on the hemisected side elsewhere (DENNY-BROWN 1964) and here wish to emphasize the progressive modification of stretch reflexes by cutaneous stimulation that then occurs. Ultimately some direct innervation of alpha neurones appears, but this also need not concern us here. The simple spasticity of the proximal extensor muscles of the limb became somewhat less, but was then easily facilitated by stretch of the long flexors of the toes, producing the general extension of the limb that is seen as the proprioceptive positive supporting reaction. Soon after this the stretch reflex of the long flexors of the toes becomes itself modified. It is facilitated by holding the animal in space with the test limb dependent. The effect is principally labyrinthine, assisted by neck reflexes. When the test limb is uppermost the positive supporting reaction is lessened, and the long flexors of the digits no longer react as briskly to stretch. The effect is such that their stretch reflex is continuously suppressed until an adequate moving tactile stimulus triggers them in intense form, resulting in clawing. When complete the clawing reaction includes active flexion of the limb. In the hemisected monkey it becomes the grasp reaction.

The first phase of the experiment we have cited here is of particular interest in that it reveals that the skin to the dorsum of the foot, though not being stimulated, nevertheless is contributing to the suppression of the stretch reflex of the long flexors. We have to predicate a continuous effect coming from skin, just as SHERRINGTON's early description (1893) of the enhancement of the knee jerk by simple section of the nerve to the hamstrings revealed that there had been a *continuous* influx from the reciprocal antagonist muscles. The last stage of the experiment showed that the integrity of the afferent endings in palmar skin likewise contributed a continuous suppressor effect on the stretch reflex as well as the potential one of contactual trigger.

In this way the full appreciation of reflex patterns requires an estimate of the tonic inhibitory effect of receptors from broad sensory fields as well as the specific drive resulting from focal receptors. There must be a widespread "polarization" of field effects in the spinal grey matter produced by random impulses in non-specific skin receptors. Tactile coarse contactual effects from the body surface become of first importance in the regulation of body posture by the higher levels of the nervous system, which we have recently discussed in relation to the mechanism of dystonia (Denny-Brown 1962 a). The type of observation reported here indicates that they also play a considerable part in determining the "set" of the spinal reflex mechanism. It is in this area that we suspect that presynaptic inhibition, so brilliantly illuminated by the recent work of Sir John Eccles and his co-workers, must play a large part.

References

Denny-Brown, D. (1962 a). *The Basal Ganglia and their Relation to Disorders of Movement*, London: Oxford University Press.

Denny-Brown, D. (1962 b). *Proc. roy. Soc. Med.* **55**, 527.

Denny-Brown, D. (1964). *The Cerebral Control of Movement* (The Sherrington Lectures, 1963). Univ. Liverpool Press (in the press).

Denny-Brown, D., Twitchell, T. E. & Saenz-Arroyo, L. (1949). *Trans. Amer. Neurol. Ass.* **74**, 108.

Sherrington, C. S. (1893). *Proc. roy. Soc.* **52**, 556.

VISCEROTOMES AND DERMATOMES: SOME COMPARISONS OF THE INNER AND OUTER SURFACES OF THE BODY

by C. B. B. Downman

"*... for the most vital parts are not the quickest of sense.*" *Francis Bacon* (1612)

"*If it should be objected that there are nerves in these viscera, my answer is, I do not alledge that they are void of all sensation, but only that it is very weak in them, viz. such as one would expect in a part which has very few nerves bestowed upon it in proportion to its bulk.*" *Allbrecht von Haller* (1755)

Bolk divided each dorsal root (DR) field into dermatome, myotome, splanchnotome and sclerotome. It is intended here to comment particularly upon the DR sensory fields of the inner surface of the body.

Dermatomes have been mapped by dissection (BOLK 1897), residual sensitivity (SHERRINGTON 1893, 1898; FOERSTER 1933), irritative lesions (HEAD 1893), antidromic vasodilation (FOERSTER 1933), and DR block (KEEGAN 1943, 1948). While the descriptions of limb dermatomes vary, there is good agreement about the fields of thoracic and upper lumbar DR. Adjacent dermatomes overlap by some 50 per cent, a place on the skin being connected to the cord via two or three DR. Mappings based on pain sensibility give narrower fields, with little overlap; any place on the skin connects through these fibres to one spinal root. Recent electrophysiological maps have been based on direct recordings of the activity in dorsal rootlets (KUHN 1953; HEKMATPANAH 1961; SUMI 1963), and confirm the older tactile maps. They show also that the dermatomes of individual rootlets each form a narrow but continuous segment of the area, with progressive overlapping of the fields of adjacent rootlets. The required spread of rootlet fibres into many cutaneous terminals depends upon plexiform arrangements within the thoracic and lumbar nerve trunks (DEMPSHER and SPRAGUE 1948).

The DR values of the intestinal sensory fields (viscerotomes, splanchnotomes of BOLK) can be defined by two approaches. Segmental entry can be projected as seen from a place on the intestine; or entry can be limited to one segment and the extent of the residual sensitivity mapped, using pupil dilation in the chloralosed cat as reflex index of volley entry. This latter method combines the techniques of SHERRINGTON and of McSWINEY. Table 1 summarises results

Table. 1. *Distribution of spinal segmental entries of intestinal afferent fibres, shown by pupil dilation in chloralosed cats* (HAZARIKA, COOTE & DOWNMAN, 1964). *Details of method in original paper*

Vagi cut. Phr. = phrenic nerves; St. = stomach; Duo. = duodenum; Jej. = jejunum; Caec. = caecum and ileocolic sphincter: G. B. = gall bladder; Pancr. = pancreas.

		Thoracic												Lumbar			
	Phr.	2	3	4	5	6	7	8	9	10	11	12	13	1	2	3	4
Oesoph			+	+	+	+	+	+	+	+	+	+					
St. body			+	+	+	+	+	+	+	+	+	+					
antrum			+	+	+	+	+	+	+	+	+	+					
Duo.			+	+	+	+	+	+	+	+	+	+					
Jej.						±	+	+	+	+	+	+	+	+	+		
Ileum						±	+	+	+	+	+	+	+	+	+		
Caec.						±	+	+	+	+	+	+	+	+	+	+	+
Colon:																	
upper										+	+	+	+	+	+	+	+
mid.																+	+
lower																+	+
G. B. & ducts	+	+	+	+	+	+	+	+	+	+							
Pancr.	+	+	+	+	+	+	+	+	+	+							

obtained in cats by such means (HAZARIKA, COOTE and DOWNMAN 1964). It shows that the DR fields are extensive and that adjacent fields overlap by up to 100 per cent. Any place on the intestine is associated with DR entry into nine or more segments. In Table 2 the results of other authors, based generally on progressive root sections, are collected. It is noted that the entries based on reflexogenic activity in acute animal experiments are much wider than those based on pain sensation in man. This reminds one of the differences in size of tactile and pain fields of the dermatome. There is not sufficient information available to discuss the relation between the reflexogenic and the pain afferents, since the methods applicable to animals and to man cannot be the same.

Table 2. *Spinal cord entry of afferent fibres from gastro-intestinal tract according to various authors. Values of* HEAD (1893), VON GAZA (1924), *and* WHITE & SMITHWICK *refer to pain sensation in man. Values of other authors refer to reflexogenic afferents in cat* (MCSWINEY & SUFFOLK 1938; UNKAUF 1939; WATANABE 1954; HAZARIKA et al. 1964) *or dog* (KAPPIS 1913; LEBEDENKO & BRJUSSOWA 1930). *Root entries based on observations in man are narrower than those based on the animal experiments.*

Abbreviations as for Table 1.

	Oes.	Stom.	Duo.	Jej.	Ileum	Caec.	Colon.	G. B.	Pancr.
HEAD (1893)	T_{5-8}	T_{7-9}	T_{9-12}	T_{9-12}	T_{9-12}	T_{9-12}		T_{8-9}	
VON GAZA (1924)	T_{5-6}	T_{6-9}	T_{9-10}	T_{9-10}	T_{9-10}		T_{12}—L_2	T_{7-8}	
KAPPIS (1913)		T_{6-9}	T_{7-13}	T_{7-13}	T_{7-13}				
LEBEDENKO & BRJUSSOWA (1930)		T_5—L_4	T_{5-12}	T_{6-11}	T_{6-11}	T_{11-12}			
MCSWINEY & SUFFOLK (1938)		T—L_2	T_6—L_1	T_{7-13}					
UNKAUF (1939)							T_7—L_2		
WHITE, SMITHWICK & SIMONE (1952)		T_{7-8}	T_{9-10}	T_{9-10}	T_{9-10}		T_{12}—L_1	T_{7-8}	
WATANABE (1954)			T_3—L_1		T_5—L_1				
HAZARIKA, COOTE & DOWNMAN (1964)	T_{3-12}	T_{3-12}	T_{3-12}	T_6—L_2	T_6—L_2	T_6—L_2	T_{10}—L_4	T_{2-10} Phr.	T_{2-10} Phr.

The DR viscerotomes do not spread equally along the intestine (Table 1). $T_{3-4-5-6}$ DR spread and overlap in the region corresponding approximately to the fore-gut; T_7—L_2 DR inclusive innervate fore- and mid-gut; while L_{3-4} DR innervate hind-gut. While the pattern of the spreading overlap remains unexplained, one suspects that it is in part determined by the embryonic moulding and final formation of the intestinal arteries (HAMILTON, BOYD and MOSSMAN 1962). Beyond the splanchnic nerve trunks, when the nerve bundles associate with intestinal arteries, the pattern of spread is limited. It can be predicted by the arterial pattern, and the overlap of adjacent intestinal fields associated with mesenteric nerves is only 20% or so (DOWNMAN 1952). This has been confirmed by fibre degeneration (SCHOFIELD 1960) and by direct recording of evoked afferent volleys in mesenteric and splanchnic nerves (HAZARIKA 1962, unpublished). Because of the wide branching of afferent fibres (NIIJIMA 1961) distant regions of the intestine connect with the same dorsal root pair. The extensive overlap of the viscerotomes is determined as far back as the splanchnic nerves, and may depend upon some plexiform arrangement of fibres within the nerve, similar to the arrangement within the somatic nerves.

We find that electrophysiological recording of dorsal root activity is complicated by the great noise of body wall afferent inflow. This obscures any activity of visceral origin, but it can be considerably reduced by cutting the intercostal nerve distal to the rami communicantes and seeking and cutting all other branches to the body wall, including somatic afferent pathways passing in the sympathetic chains and white rami (DOWNMAN and HAZARIKA 1962). It is then found that intestinal afferents are present in some dorsal rootlets but in others no evoked activity can be recorded, e. g. two or three out of seven rootlets of a whole DR. This contrasts with cutaneous afferents which are found in every rootlet. It is an expression of the relatively small number of afferent fibres from the viscera. FOLEY (1948) counted 14 000 axons in each splanchnic nerve. Of 3000 myelinated fibres, 1300 are afferent and of the 11 000 unmyelinated fibres 6500 are probably afferent. GRIFFIN, GRIFFIN and PATTON (1963) also found around 1300 myelinated afferent fibres in each splanchnic nerve. If there are branchings within the splanchnic nerves themselves, the number of parent afferent fibres may be less. Using the data of GRIFFIN et al. (1963), of AGOSTINI, CHINNOCK, DALY and MURRAY (1957) and of CRAGG and EVANS (1960) one assesses that the total of abdominal visceral afferents via vagi, splanchnics and pelvic nerves is some 55 000 fibres together with an unknown number of pelvic C fibres. Of this, the myelinated fibres total 400 in the vagi, 2600 in the splanchnics, and 3600 in the pelvic nerves. This can be compared with 10 500 myelinated fibres of T_2 DR (HJIANG 1950) or an estimated total of 1 268 000 fibres in all DR (INGBERT 1903).

These afferents are spread into a large peripheral area. WEBER (1964, unpublished) has measured the area of the serous surface of the whole gastrointestinal tract of a cat of 2.3 kg. The measured areas were: stomach 110.6 cm^2, duodenum 41.7 cm^2, jejunum and ileum 321.7 cm^2, colon and rectum 112.3 cm^2. The 2600 splanchnic afferent fibres innervate an area of stomach, intestine and upper third of colon of at least 511 cm^2. The weight: surface area formula of GREAVES (1957) gives the total skin area as 2200 cm^2. The relatively small number of visceral afferent fibres innervates a visceral area equivalent to at least a quarter of the body surface, and the histological studies of SCHOFIELD (1960), RICHARDSON (1958), IN MIN LEE (1956), OTSU (1953) and others show extensive ramifications of both myelinated and unmyelinated afferent fibres within the area.

The intestinal DR sensory field is formed, therefore, by an extensive branching of relatively few parent afferent fibres. The great overlap, up to 100 %, of the maximum fields exceeds what is found in skin. This spread depends upon frequent branchings in parallel with the vascular tree. This has been shown for the larger fibres and it remains to be determined whether the branching is as wide for all classes of afferent fibres, including the C fibres.

Hazarika's finding (1963, unpublished), that a short length of small intestine may have spots on the serosa sensitive to light pressure while a neighbouring length does not, suggests that there are differences in the extents of fibre branchings. This would form a basis for the differences of DR entries given by the clinicians and the physiologist. In spite of the few fibres of entry, afferent volleys from any place on the intestine will enter the central nervous system not only via vagi or pelvic nerves but also into thoraco-lumbar cord via many DR. Originally a consequence of embryonic shifts and mouldings, the multiple entry may ensure widespread evocation of prepotent reflexes and spatial summation in relaying nuclei.

References

Agostini, E., Chinnock, J. E., Daly, M. de B. & Murray, J. G. (1957). *J. Physiol.* (Lond.) **135**, 182.

Bolk, L. (1897). *Nederl. Tijdsch. v. Geneesk.*, Amst. 2 R. 23. ch. 1., 982.

Cragg, B. G. & Evans, D. H. L. (1960). *Exp. Neurol.* **2**, 1.

Dempsher, J. & Sprague, J. M. (1948). *Anat. Rec.* **102**, 195.

Downman, C. B. B. (1952). *J. Physiol.* (Lond.) **116**, 228.

Downman, C. B. B. & Hazarika, N. H. (1962). *J. Physiol.* (Lond.) **163**, 340.

Foerster, O. (1933). *Brain* **56**, 1.

Foley, J. O. (1948). *Anat. Rec.* **100**, 766.

Griffin, A. R., Griffin, J. T. & Patton, H. D. (1963). *Brain and Behaviour,* Vol. 2: *The Internal Environment and Alimentary Behaviour.* (ed. by Brazier). p. 192. Washington: Am. Inst. Biol. Sci.

Greaves, J. P. (1957). *J. Physiol.* (Lond.), **138**, 43—44 P.

Hamilton, W. J., Boyd, J. D. & Mossman, H. W. (1962). *Human Embryology,* 3rd ed. p. 186 et seq. Cambridge: Heffer.

Hazarika, N. H., Coote, J. & Downman, C. B. B. (1964). *J. Neurophysiol.* **27**, 107.

Head, H. (1893). *Brain* **16**, 1.

Hekmatpanah, J. (1961). *J. Neurophysiol.* **24**, 129.

Hjiang, S. H. (1950). *Acta. Anat.* **11**, 50.

In Min Lee (1956). *Arch. Jap. Chir.* **25**, 241.

Ingbert, C. *J. Comp. Neurol.* **13**, 53.

Kappis, M. (1913). *Mitt. Grenzgeb. Med. Chir.* 26, 493.

Keegan, J. J. (1943). *Arch. Neurol. Psychiatr., Chicago.* **50**, 67.

Keegan, J. J. & Garrett, F. D. (1948). *Anat. Rec.* **102**, 409.

Kuhn, R. A. (1953). *J. Neurophysiol.* **16**, 169.

Lebedenko, W. & Brjussowa, S. (1930). *Z. ges. exp. Med.* **71**, 198.

McSwiney, B. A. & Suffolk, S. F. (1938). *J. Physiol. (Lond.),* **93**, 104.

Niijima, A. (1961). *Jap. J. Physiol.* **11**, 427.

Otsu, A. (1953). *Acta. Sch. Med. Univ. Kioto. Jap.* **31**, 103.

Richardson, K. C. (1958). *Amer. J. Anat.* **103**, 99.

Schofield, G. C. (1960). *Brain* **83**, 490.

Sherrington, C. S. (1893). *Phil. Trans.* **184**, B. 641.

Sherrington, C. S. (1898). *Phil. Trans.* **190**, B. 64.

Sumi, T. (1963). *J. Neurophysiol.* **26**, 478.

Unkauf, B. M. (1939). Ph. D. Thesis. Univ. of London.

von Gaza, W. (1924). *Arch. klin. Chir.* **133**, 479.

Watanabe, Y. (1954). *Arch. g. Jap. Chir.* **23**, 38.

White, J. C., Smithwick, R. G. & Simeone, F. A. (1952). *Autonomic Nervous System.* 3rd ed. p. 136. London: Kimpton.

EXTERNAL ACTION CURRENTS AT CUT NERVE END

by Dexter M. Easton

In the usual observation of action potentials during external recording from nerve, the cut end of the nerve is expected to serve only as a stable reference, and its contribution to the record is intended to be insignificant. In fact, ordinary methods of recording tend to discourage any concern with the cut end, which is usually crushed, depolarized or otherwise maltreated.

On the other hand, the behavior of the artifactual nerve end has fascinating possibilities as a model for events at synaptic regions, (Eccles 1946). ("Artificial" is an inappropriate term here, since the severed nerve is a real nerve. The cut end is, however, artifactual, in that it is "a consequence of human intervention".)

Studies of the electrical signals from whole nerve have most often been carried out with nerves suspended on wires and exposed to air or mineral oil. That method is simple in execution, but sometimes liable to be traumatic to the preparation and subject to artifact and misinterpretation. The sucrose gap technique provides large signals, but subjects parts of the nerve to an abnormal, ion-free environment.

This paper explores some of the possibilities of a fluid electrode method of external recording from nerve. The nerve remains immersed in the appropriate physiological solution and is not subject to the deterioration accompanying exposure to oil or air or anisotonic medium. Instead it is drawn into a glass tube, and the voltage drop due to longitudinal current flowing in this tube is measured. The method is recommended for the experimental and interpretive simplicity it provides in a routine procedure of recording from whole nerve. It is particularly suitable for studying events at the cut end of a nerve.

The method is an adaptation of the suction electrode method often used in recording from heart muscle. A similar method has occasionally been used for stimulating and recording from nerve (see e. g. Larrabee 1961, Skoglund 1952).

Methods

As experimental material, sciatic nerves were taken from bullfrogs, *Rana catesbeiana*, or, in a few experiments, from leopard frogs, *Rana pipiens*. The nerve was carefully cleaned of tissue remnants, at the regions of stimulation and recording, but the sheath was left intact, aside from the cut nerve branches.

The stimuli were rectangular pulses 0.05 msec in duration, applied through an RF stimulus isolation unit proximally to the common root that yielded the largest response in the common unbranched sciatic trunk, distally.

Recording as well as stimulation was made with the aid of a suction electrode (Easton 1960, 1961, 1962) and a system of barriers. By means of the suction electrode, each end of the nerve could be drawn any desired distance into a glass tube filled with frog Ringer solution. Alternatively, a glass tube was inserted into the barrier, and the nerve drawn through the tube.

The electrode wires were silver, chlorided as necessary. The wire inside the tube containing the nerve-end was positive for stimulation. The other lead was a silver chlorided wire in another tube dipping into the Ringer bath. There was a corresponding arrangement of recording leads. Electrical activity was displayed on a cathode ray oscilloscope after DC amplification.

The schemes in Figure 1 compare the usual wire with this fluid electrode technique.

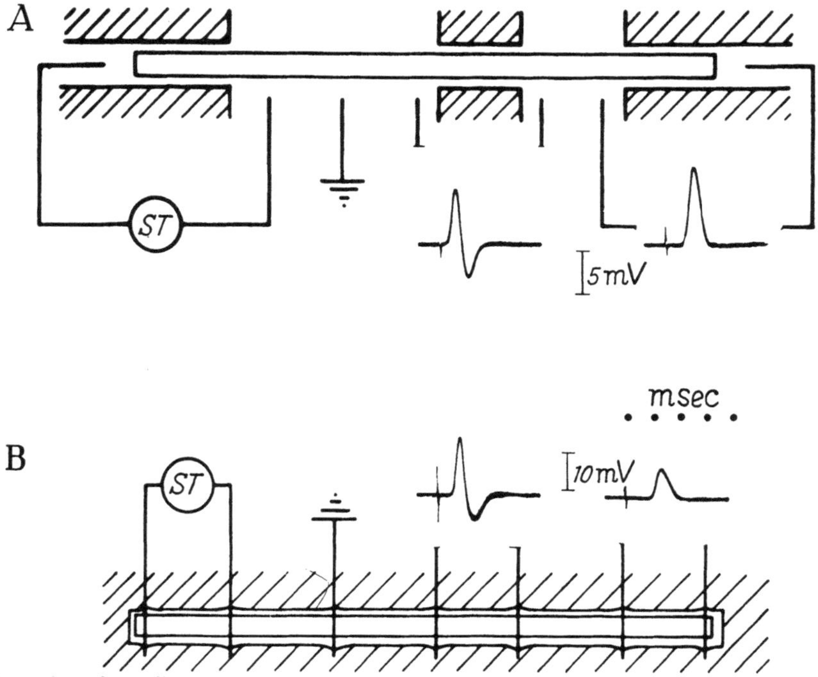

Fig. 1. Comparison of recording arrangements

A: Fluid electrode arrangement: nerve immersed in Ringer solution and pulled into tubes at each end, records *en passage* and terminating at cut end of nerve

B: Same nerve, suspended on Ag-AgCl wires, in air. Distance between wires same as lengths of tubes. Calibration: above, 5 mv., below, 10 mv. time marks 1 msec

In both situations, the recorded potential drop occurs through the high resistance of a small volume of conducting fluid along a short distance of nerve, between the electrodes. In the fluid electrode arrangement, however, there is less likelihood of artifact due to unintentional local changes in longitudinal resistance along the nerve (see e. g. Marmont 1940). The amplitude of the record depends on the snugness of fit of the nerve in the tube, and is ordinarily somewhat larger in the conventional method. Impulse velocity, as one would expect, is greater when there is a large external volume of Ringer solution.

When the nerve is drawn into a tube, the electrode within the tube is quasi-intracellular, since it looks at the cut end of the nerve. For this reason, this electrode will be considered the active electrode. This convention is consistent with that used in intracellular recording, and allows the record to be upward positive. More generally, it follows that in the observations reported here, the first of a pair of electrodes encountered by the impulses is the reference electrode. This sign convention is opposite to the usual practice in external recording, but upward positive records are in conformity with intracellular recording and with general physical practice. The convention assured that injury currents were observed downward negative, while the initial phase of an action current was upward positive. It also facilitates comparison between the two types of recording. To emphasize the role of longitudinal currents, the externally recorded nerve potentials are here described as *currents.*

Injury currents

The potential drop due to current flowing between the upstream nerve membranes and the cut end surfaces of the nerve fibers was measured as follows: The nerve was drawn into the tube the required distance, and the recording electrodes were shorted to the bath on one side of the barrier, by means of a stopcock connected in one fluid lead. Any electrode potential was balanced out at this time. The stopcock was then turned so that the electrodes were connected to opposite sides of the barrier, and the resulting deflection was due to the injury current. The DC balance of the pre-amplifier was frequently checked.

The injury current was maximal immediately after the cut had been made, and it decreased with time thereafter. When zero potential is registered under these conditions, no injury current flows and the nerve is said to be *healed.* No implication is intended by this term as to mechanism. It is a term of convenience used also by others who have observed the decrease of injury potential with time after injury (LORENTE DE NÓ 1947).

Nerves of adequate length and recently dissected, were able to heal even after several successive transections. The current was less likely to decline to zero in a short length of nerve, and in older preparations, a day or more after dissection, depending upon the temperature. In old nerves, healing, in terms of decline of injury current, was incomplete. Healing was also slowed by prolonged stimulation of the nerve, or by letting the nerve remain in the tube for many hours.

The decrease in potential difference that takes place with the passage of time can be explained by assuming that the resistance between core and outside at the cut surface increases with time after the section. When the cut is made, injury current flows to the cut end, but as time passes, the electrode sees into the interior of the fiber less effectively and the potential drop observed becomes progressively smaller. It has been suggested that core polarization develops owing to injury current (GASSER 1960). It is not unreasonable to assume that a polarization of the terminal membrane develops along with an increase in resistance. This region would then tend to share the properties of the normal membrane.

The magnitude of injury current depended upon the length of nerve in the tube. In a long uniform tube, the potential increased according to the length of nerve drawn into the tube (see Figure 2). In this instance, the cut-end potential was compared with the potential of points along the nerve at increasing distances from the cut end. When the nerve end was pulled to extend beyond the other end of the tube, the potential decreased, for in that instance the potential difference between two points along the nerve, away from the cut end, was determined.

Both the change in action current configuration (to be described) and the decline of the injury current were prevented if the nerve-end was left undisturbed for many hours in a closed tube in which it fitted moderately snugly.

Action currents

Immediately after transection of the nerve at the recording end, the action current record was monophasic positive, observed at the end of the nerve in a uniform tube. This is called a *cut-end record.* As injury current decreased with passage of time after the cut, the action current record became diphasic. The initial (positive) phase remained of constant amplitude, but the second (negative) phase increased to reach a maximum when injury current had been reduced to zero. Such a record is called a *healed-end record.* The second phase corresponds to the diphasic artifact (LORENTE DE NÓ 1947). The initial positive limb will be called the leading current, the later negative phase the *trailing* current.

The configuration of the potential change depended upon several factors: for example, the form of the tube, whether it was a narrow barrier or a long uniform tube, the elapsed time since the nerve was severed, the length of nerve in the tube, and the length of time the nerve was left in the tube. Some effects of these variables are shown in Figure 2.

Other factors such as total length of nerve, temperature, nutrition, season, Ringer composition, are significant in certain circumstances, but have not been examined systematically. Stimulus intensity and the state of the nerve end were of primary importance. The stimulus intensity was selected to be maximal for the fastest group of fibers. During stimulation, the record of the healed end was monitored while the intensity was increased until the second phase of the diphasic record was maximal. If the stimulus was excessive, the diphasic record could be made to appear monophasic as the slower fibers appeared in the trough of the second phase of the record. With passage of time after dissection, and after the manipulation involved in several passages in and out of a snug fitting tube, the capacity of the nerve end to undergo healing diminished, and the record was more likely to remain monophasic.

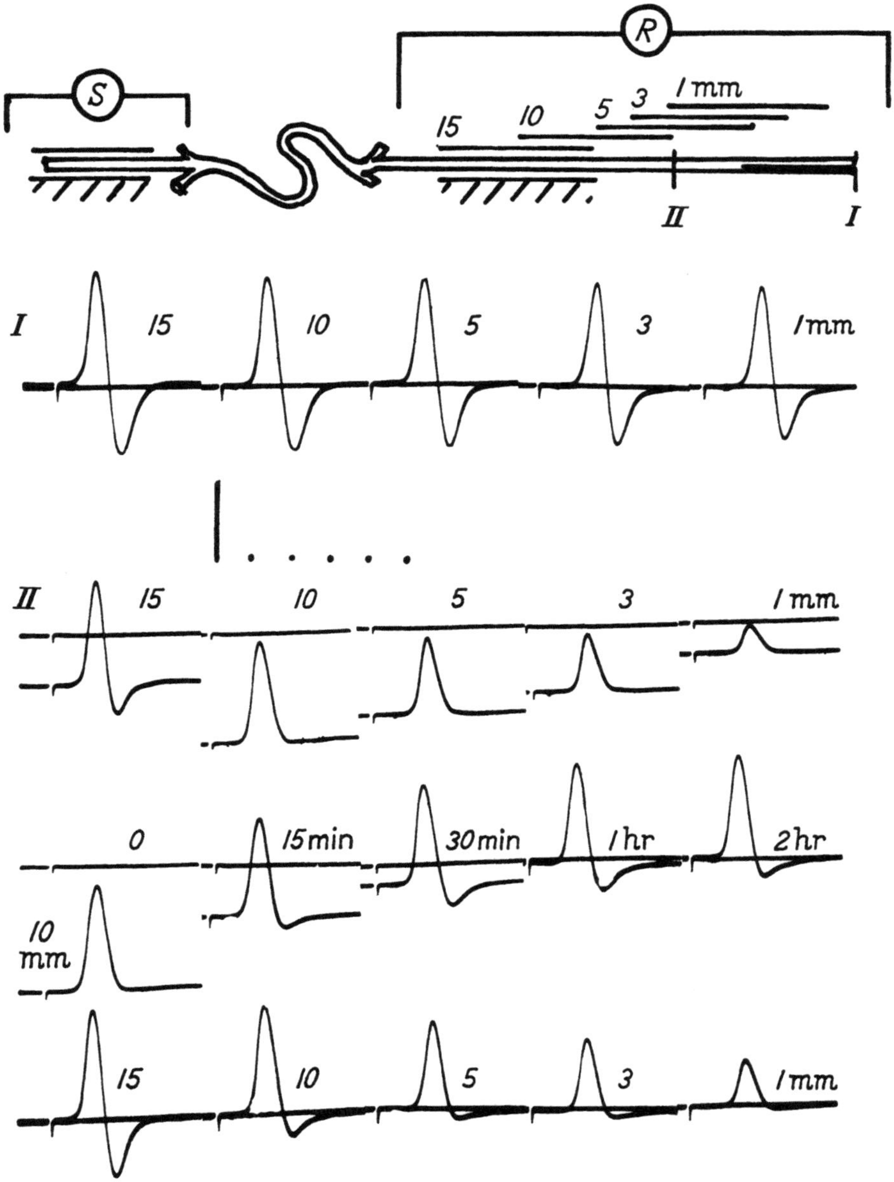

Fig. 2. Fluid electrode recordings from nerve end. Diagram shows length of distal end of nerve (relative to point II), drawn into recording tube which was 10 mm long. Row I, records when healed nerve terminated at level I in the diagram. Row II, corresponding records immediately after nerve sectioned at level II, followed by records showing time course of healing recorded at 10 mm., i. e., the distal end of cut nerve was flush with the end of the tube. Final row shows the action current configuration at distances along the end healed 2 hours. Horizontal lines: zero level for injury potential. All records from one nerve. Stimulus intensity constant throughout. Calibration: 5 mv, 1 msec

Interpretation

The records from a narrow barrier of about 1 mm. are the simplest to interpret (Figure 3).

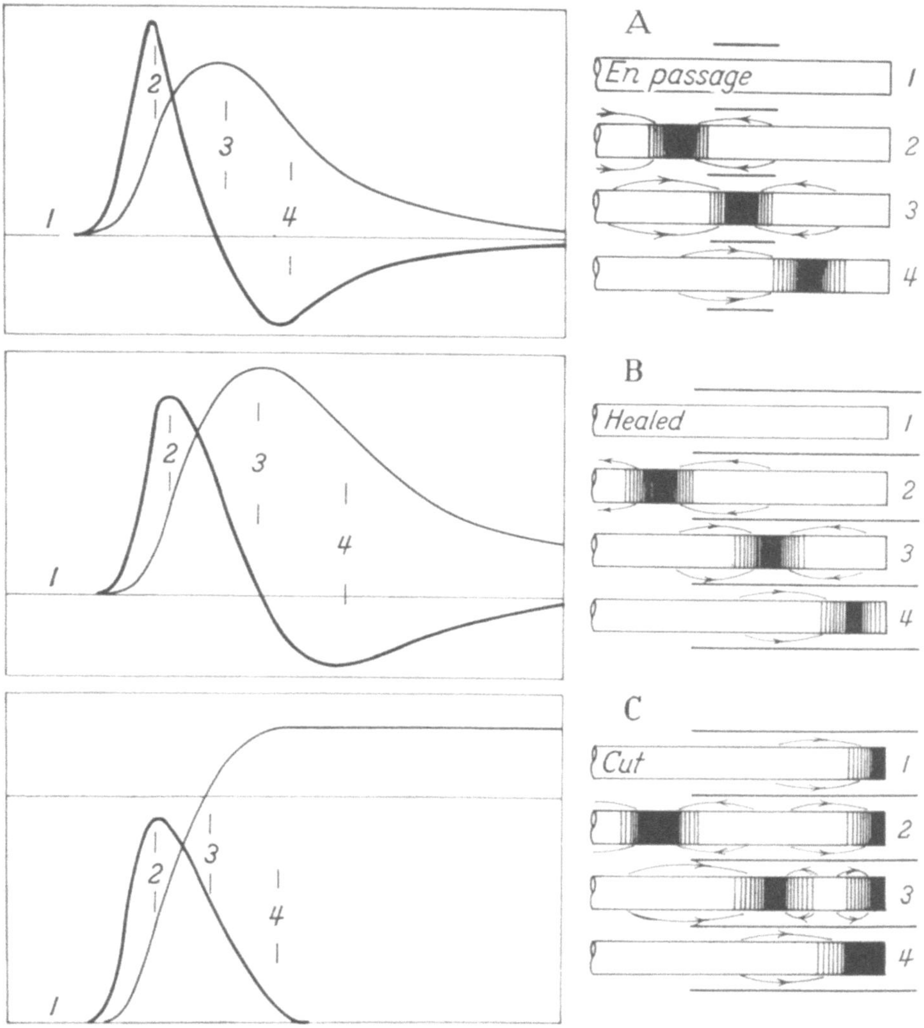

Fig. 3. Action currents at nerve end
At left: Action current record (heavy line) and superimposed integral (light line) in each instance
At right: Spatial distribution of action currents as impulse moves along region of nerve in tube. Shaded area indicates center of impulses, or depolarization at nerve end. Number indicate corresponding points during the time (at left) and in the spatial distribution (at right) of the action currents
A en passage, *B* healed end, *C* freshly cut end

Leads were connected so that a positive signal on side *B* provided an upward deflection of the cathode ray beam. Therefore the initial positive deflection indicated a flow of current from *B* to *A*. The current flowing through the aperture surrounding the nerve is directly proportional to the longitudinal potential gradient along the nerve within the aperture. The positive gradient maximum is indicated by the positive peak of the action current record. The decline of the gradient to zero indicates that the center of the sink of the longitudinal surface potential is passing through the aperture at such a point that the voltage drop due to longitudinal current at the trailing edge of the impulses comes to be equal, while opposite in direction, to the voltage drop from the longitudinal current at the leading edge of the impulses. The trough that the negative deflection reaches indicates the maximum longitudinal potential gradient at the trailing edge of the impulses.

There is a longitudinal surface potential distribution corresponding to the longitudinal current, and the time course of this potential corresponds to the integral of the longitudinal voltage drop with respect to time (KATZ and SCHMITT 1939).

The integral of the action current record with respect to time is also a statement of the quantity of charge moved as longitudinal current. The return of this integral to zero means that the charge moved in one direction is completely compensated by an equal quantity of charge moved in the opposite direction. Since the longitudinal current completes the external circuit for the membrane current, it follows that the net charge transferred in one direction through the tube equals the net charge transferred in the other direction. The diphasic record may be considered to be the vector sum of two longitudinal currents, leading and trailing with respect to the moving center of the impulse.

The record at a healed end within a long tube is also diphasic, for the net terminating current flows toward the end electrode, as it does in the *en passage* record.

When the end of the nerve has been freshly cut, the cut end serves as a sink drawing current from the nearby membrane. Impulses advancing toward the nerve end compete with the cut-end potential for current. Current at the leading edge of the impulses provides a voltage drop opposite in sign from that of the steady current, and the record is driven in a positive direction toward zero, or beyond, away from the steady negative potential level determined by the injury current.

The impulses die short of the end as they encounter the cut end sink and consequently the length of the depolarized zone is increased beyond that present prior to the invasion. However, as this zone extends back toward the tube entry, the net longitudinal gradient between the two ends becomes less. The record therefore moves down, without undershoot, toward the preexisting level, as the responding membrane repolarizes and the cut-end current reasserts itself. When a greater length of nerve is drawn into the tube, there may be an undershoot, reflecting the greater gradient then made possible by the trailing currents.

The tube recording method may also be used to observe after-potentials. The technique is also useful for recording from some muscles, and it is effective when applied to nerve ends or roots *in situ*. A special modification of the method is currently being used to provide continuous perfusion of the end so that the effect of changes in the ambient medium on the nerve end potential configuration may be evaluated.

If the artifactual nerve end is treated kindly and brought under good control, it may in the future serve as an even better model for normal nerve end behavior than it has heretofore.

Acknowledgement

Research supported in part by the National Institutes of Health and the National Science Foundation, U.S.A., the Research Council and Institute of Molecular Biophysics, the Florida State University.

References

EASTON, D. M. (1960). *Science* **132**, 1312.
EASTON, D. M. (1961). *Fed. Proc.* **20**, 347.
EASTON, D. M. (1962). *XXII. Int. Physiol. Congr. Proc.* II, 777.
ECCLES, J. C. (1946). *Ann. N. Y. Acad. Sci.* **47**, 429.
GASSER, H. S. (1960). *J. gen. Physiol.* **43**, 927.
KATZ, B. & SCHMITT, O. H. (1939). *J. Physiol.* (Lond.) **97**, 471.
LARRABEE, M. G. (1961). In: *Biophysics of Physiological and Pharmacological Actions* p. 199. Washington, D. C.: American Association for Advancement of Science.
LORENTE DE NÓ, R. (1947). *Stud. Rockefeller Inst.* **131**, 1.
MARMONT, G. (1940). *Amer. J. Physiol.* **130**, 392.
SKOGLUND, C. R. (1952). *Cold Spr. Harb. Symp. quant. Biol.* **17**, 233.

INTERNEURONES ACTIVATED BY HIGHER THRESHOLD GROUP I MUSCLE AFFERENTS

by ROSAMOND M. ECCLES

Scattered throughout the grey matter of the spinal cord are small neurones which are generally termed interneurones (HUNT and KUNO 1959). Some are found only in definite areas or nuclei. For example, the second order neurones with the lowest threshold in the cutaneous pathway from the skin occur very superficially in the dorsal horn of the spinal cord of the cat (COOMBS, CURTIS and

Landgren 1956; Eccles, Eccles and Lundberg 1960; Eccles, Kostyuk and Schmidt 1962). More ventrally lies the intermediate nucleus where many fibres from muscle and tendon receptor organs terminate (Eccles, Fatt, Landgren and Winsbury 1954; Coombs et al. 1956; Eccles et al. 1960). The interneurones on the axon collateral pathway from motoneurones may be found only in the ventral horn and are designated 'Renshaw cells' (Eccles, Fatt and Koketsu 1954). Apart from these there are numerous interneurones which are stimulated by high threshold muscle and skin, that is, by flexor reflex afferents (FRA) (Eccles et al. 1960). There seems to be no pattern in their distribution. Obviously many of these FRA interneurones are on polysynaptic pathways concerned in reflex actions onto motoneurones (Eccles and Lundberg 1959); in presynaptic inhibition (Eccles et al. 1962); in descending or ascending pathways linking the higher centres with external stimuli and effector organs (Holmqvist and Lundberg 1959).

Throughout this study on the interneurones in the intermediate nucleus, electrical stimulation of the muscle nerves was employed rather than adequate activation of the muscle receptors (Kolmodin 1957). In the intermediate nucleus there are two main groups of interneurones; one group activated monosynaptically or disynaptically by volleys in the group Ia muscle fibres, and the other by the group Ib. These two groups were designated as types A and B respectively in Eccles et al. (1960). It was further suggested, that since the interneurones could be involved in either excitation or inhibition, one might think of each type (e. g., type A) being subdivided into AE or AI indicating an excitatory or inhibitory action.

It was noticed that the potentials on the cord dorsum generated by interneuronal activity, following stimulation of a muscle nerve, varied according to the nerve stimulated. The maxima for quadriceps occurred in the fifth and sixth lumbar (L) segments; the maxima for the peroneal and deep peroneal muscles in the lower L6 to upper L7 region; for the plantaris, flexor longus digitorum and gastrocnemius lower L6 to mid L7, and L7 and S1 for the posterior biceps-semitendinosus (Eccles et al. 1954; Coombs et al. 1956; Eccles, Schmidt and Willis 1963).

Since there seemed to be a distribution of fields in the rostral-caudal direction similar to the level of entry of the related nerves in the dorsal roots (Eccles et al. 1954; Eccles et al. 1963), the dorsal-ventral distribution of interneuronal fields was investigated. In an acute spinal cat under nembutal anaesthesia the extracellularly recorded potentials evoked by group I volleys in several muscle nerves were compared at different depths. A sodium-chloride filled microelectrode (of low resistance about 1 MΩ) was inserted to a depth of 3 mm from the surface of the cord at the upper L7 region and as this electrode was carefully withdrawn records were taken at fixed depths (Coombs et al. 1956). The potential field produced by a maximum group I volley in the posterior biceps-semitendinosus nerve (PBST) had a maximum at about 2.5 mm (Fig. 1). The maxima of the fields produced by group I volleys in the gastrocnemius (GS), the plantaris (PL), and the flexor digitorum longus (FDHL) nerves were in turn slightly more superficial, the latter being located at 2.35 mm. On further withdrawal, the response to the sural nerve, reversed and reached a maximum at a depth of about 1.8 mm from the surface. Finally, as the microelectrode was withdrawn to more super-

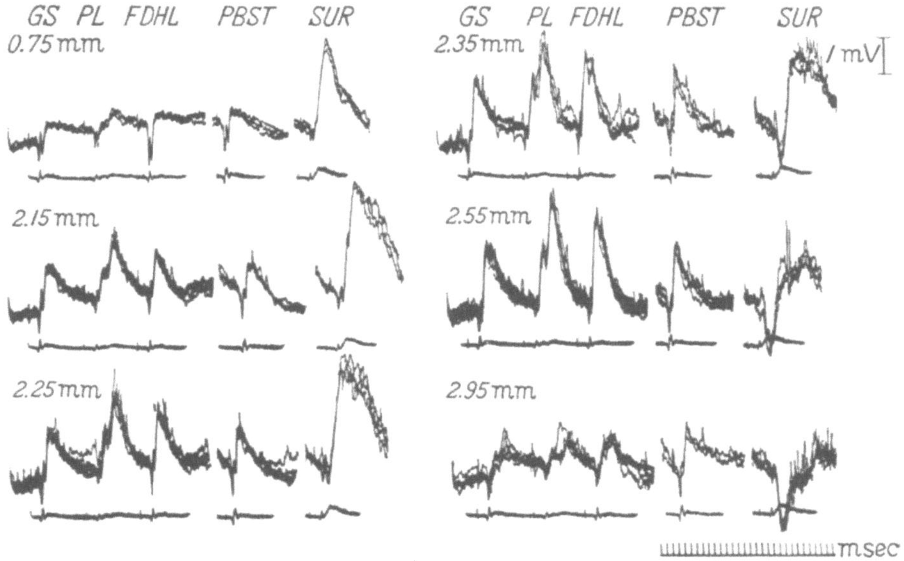

Fig. 1. A NaCl filled microelectrode was inserted into the dorsal surface of the spinal cord at a lateral angle so as to pass through the intermediate nucleus lying in the base of the dorsal horn. The microelectrode was withdrawn to the indicated depths and the potential fields (upper traces) were recorded in response to maximum group I volleys in gastrocnemius (GS), plantaris (PL), flexor digitorum brevis and longus (FDHL), posterior biceps-semitendinosus (PBST) and sural (Sur) nerves. Each record is composed of 3 or more superimposed traces. The lower traces are the cord dorsum potentials recorded by another electrode placed at the entry level of the seventh lumbar dorsal root. The potential scale applies only to the field potentials

ficial layers even the sural potential field was diminished (cf. 2.15 mm and 0.75 mm, Fig. 1). It seems as if the relationship of the group I interneuronal fields to each other is similar to the well-known patterns for the motor nuclei, both dorso-ventrally and rostral-caudally (Romanes 1951), though a certain overlap does occur in the intermediate nucleus. It should be pointed out that this distribution has been investigated only for the ankle flexors and extensors and knee flexors.

Having shown that these effects occur when maximal group I volleys were employed, it was necessary to analyse in detail the potential fields evoked by different strengths of stimulus to assess the contributions from the lowest threshold fibres of the annulo-spinal endings (Ia) and the higher threshold group I of the Golgi tendon organs (Ib) (Eccles, Eccles and Lundberg 1957 a; Laporte and Bessou 1957). In Fig. 2 the maximum potential evoked by stimulating the nerves of flexor digitorum brevis and longus (FDHL) was recorded at a depth of 2.38 mm. A volley which was nearly maximal for group Ia fibres (1.28 times threshold for the most excitable fibres) evoked a potential and when the stimulus strength was increased to 1.40 (Fig. 2), a small potential from the few Ib fibres which were activated was added to the Ia component (second arrow). If the strength of stimulus was raised to 1.60 the potential generated by the Ia com-

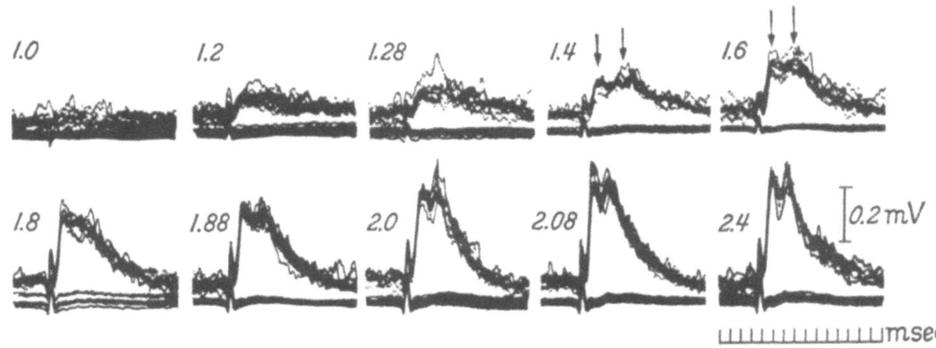

Fig. 2. The upper traces are the potential fields evoked by stimulation of the FDHL nerve and recorded at a depth of 2.38 mm from the cord dorsum where the intermediate nucleus potential field for FDHL attained a maximum. The lower traces are the records from the cord dorsum of the entering afferent volleys. Different strengths of stimulation (given on each response) were applied to the FDHL nerve. All records are composed of ten or more traces. Time scale, msec for all records. The potential scale applies only to the extracellular fields. The first arrows on the responses 1.4 and 1.6 indicate the potential fields generated by activity in group Ia fibres, the second arrows the potential field from active Ib fibres

ponent may be greater since it is difficult to judge if this was maximum at 1.40 (cf. the first arrows on 1.40 and 1.6 responses). However, there is no doubt that stimuli greater than 1.60 times threshold, which would activate most of the Ib afferent fibres, caused a large increase in the second potential up to 2.60 times threshold. The choice of FDHL for this figure was unfortunate since it has been shown that impulses in the FDHL nerve often stimulate Ia interneurones disynaptically (Eccles et al. 1960), though the very large increases when FDHL Ib fibres were stimulated indicate that disynaptic activity in group Ia fibres hardly interfered with the Ib potential fields. Graded stimulation of plantaris, gastrocnemius or posterior biceps nerves evoked similar but smaller potential fields in this track. In general it was shown that activity in Ia fibres could account for less than 10% of the activity evoked by group I volleys in the intermediate nucleus.

In Fig. 3 a composite figure shows the responses of two interneurones which were fired by only group Ib volleys. These extracellular records were obtained from the same preparation. The FDHL interneurone was located at a depth of 2.3 mm from the cord dorsum, and the plantaris at a depth of 2.45 mm. Both cells were fired only once; the FDHL interneurone by a volley 1.4 times threshold for the most excitable fibres, and the plantaris by a volley 1.45 times threshold. Volleys in other muscle or skin nerves even at thirty times threshold failed to stimulate these cells, with the exception of the plantaris interneurone (Fig. 3 B) which was fired by a volley in high threshold skin afferents (Sur).

It was found that in this series of three animals, forty-one out of forty-five neurones investigated in the intermediate nucleus could be classified as Ib interneurones. The criteria were based on the depth below the dorsal surface of the cord (1.9 to 2.5 mm), and the intensity of stimulus required to evoke a spike (from 1.3 to 2.5 times threshold of the nerve). The remaining four interneurones were activated by volleys in the lowest threshold fibres (i. e., they belonged to the A group of Eccles et al. 1960).

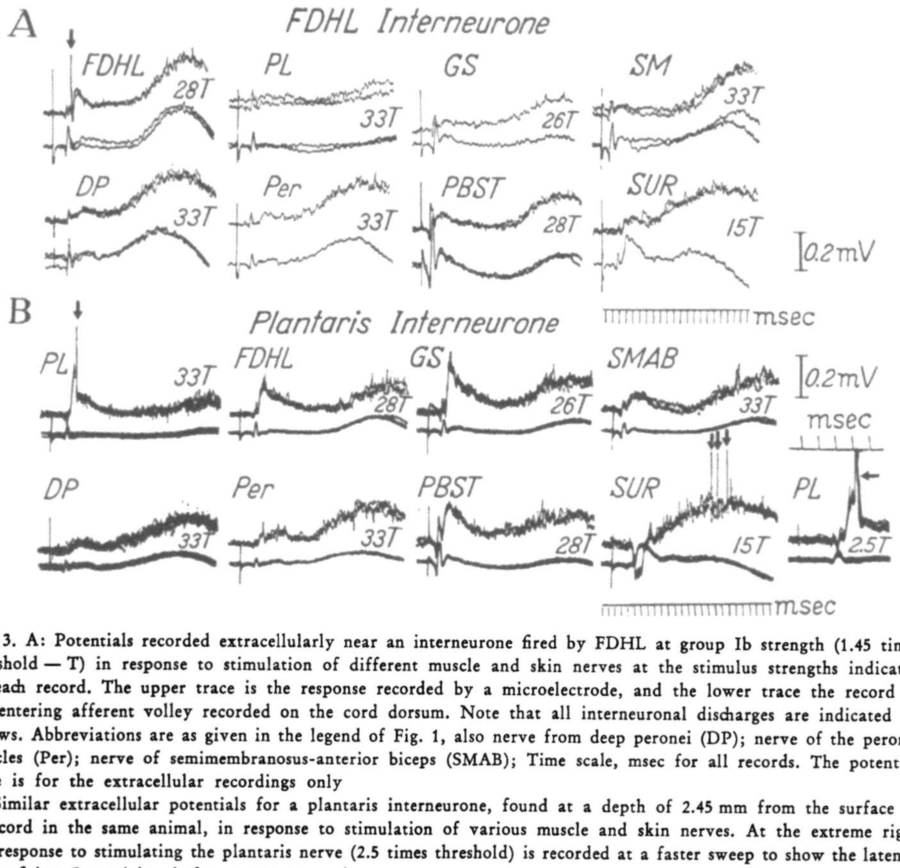

Fig. 3. A: Potentials recorded extracellularly near an interneurone fired by FDHL at group Ib strength (1.45 times threshold — T) in response to stimulation of different muscle and skin nerves at the stimulus strengths indicated on each record. The upper trace is the response recorded by a microelectrode, and the lower trace the record of the entering afferent volley recorded on the cord dorsum. Note that all interneuronal discharges are indicated by arrows. Abbreviations are as given in the legend of Fig. 1, also nerve from deep peronei (DP); nerve of the peronei muscles (Per); nerve of semimembranosus-anterior biceps (SMAB); Time scale, msec for all records. The potential scale is for the extracellular recordings only
B: Similar extracellular potentials for a plantaris interneurone, found at a depth of 2.45 mm from the surface of the cord in the same animal, in response to stimulation of various muscle and skin nerves. At the extreme right the response to stimulating the plantaris nerve (2.5 times threshold) is recorded at a faster sweep to show the latency of the firing. Potential scale for upper traces only

When the forty-one Ib interneurones were examined it was found that twenty-six were activated only once by a volley in Ib fibres of one muscle nerve, and perhaps its synergic muscle nerves. Thus these cells were in the same category as the FDHL interneurone of Fig. 3 A. Of the remainder, seven received weak excitation from flexor reflex afferents (FRA) as did the plantaris interneurone of Fig. 3 B, and eight were strongly excited by FRA. There was no sign of axons in the contralateral ventro-lateral quadrant for the twenty interneurones investigated.

In summary, these Ib interneurones in the intermediate nucleus showed a pattern of distribution not unlike that recorded for motoneurones (Eccles,

ECCLES and LUNDBERG 1957 b), the FDHL being more superficial and rostral than the PBST with the plantaris and gastrocnemius in between. The circumscribed activation of these Ib interneurones described above raises the question that perhaps these interneurones are excited by Ib fibres arising from Golgi tendon organs in only one portion of a particular muscle, and not from other areas.

Acknowledgement

The author wishes to thank Dr. M. S. DEVANANDAN for this assistance with the experiments.

References

COOMBS, J. S., CURTIS, D. R. & LANDGREN, S. (1956). *J. Neurophysiol.* **19**, 452.
ECCLES, J. C., ECCLES, R. M. & LUNDBERG, A. (1957 a). *J. Physiol. (Lond.)* **136**, 527.
ECCLES, J. C., ECCLES, R. M. & LUNDBERG, A. (1957 b). *J. Physiol. (Lond.)* **137**, 22.
ECCLES, J. C., ECCLES, R. M., & LUNDBERG, A. (1960). *J. Physiol. (Lond.)* **154**, 89.
ECCLES, J. C., FATT, P. & KOKETSU, K. (1954). *J. Physiol. (Lond.)* **126**, 524.
ECCLES, J. C., FATT, P., LANDGREN, S. & WINSBURY, G. J. (1954). *J. Physiol. (Lond.)* **125**, 590.
ECCLES, J. C., KOSTYUK, P. G. & SCHMIDT, R. F. (1962). *J. Physiol. (Lond.)* **161**, 237.
ECCLES, J. C., SCHMIDT, R. F. & WILLIS, W. D. (1963). *J. Neurophysiol.* **26**, 1.
ECCLES, R. M. & LUNDBERG, A. (1959). *Arch. ital. Biol.* **97**, 199.
HOLMQVIST, B. & LUNDBERG, A. (1959). *Arch. ital. Biol.* **97**, 340.
HUNT, C. C. & KUNO, M. (1959). *J. Physiol. (Lond.)* **147**, 346.
KOLMODIN, G. M. (1957). *Acta physiol. Scand.* **40**, suppl. 139, 1.
LAPORTE, Y. & BESSOU, P. (1957). *J. Physiol. (Paris)* **49**, 252.
ROMANES, G. J. (1951). *J. comp. Neurol.* **94**, 313.

ELECTRICAL IMPEDANCE OF STRIATED MUSCLE AND ITS RELATION TO CONTRACTION

by G. FALK and P. FATT

In an early attempt to determine the linear electrical properties of the frog striated muscle fibre, BOZLER and COLE (1935) and COLE and CURTIS (1936) obtained a capacitance between the inside and outside of 1 to 2 $\mu F/cm^2$ on the basis of measurements made at frequencies around 10 kc/s. From recent measurements carried out in a similar way to that of COLE and his co-workers, it appears that this estimate should be revised upward to about 2.5 $\mu F/cm^2$ (FATT 1964).

Using a different method, KATZ (1948) carried out measurements on the frog muscle fibre which would emphasize the behaviour at very much lower frequencies, around 10 c/s, and he obtained for the characteristics of the current path between inside and outside a capacitance of about 6 $\mu F/cm^2$ in parallel with a resistance of about 3000 $\Omega\,cm^2$. These values were confirmed in subsequent measurements in which intracellular electrodes were used (FATT and KATZ 1951). The value computed for the capacitance per unit area of muscle fibre surface seemed unreasonably high for a dielectric membrane, and the possibility was raised that this might be a result of folding of the surface membrane increasing the effective area. When, however, frog fibres were examined under the electron microscope, no such folding was evident.

Following this work on the frog muscle fibre, the linear properties of crustacean limb muscle fibres were examined with intracellular electrodes. In the crayfish the capacitance was found to be about 20 $\mu F/cm^2$ (FATT and GINSBORG 1958) and in some crabs about 40 $\mu F/cm^2$ (FATT and KATZ 1953 a). There appeared to be a correlation between muscle fibre diameter and capacitance per unit area of fibre surface, so the idea suggested itself that the large values of capacitance might arise in channels that penetrate the fibre from the surface, with a large part of the measured capacitance existing between the lumen of such channels and the myoplasm. Further interest attaches to this conception of the muscle fibre as it could help to account for the influence of the membrane potential on contraction, for it could well be that it is the potential difference across the capacitance of the walls of these channels that controls the contractile state of the myofibrils, which are closely approached by such channels.

One way in which one might hope to distinguish that part of the capacitance which is distributed along channels from that which is directly across the membrane at the surface of the fibre is by the possible presence of a resistance in series with one of these capacitances but not the other. The existance of such a combination of impedance elements could be revealed by carrying out measurements over a wide range of frequencies; the discrepancy between the determination of capacitance at 10 c/s and 10 kc/s as already described is a strong indication that this approach would yield significant results. We have therefore carried out measurements with microelectrodes to determine the impedance between a point within the muscle fibre and the surrounding extracellular fluid in the frequency range of 1 c/s to 10 kc/s. The method by which these measurements were carried out and the procedure followed in their analysis have been described in detail (FALK and FATT 1964).

Fig. 1 shows observations on a muscle fibre from the frog sartorius in *A* and on a muscle fibre from the walking leg of a crayfish in *B*. The characteristic impedance of the fibre, as observed between a point within the fibre (distant from the ends) and the outside, is here represented in the complex-plane plot, the abscissa being the resistive part of the impedance and the ordinate the reactive part. The filled circles give the experimental observations corrected for the effects of stray capacitance. If the current path between the inside and outside of the fibre consisted simply of a capacitance C_m and resistance R_m in parallel (circuit I of Fig. 1), the behaviour predicted for the characteristic impedance of the fibre would be the curve given by the dot-dashed line. As can be seen, there is

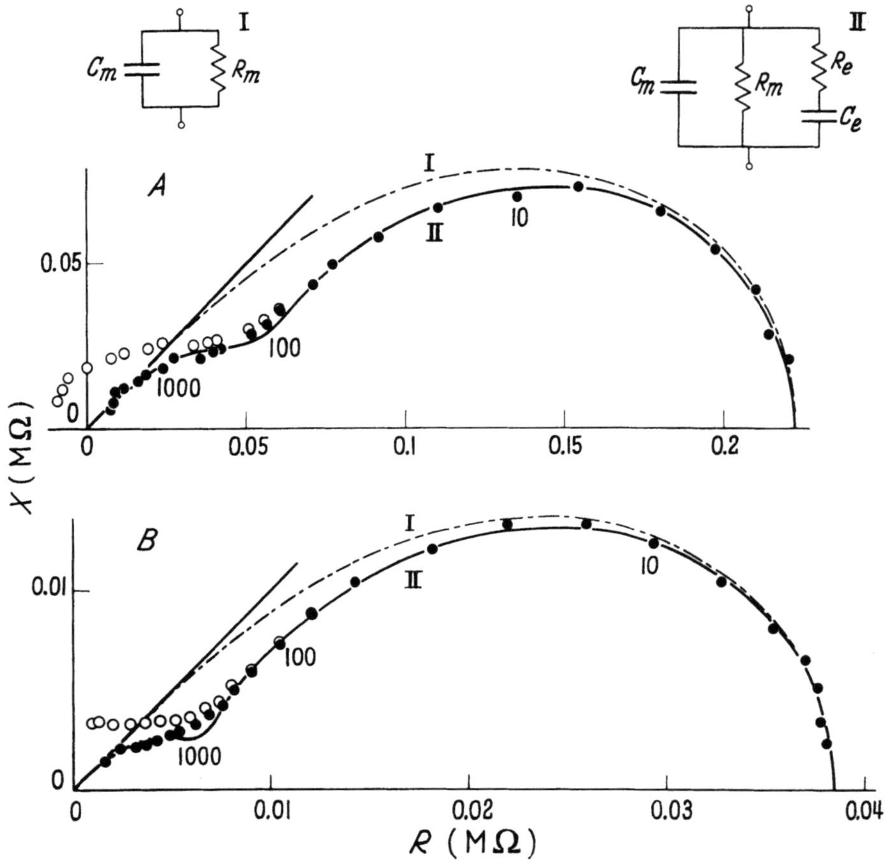

Fig. 1. Impedance-locus plots obtained with intracellular microelectrodes from a fibre of *A* frog sartorius and *B* crayfish carpopodite-extensor muscle. Current-applying and voltage-recording electrodes were placed close together. distant from the ends of the fibre. Hollow circles give uncorrected observations, filled circles after correction for stray capacitances around the microelectrodes. Measurements were made at frequencies (6 per decade) from 1 c/s to 10 kc/s; frequencies (in c/s) are labelled on a few points. The dot-dashed line gives the theoretical impedance locus for the model of inside-outside admittance shown at upper left (I), fitted to the limiting value of *R* obtained at low frequencies. The solid line gives the theoretical locus for the more complicated model of inside-outside admittance shown at upper right (II), with the additional parameters R_e/R_m and C_e/C_m adjusted to give the best fit of the observations. From FALK and FATT (1964)

a considerable divergence of the experimental points from the predicted impedance behaviour. A much better fit can be obtained if we consider the capacitance to be in two parts $C_m + C_e$ with a resistance R_e in series with C_e as shown in circuit II of Fig. 1. The theoretical curve for such a model of the inside-outside electrical connection is indicated by the solid line. Table 1 gives the average values of the impedance parameters of the two types of muscle fibres obtained by fitting this theoretical model to the experimental data.

Relevant to this revised model of the impedance of the muscle fibre are some early observations by FATT and KATZ (1953 b) on the junctional potentials set up in muscle fibres of the shore crab by stimulation of two of the excitatory nerve fibres to the muscle. As well as being used to observe the junctional potentials,

Table 1. *Electrical parameters of muscle fibres* (from FALK and FATT 1964)

	C_m (μF/cm²)	C_e (μF/cm²)	R_m (Ωcm²)	R_e (Ωcm²)	$(C_m + C_e) R_m$ (msec)	$C_e R_e$ (msec)
Frog . . .	2.6	4.1	3100	330	21	1.4
Crayfish .	3.9	17	680	35	14	0.6

microelectrodes served to carry out measurements on the electrical characteristics of the muscle fibre itself, the low-frequency behaviour being analysed in terms of an inside-outside connection consisting of a resistance and capacitance in parallel. In some fibres the junctional potentials set up by the two nerve fibres were found to differ not only in magnitude and in potentiation on repetition of nerve stimulation, which could be explained by differences in pre-junctional processes, but also in time course. The upper pair of records in Fig. 2 were obtained on a

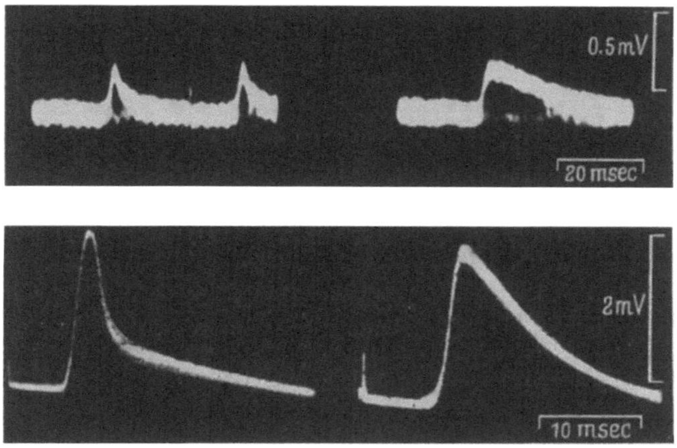

Fig. 2. Examples of excitatory junctional potentials recorded from crustacean muscle, showing responses of widely differing time course produced in the same muscle fibre by stimulation of different nerve fibres. Upper set of records: junctional potentials from a fibre of the dactylopodite-flexor muscle of the shore crab, taken from FATT and KATZ (1953 b). Note that in the record on the left the rate of stimulation was such that two responses appear on the sweep. Lower set of records: junctional potentials from a fibre of the propodite-bender muscle of the rock lobster, taken from HOYLE and WIERSMA (1958)

muscle fibre which showed this behaviour. The outstanding difference between these two responses in the same muscle fibre is in the time course of decay. From measurements of the electrical properties of the fibre, it appeared that its time constant as determined from the low-frequency behaviour would give the slower of the two rates of decay of the junctional responses. Further, the possibility that the difference in decay rates might arise from differences in the spatial spread of charge was eliminated by exploration along the length of the fibre, which showed the two junctional responses to be nearly the same everywhere. HOYLE and WIERSMA (1958) also observed differences in the rate of decay of junctional potentials on stimulation of two different excitatory nerve fibres to a muscle fibre of the rock lobster (lower set of records in Fig. 2).

On the basis of the finding that the inside-outside connection involves two time constants, a reasonable explanation is at hand for these observations on the junctional potentials. The time course of decay of the junctional potential will depend on the duration of transmitter action: if transmitter action is very brief, there will be insufficient time for the voltage across C_e to develop to the same level as across C_m, so that there will be an early rapid component of potential occurring across R_e. To illustrate this, we set up a model made up of resistors and capacitors as in Fig. 1, network II and recorded the voltage across this network as current pulses of various durations were applied to the network. If the

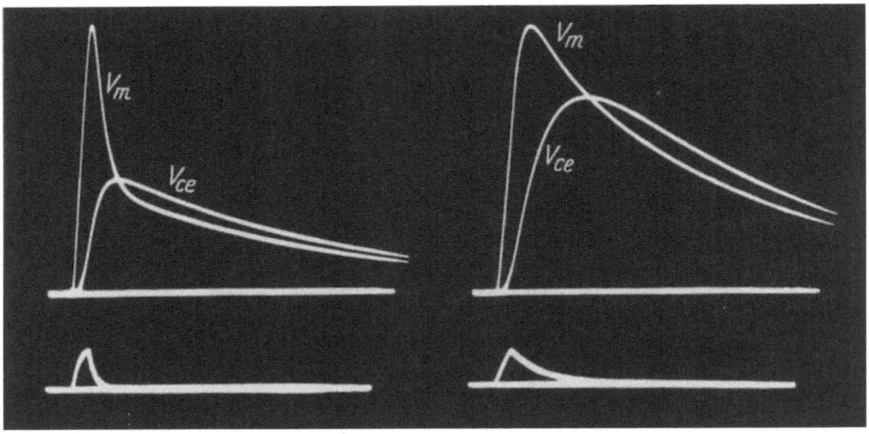

Fig. 3. Oscillograph records, illustrating voltage responses of the network II of Fig. 1 to applied current pulses of different time courses. V_m gives the voltage developed across the entire network (between top and bottom terminals) and thus across R_m and C_m individually. V_{ce} gives the voltage across the capacitance C_e. Bottom sweep gives the time course of current applied to the entire network. Voltage and time scales are arbitrary, being set by the absolute values of resistance and capacitance used in the network model

values of resistances and capacitances in the network were chosen so that $R_e/R_m = 0.2$ and $C_e/C_m = 4$, we could closely reproduce such responses as observed in crustacean muscle, by varying the time course of the current pulse applied to the network. Records obtained with the model are shown in Fig. 3, where the bottom trace gives the time course of the applied current, while the voltage response is shown as the trace labelled V_m. We have here not used rectangular current pulses, but rather pulses which have exponentially rising and falling phases, the only change in time course of the current pulse on going from the record on the left to that on the right being to lengthen the time constant controlling the rise and fall. The strength of the current pulse was adjusted so that the voltage response for the two pulses had the same peak value. Obviously, the spike-like part of the response to the shorter current pulse arises from the voltage drop across the resistance R_e in series with the larger capacitance C_e. Without such a resistance there could be no such two-stage decay as appears very prominent in the response to the shorter pulse.

As for the proposed location of the several impedance elements, C_m and R_m are taken to arise in a membrane at the surface of the muscle fibre. On the other hand, it is suggested that C_e and R_e arise in a system of channels which penetrate the fibre from the surface and extend through the entire cross-section. In frog muscle fibres, the most likely structure to give rise to these properties is the system which in longitudinal section is seen to form triads (PORTER and PALADE 1957; ANDERSSON-CEDERGREN 1959; HUXLEY 1964), R_e and C_e originating in the adjacent membranes of the triad. In the case of crayfish muscle fibres, C_e probably represents the capacitance across the walls of channels leading from the surface while R_e (when referred to unit volume) may reflect the resistivity of the channel interior.

The revised model for the electrical properties of muscle fibres may have relevance for the mechanism of initiation of contraction, for, as demonstrated by HUXLEY and TAYLOR (1958), the effectiveness of depolarization in eliciting contraction is restricted to certain regions along the surface, suggesting a special path for the inward spread of activation. It is likely that the control over contraction is effected by the potential difference across the capacitance C_e, which is presumed to exist in the walls of the sarcoplasmic tubules, rather than by the potential difference across the surface membrane. Except for steady potentials, the time course of the potential across C_e will differ from that across the surface capacitance C_m by the voltage drop in R_e, and will thus follow a slower time course. This is illustrated in the model experiment in Fig. 3 where, in addition to the voltage response V_m across the whole network, there is also recorded the voltage V_{ce} that develops across C_e. The difference between the two sweeps $V_m - V_{ce}$ is evidently the voltage drop across R_e, and includes the initial spike-like component of response to the short current pulse. From this it is seen that, although voltage responses across the entire network evoked by pulses of different time course may be of the same peak magnitude, the response to the longer current pulse involves a considerably larger potential change across the capacitance C_e. Thus, one is led to expect that junctional potentials of faster time course will be less effective in initiating contraction than slower ones of the same peak amplitude. We have here a possible explanation of the "paradox" phenomenon described by WIERSMA and VAN HARREVELD (1938) for certain types of crustacean muscles in which the larger contraction is produced by the smaller, but slower, of two junctional potentials.

One may inquire whether the delay mechanism envisaged would permit the requisite near-maximal activation of the contractile mechanism by a single action potential set up in a frog twitch fibre (HILL 1949) in which the spike may last only 2 to 3 msec. The threshold for contraction has been found by HODGKIN and

HOROWICZ (1960) to be at a depolarization of about 40 mV, with the maximum tension being reached at a depolarization only about 10 mV greater than threshold. From the results of our a. c. impedance measurements, the time constant C_eR_e by which the action potential would be distorted was, on the average, 1.4 msec and the largest value obtained was 2.1 msec. When RUSHTON's (1937) method of graphical analysis was applied to solve for the potential change which would be produced across C_e by a typical action potential from a frog sartorius fibre (peak amplitude 125 mV) when C_eR_e was 2 msec, it was found that the potential change across C_e reached 60 mV (see Fig. 18 of FALK and FATT 1964) which would be adequate for maximal activation of the contractile system.

References

ANDERSSON-CEDERGREN, E. (1959). *J. ultrastruct. Res., Suppl.* **1**, 1.
BOZLER, E. & COLE, K. S. (1935). *J. cell. comp. Physiol.* **6**, 229.
COLE, K. S. & CURTIS, H. J. (1936). *Cold Spr. Harb. Symp. quant. Biol.* **4**, 73.
FALK, G. & FATT, P. (1964). *Proc. roy. Soc. B* **160**, 69.
FATT, P. (1964). *Proc. roy. Soc.* B **159**, 606.
FATT, P. & GINSBORG, B. L. (1958). *J. Physiol.* (Lond.) **142**, 516.
FATT, P. & KATZ, B. (1951). *J. Physiol.* (Lond.) **115**, 320.
FATT, P. & KATZ, B. (1953 a). *J. Physiol.* (Lond.) **120**, 171.
FATT, P. & KATZ, B. (1953 b). *J. exp. Biol.* **30**, 433.
HILL, A. V. (1949). *Proc. roy. Soc.* B **136**, 399.
HODGKIN, A. L. & HOROWICZ, P. (1960). *J. Physiol.* (Lond.) **153**, 386.
HOYLE, G. & WIERSMA, C. A. G. (1958). *J. Physiol.* (Lond.) **143**, 403.
HUXLEY, A. F. & TAYLOR, R. E. (1958). *J. Physiol.* (Lond.) **144**, 426.
HUXLEY, H. E. (1964). *Nature* (Lond.) **202**, 1067.
KATZ, B. (1948). *Proc. roy. Soc.* B **135**, 506.
PORTER, K. R. & PALADE, G. E. (1957). *J. biophys. biochem. Cytol.* **3**, 269.
RUSHTON, W. A. H. (1937). *Proc. roy. Soc.* B **123**, 382.
WIERSMA, C. A. G. & HARREVELD, A. VAN (1938). *Physiol. Zool.* **11**, 75.

INTERPRETATION OF THE CENTRAL EXCITATORY AND INHIBITORY STATES

by RAGNAR GRANIT

In our scientific vocabulary of today Sherrington's concepts 'central excitatory state' and 'central inhibitory state' no longer play the role they used to do. This being the case it is of some interest, historical as well as theoretical, to inquire into the reasons for it. One might imagine that they finally have been replaced by more precise concepts or, again, that they simply have been forgotten. In the middle twenties these concepts were important enough to justify a belated inquiry into their fate and present interpretation. And why were they so important? The answer is simply that the two concepts drew attention to the principle of graded excitability as a function of time, and to Sherrington's view that c.e.s. and c.i.s. were capable of algebraic summation. Sherrington's deductions, as formulated in 1925, were rooted in observations over a lifetime including not only reflex work by isometric myography after the first World War, but also experiences from his earlier studies of the scratch reflex, the lengthening and shortening reactions etc.

The c.e.s. was finally analyzed by ECCLES and SHERRINTON (1931) using the technique of a conditioning shock followed by a test shock. A schematic drawing of the rise of c.e.s. as a function of time was produced which figured in most textbooks of that period. It is well known to all students of the central nervous system and need not be reproduced here.

A gradual change of emphasis took place with the introduction of monosynaptic testing by RENSHAW (1940) and LLOYD (1941) because application of the method used by ECCLES and SHERRINGTON then came to imply analysis of transmission at a specific set of synapses activated and tested by a volley in fairly homogeneous, large muscular afferents. This volley was reasonably synchronous, and so the test shock measured the magnitude of remnant excitatory or inhibitory state as a function of time. The concepts c.e.s. and c.i.s. underwent a corresponding purification and lost their general character. This work also shifted the frontier of subsequent analytical efforts to problems of transmission. This line of advance became, as it were, finally codified when ECCLES and his colleagues in 1952 introduced the intracellular approach to reflex work in general and to monosynaptic studies in particular. The monosynaptic EPSP and IPSP then became the 'purified' c.e.s. and c.i.s. respectively. In terms of membrane potential these two processes were antagonistic, and their time course suggested an explanation of the results obtained by monosynaptic testing of an assembly of extensor or flexor motoneurones.

It remained to ask whether the idea of algebraical summation would be valid also in a reflex context and this proposition was tested by GRANIT and RENKIN (1961) whose theoretical reasoning was based on opposite synaptic currents for excitation and inhibition. In their experiments excitation and inhibition cancelled out on the principle of algebraical summation, but it should be realized that the experimental conditions were carefully selected for purity.

It is no criticism of the conclusion drawn from the experiments presented above if one suggests that the original concepts of c.e.s. and c.i.s. may have lost something in the process of purification. This is simply the way science advances: it defines and re-defines its concepts until something precise emerges and in the present case this happened. One is now free to raise the question of whether or not the central excitatory and inhibitory states in the original sense are fully understood in terms of membrane potentials (restricting the inquiry to motoneurones). The rest of this discourse will be devoted to this problem. Its conclusion will be that other factors need be taken into account when considering central excitatory and inhibitory states in motoneurones.

The first process to be considered is a presynaptic one such as post-tetanic potentiation (PTP), whose importance at central synapses was first emphasized by LARRABEE and BRONK (1947). The large subsequent literature on PTP need not be reviewed here. It is well known and the actual concept is all that is needed in this connexion. Since after tetanic stimulation of the large muscular afferents excitability may be increased for up to 5 minutes without the cell becoming depolarized or having to fire a single impulse, the question arises of whether and to what extent PTP was present in the experiments on which SHERRINGTON based his idea of a subliminal c.e.s. In his days one used the old Kronecker scale for grading stimulus strength of an inductorium and also, as a first approximation, tried the effect by placing the electrodes on the tip of one's tongue. The buzzing hammer of this instrument gave stimulus frequencies much below those needed for strong PTP, but stimulus strength was not restricted to monosynaptic afferents and so a mixture of mono- and polysynaptic effects would have been obtained. When, in the decerebrate animal with its good gamma bias, muscle stretch was the stimulus (e. g., lengthening reaction), the initial frequencies of discharge in stretch would have been high enough for appreciable PTP. The surmise that a state of PTP belonged to the ingredients out of which arose the idea of graded subliminal excitability therefore seems perfectly plausible.

With regard to inhibition there is the presynaptic variety to consider which ECCLES and his coworkers have studied in such detail (cf. ECCLES 1964). The use of bared (afferent) nerves and electrical stimulation ought to have made some presynaptic inhibition possible.

On the other hand, it is clear that the concepts c.e.s. and c.i.s. could not have been defined on the basis of experiences now interpretable as presynaptic phenomena because these two central states summed algebraically by interaction at foci of convergence for excitation and inhibition — which the presynaptic processes do not do — and we know, of course, that many of SHERRINGTON's experiments were of such a nature as to be directly translatable into postsynaptic terms. The final conclusion is therefore that while there are likely to have been presynaptic components in SHERRINGTON's experiments, and while it is plausible that these served as ingredients in the intuitive approach that gradually led to more decisive experiments and to greater insight into the mode of working of the organization responsible for reflex action, the results in the end balanced out in such a way as to place the motoneurone itself in the centre of convergence for the excitatory and inhibitory synaptic influences building up the corresponding states. Since presynaptic inhibition possibly was present when SHERRINGTON studied antagonistic inhibitions, how could he save his concept of c.i.s. from the postulated error or — to put if differently — why was this error of a negligible order? The probable answer is that with the firing motoneurones used when the index was an isometric contraction, the presynaptic inhibition did not play much of a role. Inhibition at the time meant a reduction in terms of gram tension of an isometric contraction. It was not systematically studied in terms of a diminution in the size of a tendon jerk which would have corresponded to the reduction in the size of a monosynaptic EPSP. The stimuli as a rule were tetani of some duration, as pointed out above. These also led to states of excitation or inhibition of some duration.

There is in fact a gap in our knowledge of such states from the intracellular point of view which we (GRANIT, KELLERTH and WILLIAMS 1964 a, b) have tried to fill out, beginning with stretch of a number of synergistic and antagonistic muscle groups. These experiments drew attention to features of synaptic organization in motoneurones which must be considered in trying to arrive at an answer to the question of whether long-lasting c.e.s. and c.i.s. have been completely accounted for in terms of postsynaptic potentials.

The ankle antagonists *triceps surae* and *tibialis anterior* together with *extensor digitorum longus* were freed, and in addition *semimembranosus* to represent the knee joint. These muscles could be loaded by a weight, attached to the tendon across a pulley. The motoneurones were identified by their monosynaptic EPSP's to popliteal, peroneal or hamstring shocks. Monosynaptic testing during stretch was supplemented by repeating stretch while the penetrated motoneurone was fired from the inside by current injected through the tip of the microelectrode (potassium citrate). The latter test demonstrates inhibition as a reduction of firing rate, excitation as an increase of it. If other tests such as size of EPSP, average level of membrane potential or nature of synaptic activation noise are equivocal in stretch of one or all of the three muscle groups, then reduction or increase of firing rate of a firing motoneurone has the advantages of (i) being decisive and (ii) referring to postsynaptic events, because it is the membrane response itself that is influenced. A presynaptic effect is excluded by the nature of the test itself.

Consider now an assembly of motoneurones in which firing by trans-membrane current has been reduced in frequency or completely blocked by pulling upon a muscle belonging to antagonists. Quite regularly firing in the motoneurones was found to be facilitated by pull upon their synergists as defined by the monosynaptic EPSP. These basic facts are in accordance with classical reflexology. If for this assembly of motoneurones one plots the relation between size of the monosynaptic EPSP and the average membrane potential during stretch of the inhibiting muscle, then the two indices "size of EPSP" and "average membrane potential" show no systematic correlation reproducible as a curve. All kinds of combinations of the two indices are represented in such an assembly of motoneurones. There may be little if any hyperpolarization with a large reduction in size of the EPSP and, *vice versa,* definite hyperpolarization with little if any reduction of EPSP. In fact, these indices may or may not indicate inhibition in stretch while it was perfectly clear that the firing of the same motoneurones to trans-membrane current under similar circumstances always was reduced in rate or completely blocked.

Of special interest is the case of semimembranosus inhibiting motoneurones belonging to the ankle extensors. If then the EPSP is reduced by stretch of semimembranosus without an effect on the average membrane potential, it would seem reasonable to suspect the existence of presynaptic inhibition. In such cases it has always proved possible to demonstrate the postsynaptic nature of such inhibitions by using trans-membrane current to fire the cell and showing its rate of discharge to be reduced during semimembranosus pull.

The way in which average level of membrane potential and size of EPSP often fall apart as indices during stretch of a muscle need not be taken to mean that with natural stimuli the basic concepts of de- and hyperpolarization as determinants of excitability are erroneous. A more sensible conclusion is that with such modes of stimulation the instrumentation is insufficient for full information. The "field of vision" of the intracellular electrode is small relative to the size of a cell such as the motoneurone whose synapses largely are localized to its dendrites. EPSP and membrane potential may therefore well indicate events at different localities. The EPSP will be modified by leaks caused by synaptic currents around the motoneurone stimulated by muscle stretch. The instrumentation measures changes of potential only while the relevant parameter in firing is current etc.

One may actually see something of the synaptic "geography" by recording activation noise at high sensitivity. Then all three muscle groups may have effects on the motoneurone by some other index while only one of them is capable of producing strong synaptic activation noise. The synapses of the other two must then have been remote with respect to the site of the electrode tip.

The conclusion from these brief remarks on c.e.s. and c.i.s. is that we still have something to learn about the implications of these two concepts in integrative action of the nervous system.

References

ECCLES, J. C. (1964). *The physiology of synapses.* Berlin, Göttingen, Heidelberg: Springer-Verlag.
ECCLES, J. C. & SHERRINGTON, C. S. (1931). *Proc. Roy. Soc. B.* **107**, 511.
GRANIT, R., KELLERTH, J. O. & WILLIAMS, T. D. (1964a). *J. Physiol.* (Lond.) **174**, 435.
GRANIT, R., KELLERTH, J. O. & WILLIAMS, T. D. (1964b). *J. Physiol.* (Lond.) **174**, 453.
GRANIT, R. & RENKIN, B. (1961). *J. Physiol.* (Lond.) **158**, 461.
LARRABEE, M. G. & BRONK, D. W. (1947). *J. Neurophysiol.* **10**, 139.
LLOYD, D. P. C. (1941). *J. Neurophysiol.* **4**, 184.
RENSHAW, B. (1940). *J. Neurophysiol.* **3**, 373.
SHERRINGTON, C. S. (1925). *Proc. Roy. Soc. B.* **97**, 519.

THE CHOLINERGIC FACTOR IN THE MAINTENANCE OF EXPERIMENTAL ATRIAL FIBRILLATION

by H. E. HOFF and L. A. GEDDES

Introduction

The early history of atrial fibrillation has been well and fully recounted by LEWIS (1912, 1925) who gave ample credit to the pioneer studies of MACKENZIE (1902, 1908). It was MACKENZIE's analysis of its characteristic influence on the jugular and arterial pulses that gave the syndrome its clinical description in virtually its present form, but it failed to delineate the intimate cause of the failure of pulsatile contractions in the atrium. The prototype of atrial fibrillation was first observed by MCWILLIAM (1887) when he subjected the atria of dogs to faradic stimulation, and it was CUSHNY who first suggested the relationship between the events so produced in the dog's heart and the arrhythmia seen in man (1907, 1911). The validity of this suggestion was established through the use of the string galvanometer by ROTHBERGER and WINTERBERG (1909 a, 1909 b, 1910) and LEWIS (1912, 1925).

Despite these and many related studies, a large hiatus has always existed between the syndrome of atrial fibrillation in man and the conditions produced experimentally in animals. As LEWIS pointed out, the arrhythmia produced by rapid repetitive stimulation of the atrium is not genuine atrial fibrillation, although it closely resembles it, for it is not a self-maintaining condition, and rapidly ceases when the stimulating current is terminated. The condition produced by aconitine (PRINZMETAL 1952) is apparently akin to that evoked by electrical stimulation, for this drug sets up a localized pacemaking area that discharges the rest of the atrial musculature in an irregular and disorderly manner. There has been, then, no effective manner of evoking atrial fibrillation for long periods in experimental animals in any way comparable with the days, weeks, and even years which the disorder may persist in man, and even the most recent studies on the arrhythmia have had to depend on MCWILLIAM's method of repetitive electrical stimulation (MOE and ABILDSKOV 1964). Nor is atrial fibrillation at all common as a spontaneous syndrome in the dog. ROOS, who reported three cases in the dog, found that "in dogs the affection is a rare one indeed" (1924), and DETWEILER does not mention it in his survey of cardiovascular disease in dogs (1960).

One of the more physiological approaches to the factors involved in maintenance of atrial fibrillation is its relation to the vagus nerve, reviewed in 1920 by LEWIS, DRURY, and BULGER (1920). The effect of vagal stimulation to increase the rate of atrial fibrillation, and to convert flutter into fibrillation was related to the influence of the vagi to shorten the atrial refractory period, according to an earlier suggestion by ROTHBERGER and WINTERBERG (1914—1915) who had noticed that although atrial fibrillation never persisted beyond a few seconds after cessation of atrial stimulation, it could be made to persist for several minutes by physostigmine or muscarine. WINTERBERG (1907) had some years earlier noted that concurrent vagal stimulation would also maintain fibrillation after cessation of atrial stimulation, although no data were given of the duration of fibrillation in these experiments.

In 1940 NAHUM and HOFF (1940) reported the production of auricular fibrillation in the dog and cat following the local application to the atria of acetyl-β-methylcholine in concentrations of 1 : 500 to 1 : 2000. This substance, closely related to acetylcholine, possesses a muscarine action of high potency, along with a considerable resistance to hydrolysis by acetylcholine esterase, and thus reproduces the general effects of vagal stimulation in the atrium. After treatment with the drug, fibrillation appeared spontaneously, or followed a normal atrial systole, or could be readily precipitated by one or more electrical or mechanical stimuli. The possible role of cholinergic factors in the precipitation of atrial fibrillation in hyperthyroidism was suggested by the production of the arrhythmia by injection of acetyl-β-methyl choline in hyperthyroid patients (NAHUM and HOFF 1935), and there is evidence implicating vagal activity in the production of atrial fibrillation in other circumstances (ALTSCHULE 1945). In these and in a later study by HOFF and GEDDES (1955) emphasis was placed on the role of cholinergic factors in the precipitation of atrial fibrillation. In the present report the cholinergic factor is considered more particularly from the standpoint of the maintenance of the fibrillating state in the atria, once it has been initiated.

Methods

These experiments were carried out on mongrel street dogs of from 10—15 kg body weight. Anesthesia consisted of morphine sulphate in subcutaneous doses of 10—20 mg/kg, at times in combination with 5—15 mg/kg pentobarbital, intravenously. Recordings of atrial activity and stimulation of the atrium were effected via a bipolar catheter inserted into the atrium through the jugular vein. Proper placement of the catheter was determined by the amplitude of the atrial component as compared with the ventricular contribution to the electrogram so recorded, and this comparison was facilitated by the concurrent registration of lead II of the standard electrocardiogram. Preliminary single testing shocks were then delivered, and in the absence of direct ventricular responses to these shocks as delivered via the atrial leads, it was considered safe to proceed with repetitive stimuli at 25—100/sec. Because of these precautions, ventricular ectopic arrhythmias or ventricular fibrillation were never precipitated by the stimuli designed to fibrillate the atria. Moving the catheter rapidly in and out was often employed to precipitate atrial fibrillation after cholinergic treatment of the atrium, but when the catheter was left in place and was not moved, it did not stimulate the atrium in which it lay, and there was nothing to suggest that it formed a source of stimulus to maintain the fibrillating state.

Results

Duration of atrial fibrillation in the unanesthetized dog

All previous studies on experimental atrial fibrillation have involved anaesthesia of some kind, more often morphine in some combination in earlier studies, and barbiturates in later. The method introduced in these studies of catheterization of

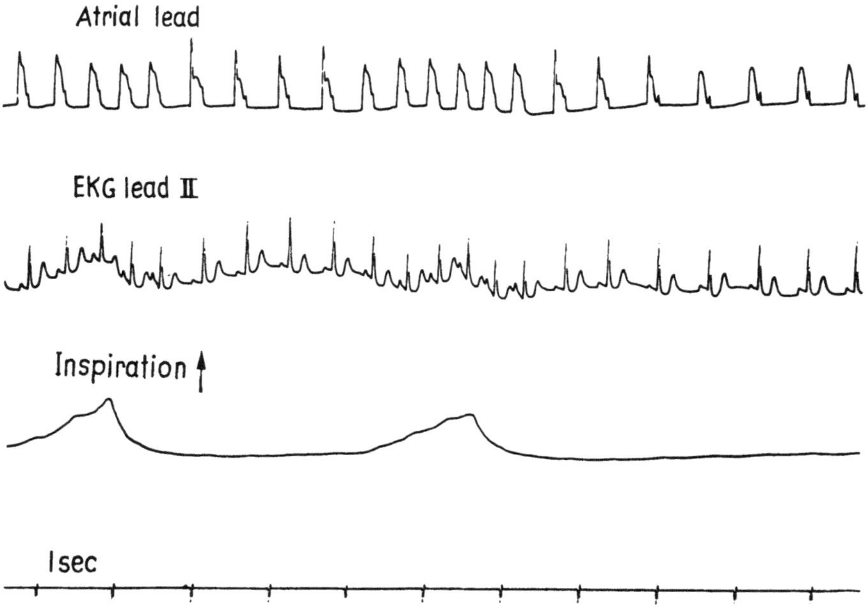

Fig. 1. This 13.0 kg dog was anesthetized by a combination of morphine sulfate (10 mg/kg) subcutaneously, and 10 mg/kg pentobarbital intravenously. An indwelling intra-atrial catheter for recording and an extra-atrial catheter fo injection were installed and the thoracic wall reconstructed. The atrial lead in the upper channel shows a characteristic monophasic action potential, interrupted near the baseline by a ventricular component. The animal shows a moderate respiratory heart-rate reflex, and during the slowing phase in expiration the monophasic component is reduced in amplitude and curtailed in duration, as demonstrated by the clear emergence of the ventricular component in the last complexes. During inspiratory acceleration the classical cardiac monophasic wave is restored

the right atrium via the jugular vein makes possible a completely painless means of initiating and monitoring the course of the arrhythmia in unanesthetized and unrestrained animals, providing that the area of incision over the jugular in the neck is well anesthetized locally with procaine. Three of such animals were quiet and unexcited by the initial restraint and maintained throughout the experiment a normal respiratory heart-rate response. Atrial fibrillation was precipitated in these animals repeatedly, and its persistence after cessation of the stimulus was recorded. In the first there were 28 trials, and fibrillation lasted from 2—45 secs, with average duration of 9.0 secs. In the second, the arrhythmia was precipitated 12 times and lasted from 1—58 secs with an average of 6.3 secs. In the third there were 76 trials, and the resulting fibrillation lasted from 1—58 sec, with an average duration of 16.5 sec. This animal was later given 0.5 mg atropine sulphate per kilogram of body weight, and thereafter the arrhythmia did not persist beyond the cessation of the electrical stimulus. It is thus apparent that in the normal unanesthetized dog, some vagal activity, either present normally, or elicited or added to by concurrent stimulation of vagal endings in the atrium, is essential for the presistence of atrial fibrillation to any significant degree in the absence of the electrical stimulus that initiates it.

Duration of atrial fibrillation induced by local application of acetyl-β-methyl choline

Ten dogs were anesthetized by intravenous pentobarbital (30 mg./kg.), artificial respiration was maintained by insufflation, and the right atrium exposed through the third interspace. The electrocardiogram was recorded from lead II, and the atrium was not traumatized by implacement of leads. A small quantity of the drug (0.1 to 0.2 cc) in a dilution of 1 : 2000 was applied to the surface of the right atrium by dropper, and atrial fibrillation induced by mechanical stimulus. After the fibrillation had spontaneously terminated, the atrium was irrigated with physiological saline, and after an interval of 5—10 min the procedure was repeated. Thirty-seven periods of fibrillation were recorded, varying from 2 min 34 sec to 16 min 14 sec, with an average duration of 6 min 37 sec.

After arrest occurred, fibrillation could almost always be restarted by faradic stimulation of the atrium, but in this case the resulting fibrillation rarely lasted more than a minute. After this episode stopped, further stimulation might again precipitate fibrillation, but for still shorter periods.

The time course of these experiments approximates generally that of the other muscarinic actions of acetyl-β-methyl choline, and suggests that the duration of its action in reducing the refractory period of the atrial musculature is the critical factor determining the persistence of the arrhythmia. This view is generally supported by the observation that when electrodes are placed on the atrium and monophasic records are obtained, the atrial records that emerge after cessation of the fibrillation show a return of the classically prolonged cardiac monophasic action current, in contrast with the brief monophasic wave produced by the application of the drug.

Direct leads on the right atrium generally recorded rates of 1600—1800, and occasionally as high as 3000 per min during the early stages of acetyl-β-methyl choline induced fibrillation, and the rate progressively dropped to the region of

800—900 min immediately before reversion to the normal rhythm. When the fibrillation was temporarily restarted by further stimulation without further cholinergic treatment, the lower, prereversion rates prevailed. When additional cholinergic drug was applied, the faster rates reappeared.

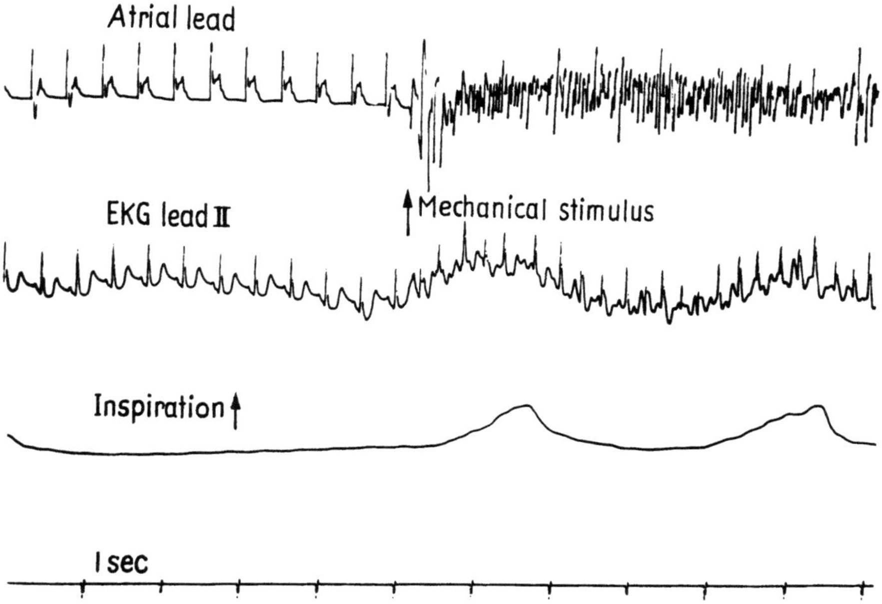

Fig. 2. This record was made a few seconds after Fig. 1., and after injection of 0.5 cc. of 1 : 1000 acetyl-β-methylcholine onto the external surface of the right atrium. The slow component of the atrial monophasic wave has almost completely disappeared, especially in the first three complexes, and the ventricular component is prominent. The jugular electrode was rapidly pushed in and out a few millimeters, producing a series of ectopic atrial beats and precipitating atrial fibrillation, which then continued for over four hours as additional acetyl-β-methylcholine was instilled. This figure illustrates the influence of normally released acetylcholine to shorten the monophasic action current and hence reduce the refractory period, and the precipitation of the arrhythmia by a few closely-placed atrial extrasystoles, mechanically elicited in this case

This observation that fibrillation appeared to persist as long as a critical level of cholinergic activity could be maintained in the atrial musculature suggested that the arrhythmia could be prolonged by repeated administration of the cholinergic agent. The reputed anti-cholinergic action of pentobarbital was in part avoided by use of morphine as the basal anesthetic with sufficient pentobarbital to insure anesthesia during the surgical procedures. Usually 10 mg/kg morphine and 5—10 mg/kg pentobarbital proved sufficient, and the lack of cholinergic blockade was demonstrated by the generally slow heart rate and marked respiratory heart-rate response exhibited with this combination. Recordings of atrial activity were made via jugular catheterization, the cholinergic drug was introduced via a polyethylene tube introduced into the pericardium and stitched to the atrium, and the air-tight integrity of the thorax was restored. In a series of ten trials of fibrillation induced by the usual combination of acetyl-β-methyl choline and mechanical or electrical stimulation of the atrium, in a single animal, durations of 2 min 30 sec to 16 min were recorded, with an average of 6 min 25 sec, not essentially different from the previous series.

In two experiments it was then attempted to maintain fibrillation by repeated administration of the cholinergic drug. The individual trials mentioned above made it apparent that administration could not be governed entirely by time, so it was regulated by the rate of atrial activity. While the electrical activity in atrial fibrillation is never perfectly regular, and difficulties in counting are apparent, it was found that when the overall atrial rate, as recorded via the intra-atrial lead, was maintained above 900/min by instillation of doses of 0.5 to 1.0 cc. of 1 : 1000 acetyl-β-methyl choline, atrial fibrillation would persist as long as desired. In one experiment it was continued for 1 hr 53 min; in the other for 4 hrs 10 min. By not attempting to produce atrial rates above 1500—1700/min excessive A-V block was avoided and the circulation was well-maintained; particularly in the second experiment there seemed no obvious obstacle to indefinite maintenance of the fibrillation as long as a cholinergic effect could be maintained (Figs. 1 and 2).

The influence of vagal stimulation on the duration of atrial fibrillation

These animals were studied under a combination of morphine and pentobarbital anesthesia. By potentiating the vagus, morphine alone increases the duration of atrial fibrillation; in 32 trials in one animal the arrhythmia persisted for from 5 to 127 secs with an average of 34.3 secs. With the addition of pentobarbital this is greatly reduced; in one animal the average of four trials was 11 secs., very close to the average in the nonanesthetized animals. Stimulation was via bipolar shielded electrodes placed alongside the vagus nerves, either after ligation to preclude central effects, or in continuity. Stimuli were 0.1 msec square-wave pulses, at 5—10 volts, and a freqency of 10 to 50 per sec. Rate and intensity were generally chosen so that the fibrillating rate remained above 1000, and A-V conduction continued, with a supraventricular rhythm. At times the necessity to maintain an adequate atrial rate required stimulus intensities and frequencies that produced total A-V block and permitted the establishment of an idioventricular rhythm. In such cases the vagal intensity could be freely increased without altering the ventricular rate. Only one vagus nerve was stimulated at a time, and as fatigue began to set in, as it did after several hours, recourse was had to the other nerve. The right nerve appeared to be somewhat more effective than the left.

Once these simple details were attended to it was possible to maintain atrial fibrillation for comparatively long periods, as far as anything reported to date is concerned, and in five successive trials in four animals, durations of 2 hrs 17 min, 4 hrs, 6 hrs 5 min, 6 hrs, and 8 hrs 12 min were recorded (Fig. 3). Most experiments were stopped arbitrarily, or terminated accidentally by displacement of the vagal electrode, and there seems to have been no intrinsic reason why the arryhthmia could not have been maintained indefinitely. In all cases the arrhythmia ceased within a few seconds of discontinuance of vagal stimulation, and in subsequent trials fibrillation induced without vagal stimulation reverted within a few seconds. There were thus no changes induced by the presence of fibrillation that tended to maintain the arrhythmia.

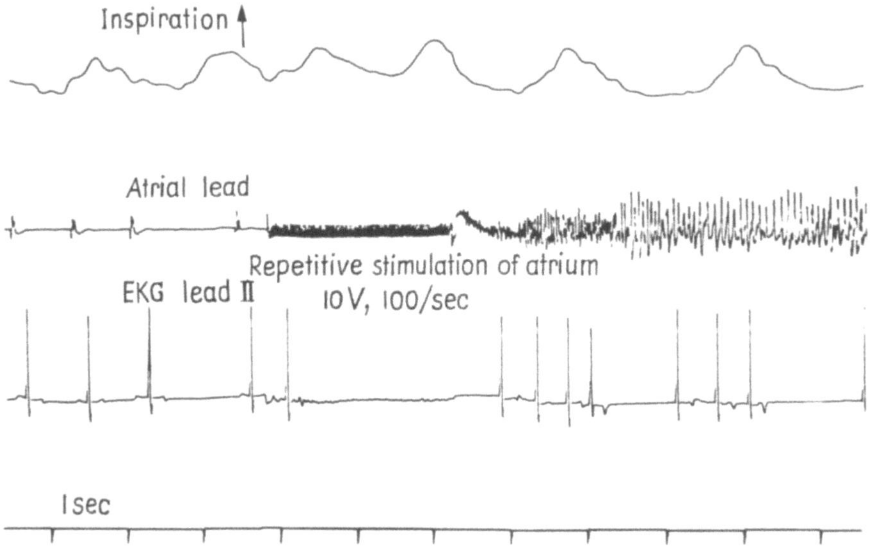

Fig. 3. This 20 kg dog received 240 mg morphine subcutaneously. After careful procaine infiltration of the neck over the right external jugular vein, the right atrium was catheterized with the stimulating-recording electrodes, whose localization is demonstrated by the large atrial component in channel II, coincidental with P wave in lead II (channel III), and the absence of any discernable ventricular component. There was a marked respiratory heart-rate reflex, with acceleration during inspiration and slowing during expiration.
The right atrium was stimulated at approximately 100/sec at 10 V. for approximately 2.5 sec and during this period there were no ventricular beats, presumably because the stimulus excited vagal fibres en route to the A-V node and thereby inhibited conduction over this pathway. When the stimulus is stopped and atrial fibrillation continues spontaneously, ventricular beats very shortly appear, but again there is slowing of the heart during expiration, this time because of reflex vagal inhibition of A-V conduction. The rate of atrial fibrillation is close to 3000 per min in the first second; it falls quickly to 1800 in the second, 1020 in the fourth, and 720 in the sixth. It stops within 20 seconds after cessation of the stimulus. This figure illustrates that there is indeed stimulation of vagal fibers in the atria and supports the contention that the maintenance of atrial fibrillation after cessation of the exciting stimulation was contributed to by the concomitant stimulation of vagal fibers to the atrial musculature. This view is supported by the rapid rate of fibrillation immediately after cessation of the stimulation, and its rapid decline thereafter, as the acetylcholine so released is removed and its influence to shorten refractoriness disappears

Discussion

These experiments make it amply clear that in the dog vagal activity is essential for the maintenance of atrial fibrillation beyond the period of active repetitive atrial stimulation, and in essence for the development of atrial fibrillation as a self-perpetuating arrhythmia within the atrium. It is equally clear that once the disorder has begun it can be maintained in being at will by the continuation of vagal stimulation or pharmacological, cholinergic treatment of the fibrillating auricle. While the longest period of persistence of the arrhythmia was 8 hours, obviously immeasurably short in relation to the years that spontaneous atrial fibrillation may persist in man, it is significantly long in relation to the duration of independent atrial fibrillation hitherto reported, and there appear to be no theoretical obstacles to much longer maintenance of the arrhythmia by the same methods.

These results are most simply explained on the basis of the original concept of circus motion, which invokes a circus pathway, or feed-back loop of finite length, largely determined by the size of the auricle, which describes the pathway of the re-entrant impulse, a rate of conduction within a given range, which limits the rate at which the impulse can traverse the loop, and an absolute refractory period which determines whether the re-entrant impulse will find excitable or refractory tissue when it has traversed its circuit.

Presumably, according to this formulation, the atrium of the dog is below the critical size for the maintenance of re-entrant excitation at the conduction rates and refractory periods generally prevailing. As soon, therefore, as the stimulation ceases that has maintained the appearance of atrial fibrillation by establishing an area of rapid repetitive discharge, the disordered rapid beat will cease because a true re-entrant excitation cannot be maintained. Understandably the current applied to the atrium to produce the disordered rhythm also discharges acetylcholine from the endings of vagal postganglionic nerve fibers stimulated along with the atrial musculature. This reduces atrial refractoriness sufficiently so that when the re-entrant wave of excitation returns — its rate of travel in the atrium is not proportionally altered — the refractory period is over and the re-entrant impulse can continue in its circuitous path. Thus, until the acetylcholine so released is hydrolyzed, genuine atrial fibrillation is maintained, however briefly, after cessation of the stimulus as in Fig. 3. This persistence is prolonged by the presence of choline esterase inhibitors, by the state of vagal tone in the animal, by the vagal discharge accompanying the expiratory phase of the respiratory heart-rate response, or other vagal reflexes, or by the administration of drugs that otherwise stimulate, or potentiate, or reproduce the muscarinic effects of vagal stimulation and acetylcholine release. The success of these experiments in so maintaining atrial fibrillation for many hours suggests strongly that by proper regulation of these or similar cholinergic factors, atrial fibrillation might be maintained almost indefinitely in the dog.

It must be admitted, however, that these experiments throw little light upon the alternate possibility that the cholinergic factor might itself create a local rapidly firing focus, which maintains the appearance of atrial fibrillation much as does a repetitive electrical stimulation or the local instillation of aconitine. This possibility is inconsistent with the known actions of acetylcholine in atrial fibers, where it is known to hyperpolarize the membrane, and suppress the usual pacemaker potentials. Nevertheless the original observation of Nahum and Hoff that quite localized application of acetyl-β-methylcholine is capable of generating atrial fibrillation suggests the possibility of a local and not a generalized phenomenon, more consistent with pacemaking than with a feedback loop. Careful restudy of the influence of localized application is certainly indicated.

The critical factor of the rate of atrial discharge, as observed in these experiments from the aspect of bipolar surface electrodes some two or three millimeters apart, and therefore not well relatable to local fiber discharge rate, suggests nevertheless that fibrillation cannot maintain itself when this rate falls below the range of 800 to 900 a minute. If rapid pacemaker firing were the fundamental driving force in atrial fibrillation it would seem that some form of rapid atrial discharge might continue throughout the whole range of rates as the pacemaker decelerated, while according to the circus motion theory fibrillation ought to cease abruptly at a rate representing the transit time through the largest possible anatomical loop. With a maximal cholinergic effect, and a consequent minimal refractoriness, small loops might be traversed to give the observable rates approaching the maximum we have observed in such experiments of 3000 a minute. As the cholinergic effect diminishes, growing refractoriness would force the circulating impulses through larger and larger loops until the upper limit determined by the size of the atrium is reached. Any further increase in refractoriness would then abruptly stop the fibrillation. It has, however, been pointed out in nerve fibers that rhythmic discharge can develop from rebounds of the several phases of the action potentials. Very rapid rhythms might develop from rebounds of the spike potential while slower rhythms can develop from oscillations of the negative and positive after-potentials. By drastically accelerating repolarization and in this way shortening the atrial monophasic action potential it might well be that acetylcholine could potentiate the development of rapid local pacemaking, in some way critically linked to a certain degree of influence on the monophasic action potential and not related to the cholinergic influence on the ordinary pacemaker potentials.

These experiments make it unlikely that the maintenance of the long-lasting atrial fibrillation encountered in man is based primarily upon the cholinergic factor described here, although in the larger human atria a lesser degree of cholinergic activity might well be adequate and even at times critical. Episodes of paroxysmal atrial fibrillation, and the atrial fibrillation of rheumatic fever where potentiation of vagal activity has been reported to exist might well involve an element of vagal activity. The well recognized influence of digitalis to perpetuate atrial fibrillation might also be linked with the increased vagal activity associated with the administration of the drug. However, atropinization quickly abolished atrial fibrillation in these experiments whenever it was administered; it has no such influence in the human syndrome. On the whole, therefore, some more important factor must be involved in the maintenance of atrial fibrillation in man.

Summary and conclusions

1. In the normal unanesthetized dog, atrial fibrillation persists no more than a few seconds, averaging about 10, after cessation of adequate repetitive electrical stimulation of the right atrium. After atropinization, it does not persist at all, suggesting the necessity of cholinergic activity, presumably linked to the shortening of the myocardial refractory period, as the requisite factor in the establishment of atrial fibrillation as a self-perpetuating arrhythmia in the dog.

2. Instillation of acetyl-β-methylcholine onto the surface of the atrium permits the establishment of atrial fibrillation for longer periods, averaging some 7 minutes after a single application. If the drug is instilled periodically, as required to maintain the all-over rate of atrial discharge safely above the critical level of 800—900 per min., the arrhythmia has been maintained virtually as long as desired. When instillation was terminated, the rhythm reverted to normal within a few minutes.

3. By stimulating the vagus appropriately to maintain the rate of atrial discharge in the range of 1000 to 1500 the arrhythmia was maintained as long as 8 hours. Here also there appeared to be no theoretical limit to the potential duration of atrial fibrillation so maintained. When the vagal stimulus was interrupted, the atria reverted to a normal rhythm within a few seconds.

4. These studies do not offer decisive evidence in favor of either one of the two most debated current theories of the nature of atrial fibrillation, although they appear to be more compatible with the concept of circus motion.

5. It cannot be suggested from these experiments that cholinergic activity is the principal factor operating to maintain persistent atrial fibrillation in man.

Acknowledgement

This investigation was supported in part by USPH Grant 5-Tl-HE-05125.07.

References

Altschule, M. D. (1945). *New Engl. J. Med.* **233**, 265.

Cushny, A. R. (1911). *Amer. J. med. Sci.* **141**, 826.

Cushny, A. R. & Edmunds, C. W. (1907). *Amer. J. med. Sci.* **133**, 66.

Detweiler, D. K., Hubben, K. & Patterson, D. F. (1960). *Amer. J. Vet. Res.* **21**, 329.

Hoff, H. E. & Geddes, L. A. (1955). *J. appl. Physiol.* **8**, 177.

Lewis, T. (1912). *Brit. med. J.* **1**, 57.

Lewis, T. (1925). *The mechanism and graphic registration of the heart beat, 3rd ed.* London: Shaw and Sons.

Lewis, T., Drury, A. N. & Bulger, H. H. (1920). *Heart* **8**, 141.

Mackenzie, J. (1902). *The study of the pulse, arterial, venous, and hepatic, and of the movements of the heart.* Edinburgh and London: Pentland.

Mackenzie, J. (1908). *Diseases of the Heart.* London: Oxford Univ. Press.

McWilliam, J. A. (1887). *J. Physiol.* **7**, 296.

MOE, G. K. & ABILDSKOV, J. A. (1964). *Circulat. Res.* **14**, 447.
NAHUM, L. H. & HOFF, H. E. (1935). *J. Amer. med. Ass.* **105**, 254.
NAHUM, L. H. & HOFF, H. D. (1940). *Amer. J. Physiol.* **129**, 428.
PRINZMETAL, M. (1952). *The auricular arrhythmias.* Springfield, Illinois: Thomas.
ROOS, J. (1924). *Heart* **11**, 1.
ROTHBERGER, C. J. & WINTERBERG, H. (1909 a). *Wien. klin. Wschr.* **22**, 839.
ROTHBERGER, C. J. & WINTERBERG, H. (1909 b). *Wien. klin. Wschr.* **22**, 1792.
ROTHBERGER, C. J. & WINTERBERG, H. (1910). *Arch. ges. Physiol.* **131**, 387.
ROTHBERGER, C. J. & WINTERBERG, H. (1914—15). *Arch. ges. Physiol.* **161**, 42.
WINTERBERG, H. (1907). *Arch. ges. Physiol.* **117**, 233.

THE ORIGIN AND SIGNIFICANCE OF ANTIDROMIC ACTIVITY IN MOTOR NERVES

by J. I. HUBBARD

One of the most important contributions made by Sir JOHN ECCLES to physiology has been a new appreciation of the importance of nerve terminals in synaptic function. It is therefore appropriate to consider afresh the properties of motor nerve terminals, for many years a most controversial subject and characteristically one in which Sir JOHN was, and is, still interested.

The phenomenon which has led to so much speculation and controversy is of course the generation of nerve impulses, apparently in the region of nerve terminals, when mammalian nerve muscle preparations are exposed to acetylcholine (ACh) or anticholinesterases such as prostigmine bromide (PrBr) and eserine. When first discovered (MASLAND and WIGTON 1940; FENG and LI 1941) this generation of antidromic nerve impulses was attributed to nerve terminal depolarization by the drugs. Later workers however laid stress on the muscle twitching that also occurred during drug action. In particular, ECCLES, KATZ and KUFFLER (1942) and LLOYD (1942) soon showed that nerve terminals could be excited by muscle action potentials even in the absence of drugs, and that much of the effect of eserine in setting up antidromic nerve impulses could be attributed to this ephaptic action.

A new impetus was given to this controversy when WERNER (1960 a, b, 1961) showed that the ephaptic action and the setting up of impulses by drugs could be differentiated on several grounds such as latency, effect of paired and tetanic stimulation, and differential sensitivity to curare. BARSTAD (1962), who showed that ACh and PrBr could set up nerve impulses when the muscle was so damaged (by cutting) that it could not generate action potentials, also demonstrated the independence of the generation of antidromic impulses from the generation of muscle action potentials. A further postsynaptic type of origin of the antidromic potentials has however been suggested (DIAMOND 1959; KATZ 1962), based on the release of potassium from muscle fibres which follows the reaction of ACh with the endplate receptors (FATT and KATZ 1951; TAKEUCHI and TAKEUCHI 1960). If this process was intense, the potassium might diffuse across the synaptic cleft and depolarize nerve terminals (LILEY 1956 b; TAKEUCHI and TAKEUCHI 1961).

The lack of unequivocal evidence for a direct action of ACh and related drugs on motor nerve terminals has not prevented much speculation about the possible physiological significance of the antidromic impulses. Thus RIKER, WERNER, ROBERTS and KUPERMAN (1959) suggested that the presence of antidromic nerve discharges reveals the presence of a generator potential set up in the (presumably) decrementally conducting nerve terminals (RIKER 1960). This generator potential was thought to release ACh from a bound state. This ACh then reacted with further presynaptic receptors (RIKER, ROBERTS, STANDAERT and FUJIMORI 1957) to release more ACh which then acted upon the postsynaptic membrane (KOELLE 1962). The generator potential which was thought to be augmented or depressed by suitable drug treatment was thus given a regulating role in neuromuscular transmission, and ACh was given an essential role in transmitter release. A less extreme view (ECCLES 1964) would be that there is a cholinoceptive system in the nerve terminals but that this is only of pharmacological, and perhaps developmental, interest.

Recently new techniques have been described which enable the placing of a microelectrode close to nerve terminals in a nerve-muscle preparation (HUBBARD and SCHMIDT 1961, 1963). It has thus been possible to look for the postulated actions of ACh and related drugs upon nerve terminal excitability (HUBBARD, SCHMIDT and YOKOTA 1965). Fig. 1 illustrates the essential details of the technique applied to the rat diaphragm-phrenic nerve preparation *in vitro* bathed in a modified Krebs-Ringer solution (HUBBARD 1961). The microelectrode (Fig. 1) was filled with 4 M NaCl and had a 1—3 μ tip and a resistance of 1—3 MΩ. It was placed in the vicinity of nerve terminals by using as a criterion (DEL CASTILLO and KATZ 1956) the ability to record extracellular miniature endplate potentials (m. e. p. p. s.) in Mg-paralysed preparations, or focal endplate potentials preceded by a presynaptic spike potential in curarised preparations. Once in position, as Fig. 1 shows, the electrode was connected to a stimulating circuit, a 0.1 msec cathodal pulse from the isolation unit of a Grass stimulator being monitored by an amplifier and a 50 Ω resistance. On stimulation at 1 sec intervals all or nothing antidromic potentials were set up in single fibres in the phrenic nerve and were recorded by two platinum electrodes connected to a high gain amplifier system. On an oscilloscope the pulse and the antidromic response were simultaneously monitored, and the terminal threshold was measured by finding the pulse amplitude at which the antidromic response was set up in 2—4 of 6 trials.

Using this technique two mechanisms have been demonstrated by which ACh and related drugs can initiate presynaptic impulses. The first, a direct depolarizing action, is illustrated in Fig. 2, which shows the threshold changes as the terminals were exposed to solutions containing 4.4×10^{-6} M ACh in the absence and presence of 2.54×10^{-6} M d-tubocurarine chloride (dTC). In the graph the

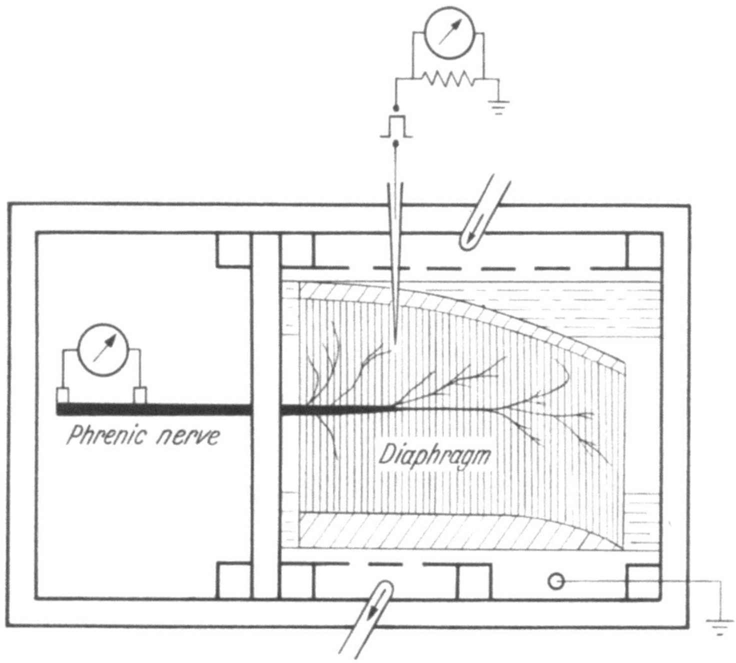

Fig. 1. Stimulation of motor nerve terminals in a rat diaphragm-phrenic nerve preparation. The diagram shows a two chambered clear plastic muscle bath with the rat diaphragm strip bathed in Krebs-Ringer solution on the right, and the phrenic nerve in paraffin oil on the left. The microelectrode tip has been placed in the vicinity of a nerve terminal and is shown connected to a stimulating circuit. Antidromic potentials set up by stimulation are recorded by the platinum electrodes shown on the whole nerve in the adjoining chamber. Further description in text

average control pulse amplitude is taken as 100% and later changes are expressed as a percentage of the control threshold. The preparation was paralysed by raising the $MgCl_2$ concentration to 14 m-moles/l. The threshold was steady under these conditions but when the solution was changed (Fig. 2, first vertical dotted line) to the ACh containing solution the threshold fell after about 2 min to 50% of the control threshold. When the threshold was stable at this low level the solution again changed, the new solution containing dTC in addition to the ACh (Fig. 2, arrow and hatched bar). After a delay of 10 min, partially due to lag time in the fluid exchange system, the threshold rose eventually reaching the control level where it was stable. After 23 min (arrow and end of hatched bar) the dTC was washed out, but the ACh remained. The threshold fell again under the unopposed influence of ACh but reverted to the control level when the ACh was washed out (Fig. 2, second vertical dotted line).

If, as seems reasonable (WALL 1958), the fall in threshold produced by ACh (Fig. 2) is attributed to a depolarization of the terminals, a mechanism is revealed which explains the generation of antidromic impulses by ACh and the blocking of this action by dTC (MASLAND and WIGTON 1940; FENG and LI 1941; BARSTAD 1962). In other experiments (HUBBARD et al. 1965) it has been found that the ACh action is prolonged by PrBr and presumably by other anticholinesterases. In higher concentrations PrBr will depolarize the terminals in the same way as ACh. The setting up of impulses by such drugs (MASLAND and WIGTON 1940; FENG and LI 1941; VAN DER MEER and MEETER 1956) can thus also be partly attributed to a direct action on terminals.

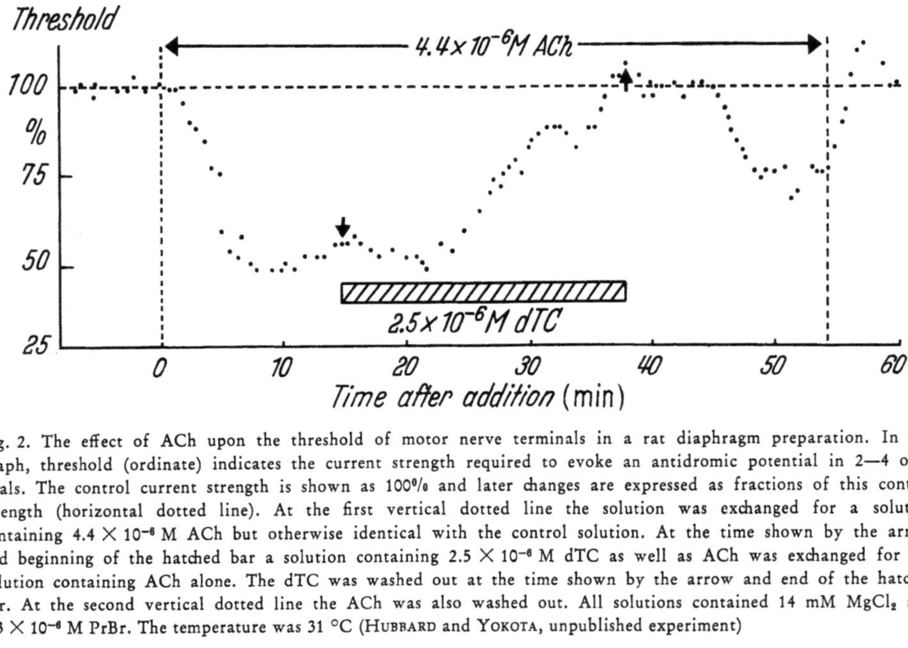

Fig. 2. The effect of ACh upon the threshold of motor nerve terminals in a rat diaphragm preparation. In the graph, threshold (ordinate) indicates the current strength required to evoke an antidromic potential in 2—4 of 6 trials. The control current strength is shown as 100% and later changes are expressed as fractions of this control strength (horizontal dotted line). At the first vertical dotted line the solution was exchanged for a solution containing 4.4×10^{-6} M ACh but otherwise identical with the control solution. At the time shown by the arrow and beginning of the hatched bar a solution containing 2.5×10^{-6} M dTC as well as ACh was exchanged for the solution containing ACh alone. The dTC was washed out at the time shown by the arrow and end of the hatched bar. At the second vertical dotted line the ACh was also washed out. All solutions contained 14 mM $MgCl_2$ and 3.3×10^{-6} M PrBr. The temperature was 31 °C (HUBBARD and YOKOTA, unpublished experiment)

The findings illustrated in Fig. 2 leave unexplained experiments in which the drug conditioned antidromic discharges were only developed after an orthodromic nerve volley (RIKER et al. 1957, 1959; WERNER 1960 a, b, 1961). These experiments may be fully explained (as postulated by WERNER 1960 a) by a second mechanism; a drug induced increase in the amplitude and duration of the negative after potential recorded in the terminals. Such an increased after potential has been found by HUBBARD et al. (1965) in preparations paralysed with $MgCl_2$ or dTC, exposed to as little as 3.14×10^{-6} M PrBr. Conversely dTC was found to reduce the duration and magnitude of negative after potentials. An experiment which illustrates this action of dTC, and also the effect of a slight temperature reduction which mimics PrBr action, is shown in Fig. 3. In these experiments the after potentials in the terminals were explored by determining the related excitability changes (GASSER and ERLANGER 1930) by a two-shock technique, that is, the threshold was determined in the absence of, and when preceded by, a supramaximal conditioning shock at an interval of 2—80 msec. In the graphs the ordinate indicates the ratio of the conditioned to the unconditioned threshold at each interval. In Fig. 3 A the curve (open circles) at

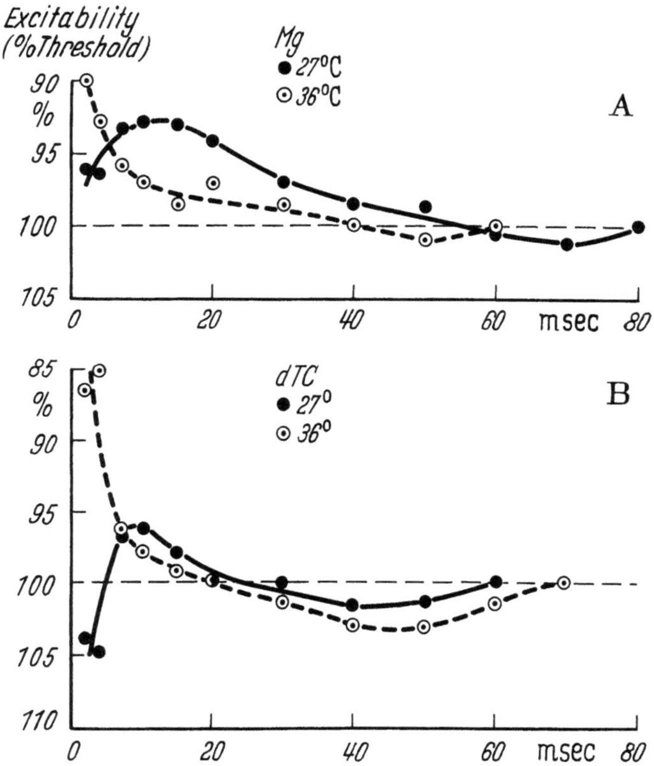

Fig. 3. The effect of temperature changes and dTC upon the after potentials in nerve terminals in a rat diaphragm preparation. In the graphs the excitability changes after a single conditioning stimulus are plotted at intervals (abscissa) from 2—80 msec. At each interval the threshold current required to set up antidromic impulses was measured in the absence and presence of the conditioning impulse. The ratio of the conditioned threshold current strength (μA) to the unconditioned threshold current strength is expressed as a percentage (ordinate). In this experiment the excitability changes were first measured in a solution containing 11 mM $MgCl_2$ at 36 °C (A, open circles). The temperature was then lowered to 27 °C and the procedure repeated (A, filled circles). The solution was then exchanged for one containing 1 mM $MgCl_2$ and 5.05×10^{-6} M dTC. 20 min later the excitability changes were determined in the presence of dTC, firstly at 27 °C (B, filled circles) and then at 36 °C (B, open circles) (HUBBARD and SCHMIDT, unpublished experiment)

36° C in 13 mM $MgCl_2$ shows a supernormal period lasting for 40 msec followed by a small subnormal period which became undetectable at an interval of 60 msec. Lowering the temperature to 27° C (Fig. 3 A, filled circles) increased the supernormal period so that it lasted almost 60 msec. The bathing solution was now changed for a solution containing 5.09×10^{-6} M dTC. Investigation 20 min later (Fig. 3 B, filled circles) showed that the supernormal period, despite the low temperature, had shortened to 25—30 msec, revealing a longer subnormal period. Raising the temperature to 36° C once more (Fig. 3 B, open circles) further shortened the supernormal period which could now be detected for only 15—20 msec after the conditioning impulse.

These observations could well be explained if ACh released by nerve impulses depolarized nerve terminals, the action being manifest as a prolongation and enlargement of the negative after potential, and detected by the resultant changes in the supernormal period (Gasser and Erlanger 1930). The accentuation by PrBr and the reduction by dTC are then similar to the effects of these drugs upon endplate potentials. It seems probable that drug enhanced negative after potentials would be large enough to generate impulses by local circuit action involving the adjacent nodes of Ranvier. The observations of Riker et al. (1957, 1959), of Werner (1960 a, b, 1961), of Blaber and Bowman (1963 a, b) and of Standaert (1964) of drug conditioned antidromic discharges following orthodromic volleys and their suppression by dTC would then be fully explained. The resemblance between tetanic stimulation and drug administration in facilitating the generation of antidromic discharges by an orthodromic volley (Feng and Li 1941; Blaber and Bowman 1963 a) and the facilitation of drug action by previous tetanic stimulation (Kuperman and Werner 1960; Blaber and Bowman 1963 a) may be explained by the enhancement of negative after potentials to be expected (Graham 1942) during the post-tetanic hyperpolarization of the terminals.

With this newer evidence to hand it is appropriate to reconsider the site of ACh action. In the opinion of the present author it must be considered a direct reaction with cholinoceptive sites on, or adjacent to, nerve terminals. The alternative hypothesis (Katz 1962) that it is secondary to postsynaptic potassium release must be dismissed for three reasons: firstly, intra-arterially injected KCl, although it causes muscle contraction and twitching, usually does not cause antidromic discharges (Feng and Li 1941); secondly, ACh does not influence the frequency of miniature endplate potentials (Fatt and Katz 1952; del Castillo and Katz 1955; Hubbard et al. 1965), whereas this frequency increases markedly when the KCl concentration of the bathing medium is raised (Liley 1956 a; Takeuchi and Takeuchi 1961); thirdly, ACh decreases the quantal content of endplate potentials in frog and rat muscle (Ciani and Edwards 1963; Hubbard et al. 1965). In contrast increasing the KCl concentration of the bathing solution increases the quantal content of endplate potentials at frog and rat neuromuscular junctions (Takeuchi and Takeuchi 1961; Liley 1956 a).

It remains to consider the physiological significance of the presynaptic cholinoceptive system. Riker's hypothesis (1960) that it is involved in an intermediary event — a generator potential — controlling neuromuscular transmission must now be discarded. Recent experiments with frog and rat neuromuscular junctions *in vitro* indicate that action potentials invade nerve terminals (Hubbard and Schmidt 1963; Katz and Miledi 1964) and that negative after potentials, enhanced by the depolarizing action of the ACh released by nerve impulses, fulfil all the functions postulated (RIKER 1960) for a generator potential (HUBBARD et al. 1965). Again, contrary to KOELLE's (1962) hypothesis, ACh does not appear to have any essential role in transmitter release at the neuromuscular junction. As previously mentioned, ACh depresses the amount of transmitter released by nerve impulses. It is further particularly significant that ACh applied electrophoretically to nerve terminals does not alter the miniature endplate potential frequency, monitored simultaneously by an intracellular electrode (DEL CASTILLO and KATZ 1955; CURTIS and HUBBARD, unpublished experiments). It must be concluded that at present no physiological function can be assigned to the presynaptic ACh receptors.

The ability to react with ACh may well be a general property of unmyelinated membranes for such receptors have been reported in mammalian C fibres (ARMETT and RITCHIE 1960, 1961, 1962), lobster nerves (DETTBARN and DAVIS 1963), sensory receptors (see GRAY 1959), regenerating sensory nerves (DIAMOND 1959) and spinal interneurones (CURTIS and RYALL, personal communication). Like the ACh receptors in motor nerve terminals, the receptors in sensory receptors and C fibres have no detectable physiological role (GRAY 1959; RITCHIE and ARMETT 1963). The particular pharmacological interest of the ACh receptors in nerve terminals is then only a property of their position at the site of ACh release.

Acknowledgements

It is a pleasure to acknowledge the assistance of Drs. R. F. SCHMIDT and T. YOKOTA in these experiments.

References

ARMETT, C. J. & RITCHIE, J. M. (1960). *J. Physiol. (Lond.)*, **152**, 141.
ARMETT, C. J. & RITCHIE, J. M. (1961). *J. Physiol. (Lond.)*, **155**, 372.
ARMETT, C. J. & RITCHIE, J. M. (1962). *J. Physiol. (Lond.)*, **164**, 18 P.
BARSTAD, J. A. B. (1962). *Experientia (Basel)*, **18**, 579.
BLABER, L. C. & BOWMAN, W. C. (1963 a). *Int. J. Neuropharmacol.*, **2**, 1.
BLABER, L. C. & BOWMAN, W. C. (1963 b). *Brit. J. Pharmacol.*, **20**, 326.
CIANI, S. & EDWARDS, C. (1963). *J. Pharmacol.*, **142**, 21.
DEL CASTILLO, J. & KATZ, B. (1955). *J. Physiol. (Lond.)*, **128**, 157.
DEL CASTILLO, J. & KATZ, B. (1956). *J. Physiol. (Lond.)*, **132**, 630.
DETTBARN, W. D. & DAVIS, F. A. (1963). *Biochem. biophys. acta* **66**, 397.
DIAMOND, J. (1959). *J. Physiol. (Lond.)*, **145**, 611.
ECCLES, J. C. (1964). In *The Physiology of Synapses*, p. 130. Berlin: Springer Verlag.
ECCLES, J. C., KATZ, B. & KUFFLER, S. W. (1942). *J. Neurophysiol.*, **5**, 211.
FATT, P. & KATZ, B. (1951). *J. Physiol. (Lond.)*, **115**, 320.
FATT, P. & KATZ, B. (1952). *J. Physiol. (Lond.)*, **117**, 109.
FENG, T. P. & LI, T. H. (1941). *Chin. J. Physiol.*, **16**, 37.

Gasser, H. S. & Erlanger, J. (1930). *Amer. J. Physiol.*, **94**, 247.
Graham, H. T. (1942). *J. Neurophysiol.*, **5**, 137.
Gray, J. A. B. (1959). In *Handbook of Physiology*, vol. 1, (ed. by Field, F., Magoun, H. W. & Hall, V. E.) p. 123. Washington D. C.: American Physiological Society.
Hubbard, J. I. (1961). *J. Physiol. (Lond.)*, **159**, 507.
Hubbard, J. I. & Schmidt, R. F. (1961). *Nature (Lond.)*, **191**, 1003.
Hubbard, J. I. & Schmidt, R. F. (1963). *J. Physiol. (Lond.)*, **166**, 145.
Hubbard, J. I., Schmidt, R. F. & Yokota, T. (1965). *J. Physiol. (Lond.)*, submitted for sublication.
Katz, B. (1962). *Proc. Roy. Soc. B*, **155**, 455.
Katz, B. & Miledi, R. (1964). *J. Physiol. (Lond.)*, **172**, 26 P.
Koelle, G. B. (1962). *J. Pharm. Pharmacol.*, **14**, 65.
Kuperman, A. S. & Werner, G. (1960). *Nature (Lond.)*, **188**, 1032.
Liley, A. W. (1956 a). *Transmission at the mammalian neuromuscular junction.* Ph. D. Thesis. Australian National University.
Liley, A. W. (1956 b). *J. Physiol. (Lond.)*, **134**, 427.
Lloyd, D. P. C. (1942). *J. Neurophysiol.*, **5**, 153.
Masland, R. L. & Wigton, R. S. (1940). *J. Neurophysiol.*, **3**, 269.
Ritchie, J. M. & Armett, C. J. (1963). *J. Pharmacol.*, **129**, 201.
Riker, W. F. Jr. (1960). *Arch. Neurol.*, **3**, 488.
Riker, W. F. Jr., Roberts, J., Standaert, F. G. & Fujimori, H. (1957). *J. Pharmacol.*, **121**, 286.
Riker, W. F. Jr., Werner, G., Roberts, J. & Kuperman, A. (1959). *J. Pharmacol.*, **125**, 150.
Standaert, F. G. (1964). *J. Pharmacol.*, **143**, 181.
Takeuchi, A. & Takeuchi, N. (1960). *J. Physiol. (Lond.)*, **154**, 52.
Takeuchi, A. & Takeuchi, N. (1961). *J. Physiol. (Lond.)*, **155**, 46.
Van der Meer, C. & Meeter, E. (1956). *Acta physiol. pharm. neerl.* **4**, 454.
Wall, P. D. (1958). *J. Physiol. (Lond.)*, **142**, 1.
Werner, G. (1960 a). *J. Neurophysiol.*, **23**, 171.
Werner, G. (1960 b). *J. Neurophysiol.*, **23**, 453.
Werner, G. (1961). *J. Neurophysiol.*, **24**, 401.

THE PERIPHERAL MECHANISMS OF CUTANEOUS SENSATION

by A. Iggo

The peripheral mechanisms of cutaneous sensation

Since Adrian in 1926 first established the general principles of afferent nervous activity, there have been numerous studies of the peripheral cutaneous sensory mechanisms. Recent interest has turned on more detailed and quantitative studies of the mechanisms by which the peripheral afferent nerve fibres encode the changing conditions at the surface of the body. There are at present two principle hypotheses to account for the phenomenon. One states that the incoming information is coded in a spatial and temporal pattern of impulses in a population of afferent fibres with nerve-endings which are individually similar, if not identical, but which because of their morphological disposition may respond differently to identical stimuli. This hypothesis has been expressed in several forms by CAUNA (1961), CATTON (1964) and WEDDELL and his colleagues (WEDDELL 1961). The alternative classical hypothesis states that the coding, although depending in part on the temporal pattern of impulses, arises in peripheral receptors that display a considerable degree of dissimilarity, so that the receptors have played a significant part in determining or selecting the afferent fibres that are active in any particular situation.

This article sets out to examine critically the evidence for the two hypotheses and takes as its starting point the kind of discharge that occurs in cutaneous afferent nerve fibres.

Cutaneous afferent discharge

ADRIAN and ZOTTERMAN (1926) established that the activity in an individual myelinated axon, excited through its receptor, is coded in an impulse frequency pattern, intensity being signalled by the frequency as modified by the rate of adaptation of the afferent unit. Subsequent work on myelinated axons has confirmed these original conclusions. Recently studies of the skin of monkeys, cats, dogs and rats have extended the analysis to the non-myelinated axons and yielded new information on the various kinds of non-myelinated afferent unit that exist. Electrophysiological evidence that it was technically possible to record impulses in individual C fibres was first obtained in 1956 (IGGO 1958), and BOWER (1959) has recorded spike potentials from mammalian nerves containing only non-myelinated axons. The possibility of actually dissecting a single non-myelinated fibre from a peripheral nerve trunk was made unlikely by GASSER's discovery (1955) that the non-myelinated axons in a peripheral nerve trunk are enclosed in a Schwann cell complex, with as many as 20 axons in a single Schwann bundle in a peripheral nerve. The evidence that within a bundle the axons are behaving independently rests largely on the unitary behaviour of the action potentials. Within each Schwann bundle in a peripheral nerve there may be a four-fold range in diameter of the non-myelinated axons (Fig. 1). Since there is a linear relation between diameter and conduction velocity (GASSER 1955), the individual axons would be expected to conduct impulses at different velocities. Even if an impulse started simultaneously in all the axons the impulses would quickly disperse so that a unitary action potential would not be present 100 mm or more proximal to the receptor. Unitary potentials, however, are always recorded. Records from thin strands dissected from a peripheral nerve often show several distinctive action potentials with C conduction velocities (i. e., less than 2.5 m/sec). When a "natural" stimulus is applied, the discharges in the individual units are quite distinct (Fig. 2), and an interaction among the various action potentials is found only exceptionally. The action potentials bear no temporal relation to each other. This independence of operation could only arise if the impulses in each axon were free of suprathreshold interactions with the other. In the experiments illustrated in Fig. 2 each spike could be

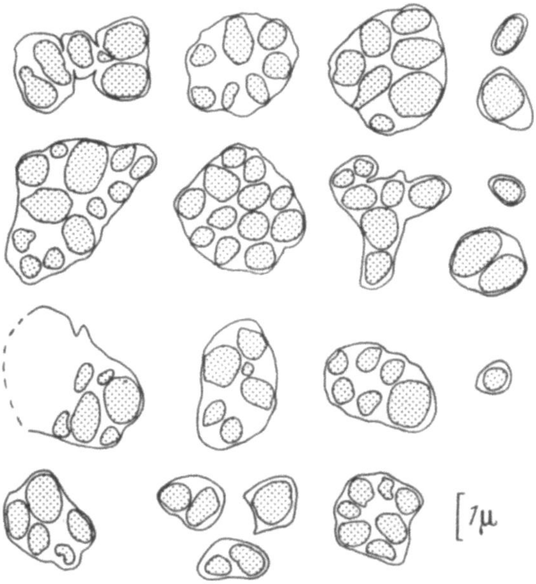

Fig. 1. Non-myelinated axons in the sural nerve of the cat. Each group of axons is associated with one Schwann cell. Within each group there is a wide range of axon diameters (after Gasser 1955)

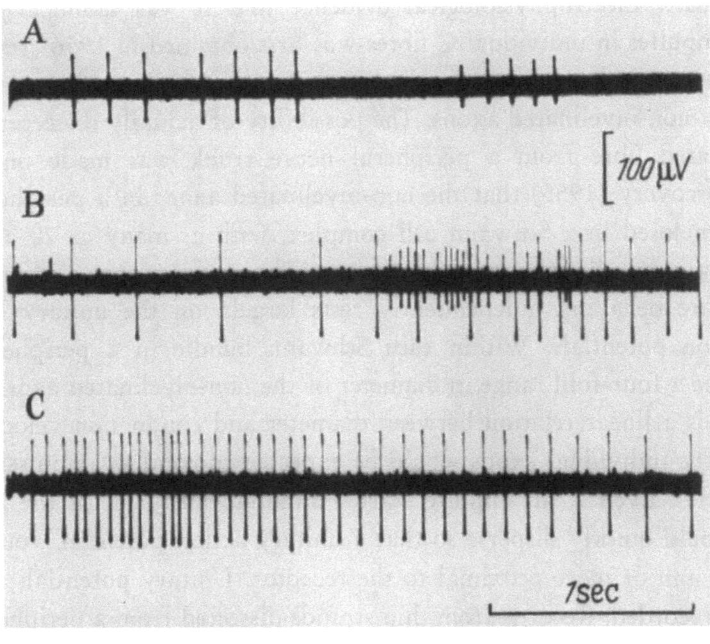

Fig. 2. Independence of conduction in C fibres. The records were made by leading from a thin strand dissected from the saphenous nerve of a cat, and show the responses to mechanical stimuli applied to adjacent areas of skin. In A) each receptive field was stimulated separately; in B) the field of the large spike was stimulated and an intercurrent stimulus was given to the field of the small-spike. In C) the stimulus was restricted to the field of the large spike alone at higher intensity. The conduction velocities were 1 m/sec for the large spike and 0.8 m/sec for the small spike

blocked by the antidromic conduction from a pair of stimulating electrodes of an action potential that was identical in shape and size with the respective spikes. The antidromic spikes appeared in an all-or-none manner when the electrical stimulus was adjusted to play about threshold, and each orthodromic spike also behaved as a unit.

Further evidence is provided by the effect on the amplitude of the action potential of a period of repetitive activity. It is now firmly established that the compound action potential of non-myelinated axons in a peripheral nerve trunk becomes larger during a period of repetitive activity (Brown and Holmes 1956) due to a progressive enhancement of the negative after-potential (Ritchie and Straub 1956). The records in Fig. 2 and other published records illustrating a repetitive discharge in C fibres at frequencies of 50/sec or greater do not reveal any progressive enhancement in the size of the action potential spikes, a result which is consistent with the hypothesis that the single unit record is generated by activity in a single axon.

The general conclusion is, therefore, that the action potential spikes with conduction velocities in the C range (below 2.5 m/sec), recorded from strands of mammalian peripheral nerve, are generated by single axons which are non-myelinated at the recording electrodes. One possibility is that these spikes are carried *at the recording electrodes* by thinly myelinated axons. This point has been tested by measuring the conduction velocities at several points along the nerve (Iggo 1958). The majority of the axons with low conduction velocities (<2.5 m/sec) in the peripheral nerve trunks also had similar conduction velocities at the recording electrodes, i. e., they were non-myelinated at the recording electrodes.

The conclusion reached, therefore, is that the single unit action potentials with C conduction velocities are carried by single non-myelinated axons. An exception may be provided by those axons which lose their myelin sheaths in the nerve trunks as they travel distally. Afferent discharge in non-myelinated axons follows the same general rules as in the myelinated fibres.

Cutaneous afferent units

The results obtained by single fibre recording from both non-myelinated and myelinated cutaneous afferent nerves establish unequivocably that the individual afferent fibre-receptor units have distinctive properties.

Non-myelinated units

The C fibre group in monkeys, cats, dogs and rats (Iggo 1959, 1960, 1963 b; Hensel, Iggo and Witt 1960; Iriuchijima and Zotterman 1960) includes units *selectively sensitive* to either mechanical or thermal stimuli. Within these two broad groups were found units with widely differing thresholds. The most sensitive mechanoreceptors with afferent C fibres could be excited by movements of the tips of the hairs, but these units were less sensitive than similar receptors supplied by thin myelinated axons which have recently been found to innervate the fine down hairs (see below) which can be excited by hair displacements as small as 1 μ. At the other extreme of the range there are non-myelinated mechanoreceptors that require a force of several grams before a discharge can be provoked. There is a similar wide range of sensitivity to temperature changes among the thermoreceptors (Hensel et al. 1960). In some species (cat, dog and rat) the sensitive cutaneous thermoreceptors are innervated by non-myelinated axons and probably for this reason were not discovered until techniques were developed for examining the responses of single C fibres. In the monkey, on the other hand, the sensitive thermoreceptors are innervated either by small myelinated axons or by non-myelinated axons (Iggo 1964). Human cutaneous thermoreceptors are probably similar in innervation to monkey temperature receptors (Hensel and Boman 1960), a fact which may account for the earlier loss of thermal sensations than "pain" in differential block of cutaneous nerves in some human experiments. In addition to being innervated by larger axons the cutaneous thermoreceptors in monkey skin displayed an enhanced "differential sensitivity" when compared with cat thermoreceptors.

The myelinated axons supplying the skin also innervate receptors with distinctive properties. The original electrophysiological work of Adrian and his colleagues (Adrian 1931) established that some axons were excited by hair movements and others by pressure on the skin. Since then every investigation using single fibre methods has confirmed these differences and added to them. Recent detailed re-investigations of the myelinated afferent units in the monkey (Iggo 1963 b), cat (Hunt and McIntyre 1960 a, b; Iggo 1963 a) and the rabbit (Brown and Iggo 1964) have clearly established the existence of several main groups of sensitive mechanoreceptors in hairy skin. The *hair follicle receptors,* which are identified by the particular sensitivity of the afferent fibres to hair movements, are innervated by axons with conduction velocities ranging from 5 to 100 m/sec (Hunt and McIntyre 1960 b). Recently it has been found that at least 4 subgroups can be made in the rabbit, according to the kind of hair follicle and the properties of the receptor and conduction velocities of the afferent fibre (Brown and Iggo 1964). These are *type D* — afferent fibres conducting at less than 20 m/sec with receptors of high mechanical sensitivity that adapt relatively slowly. The receptive fields are large and the hairs innervated include the fine down hairs and also the guard hairs. *Type G:* afferent fibres conducting at velocities greater than 20 m/sec. The discharge to hair movement is phasic, the mechanical thresholds are higher and the discharge adapts more rapidly than for the type D. The receptive fields are small and the only hairs innervated, judged by threshold sensitivity to displacement, were the guard hairs, excluding those described below in type T. The *type T* units also had afferent fibres with conduction velocities greater than 20 m/sec, and covered the same range of velocities (20—100 m/sec) as the type G units. The units had a phasic discharge to hair movement and a mechanical threshold higher than type D. An individual axon innervated only a few specialised guard hair follicles, called tylotrich follicles by STRAILE (1960). The fourth subgroup contains the vibrissa follicles, which are known to have an elaborate innervation and which FITZGERALD (1940) has found to contain slowly-adapting receptors with a directional sensitivity. The types D, G and T are all distinguishable in the body skin of the rabbit and so far types D and G have been found in the cat. These latter types probably correspond to the slow large receptive field and fast small receptive field units, respectively, that HUNT and MCINTYRE (1960 b) mention in their examination of the relation between the size of afferent fibres and area of receptive fields of hair receptors in the cat.

Two other receptors, one dermal and the other epidermal and innervated by myelinated fibres, have been examined in detail by electrophysiological and histological techniques. The best known is the Pacinian corpuscle which lies in the dermis. This receptor is innervated by a large myelinated axon and adapts very rapidly to mechanical stimulation (GRAY and MATTHEWS 1951). Detailed electronmicroscopic studies have established that at the centre of the receptor there is a non-myelinated nervous element, completely enclosed by the lamellated outer coats. LOEWENSTEIN (1961 a) has established that the mechanical sensitivity is restricted to the non-myelinated terminal. The receptor in the epidermis of hairy skin that has been given detailed electrophysiological and histological examination is the "touch-spot" receptor. This afferent unit is supplied by a large myelinated axon, adapts slowly to a steadily maintained mechanical deformation and has a spot-like receptive field in the rabbit (FRANKENHAEUSER 1959), the cat (HUNT and MCINTYRE 1960 a; IGGO 1963 b), and the monkey (IGGO 1963 b). The receptive fields of these afferent units are discrete roughly circular elevations of the skin surface about 250 μ in diameter (IGGO 1963 a; IGGO and MUIR 1963). The epidermis of the elevation is modified and thicker than the adjacent epithelium. The sensory endings consist of MERKEL's corpuscles which hang along the lower border of the stratum basalis, being attached to this layer by many desmosomes, and are enclosed by the basement membrane of the epidermis (Fig. 3). Similar MERKEL's corpuscles have been described by CAUNA (1961) in hairless human skin. Each "touch spot" is innervated by a single thick myelinated axon that branches profusely at the base of the elevation. These receptors have been found in the cat, monkey and rabbit and in the original histological description by PINKUS (1904), who called them Haarscheibe, they are described in human skin.

The recent quantitative search for cutaneous thermoreceptors has revealed that many of the sensitive mechanoreceptors respond to change of temperature with a phasic discharge and that some display, in addition, a static sensitivity

(HENSEL and ZOTTERMAN 1951; WITT and HENSEL 1959; IGGO 1963 a, b). The most conspicuous features of the results are that the more slowly an afferent unit adapts to a mechanical stimulus, the more strongly it can be influenced by a fall of temperature (IGGO 1963 b), and that the afferent units are excited only by a *fall* of temperature. LOEWENSTEIN (1961 b) reports that Pacinian corpuscle which adapts very rapidly could not be excited by a sudden, rapid fall of temperature. Most of the G type and T type hair follicle receptors, which also adapt

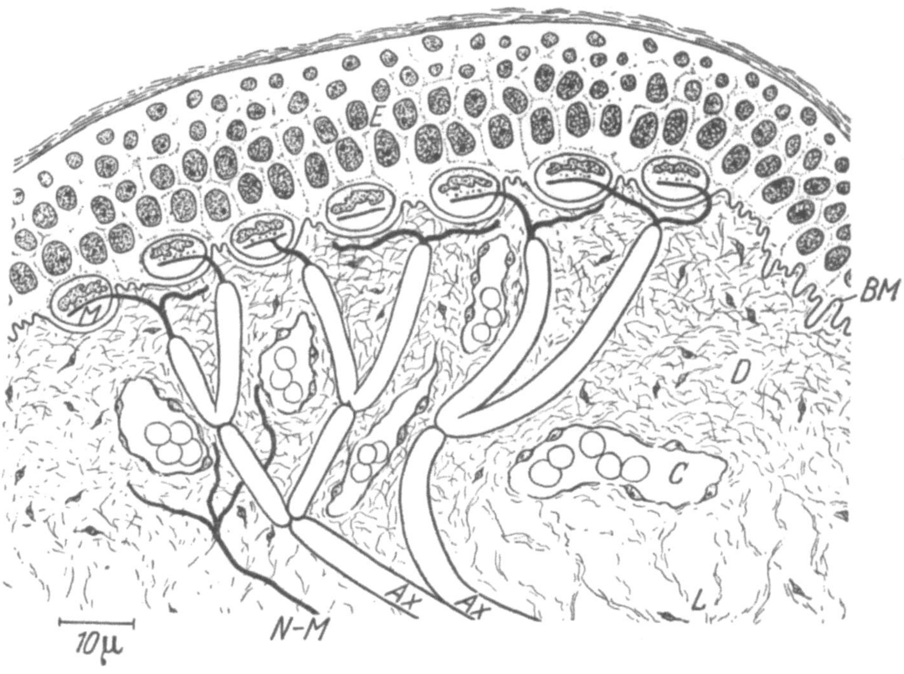

Fig. 3. Diagram of a slowly-adapting 'touch spot' from the hairy skin of a cat, based on light and electron-microscopic studies (IGGO & MUIR 1963). The branches (AX) of a single myelinated axon divide and end in Merkel's corpuscles (M) inside the basement membrane (B M) beneath a thickened epidermis (E). The dermal plug contains numerous capillaries (C) embedded in dense collagen (D). The deeper layers of the dermis (L) contain loosely packed collagen. A bundle of non-myelinated axons (N—M) also enters the dermal plug

rapidly, fail to be excited, although some may discharge a few impulses for a 10° C fall. The D type hair follicle receptors which adapt more slowly are excited weakly by a fall of temperature. The effect of a temperature change is more conspicuous with the "touch spot" units (HUNT and MCINTYRE 1960 a; IGGO 1963 a) which may be excited for several seconds by a fall in temperature. The most striking effects are seen in the very slowly-adapting mechanoreceptors in monkey skin (IGGO 1963 b) where a discharge may persist for 10 seconds or longer. These last units also displayed clearly the converse effect evident with all slowly-adapting units, the disappearance of the discharge when the skin temperature was raised (the "pause" of WITT and HENSEL 1959). A similar excitation has been found for the non-myelinated cutaneous afferent mechanoreceptor units (DOUGLAS, RITCHIE and STRAUB 1957; IGGO 1959, 1960; HENSEL, IGGO and

WITT 1960). Originally this effect was interpreted by DOUGLAS et al. as implying a multi-specific type of ending, but later work with single C fibres established significant differences among the C non-myelinated afferent units based on selective sensitivity to mechanical and thermal stimuli. The thermal effects just described have been reported for other slowly-adapting mechanoreceptors, such as the muscle spindle (LIPPOLD et al. 1960). The significance of the phenomenon awaits elucidation. It appears to be a very general property of mechanoreceptors and may be caused by some action on the receptor/generator potential mechanism that only leads to excitation when the permeability changes that underly the generator potential last a sufficient time.

From these results it is clear that axons of all diameters supply many different kinds of afferent unit in normal skin, and that several of them have specialised structures (receptors) associated with their terminals. There is no evidence that an individual axon supplies several different kinds of receptor. The significance of at least one of these receptors in conferring special properties on the afferent unit has been examined (BROWN and IGGO 1963). The "touch spot" receptors degenerate when the afferent nerve is cut and when the axon grows back into the skin the selective sensitivity of the afferent unit reappears at the same time as the MERKEL's discs reappear. The receptor complex thus appears to be required for the normal selective sensitivity of the afferent unit.

The conclusion reached, therefore, is that the second of the two hypotheses outlined at the start of this article is consistent with the results reviewed. It thus becomes the appropriate starting point for any investigation of cutaneous sensory mechanisms in the central nervous system.

References

ADRIAN, E. D. (1926). *J. Physiol.* (Lond.) **61**, 49.
ADRIAN, E. D. (1931). *Proc. Roy. Soc.* B. **109**, 1.
ADRIAN, E. D. & ZOTTERMAN, Y. (1926). *J. Physiol.* (Lond.) **61**, 151.
BOWER, E. A. (1959). *J. Physiol.* (Lond.) **148**, 2 P.
BROWN, A. G. & IGGO, A. (1963). *J. Physiol.* (Lond.) **165**, 28 P.
BROWN, A. G. & IGGO, A. (1964). *J. Physiol.* (Lond.) **172**, 33 P.
BROWN, G. L. & HOLMES, O. (1956). *Proc. Roy. Soc.* B. **145**, 1.
CATTON, W. T. & PE TOE, N. (1964). *J. Physiol.* (Lond.) **172**, 21 P.
CAUNA, N. (1961). In: *Cytology of Nervous Tissue*, London: Taylor and Francis.
DOUGLAS, W. W., RITCHIE, J. M. & STRAUB, R. W. (1957). *J. Physiol.* (Lond.) **150**, 266.
FITZGERALD, O. (1940). *J. Physiol.* (Lond.) **98**, 163.
FRANKENHAEUSER, B. (1949). *Acta physiol. scand.* **18**, 68.
GASSER, H. (1955). *J. gen. Physiol.* **38**, 709.
GRAY, J. A. B. & MATTHEWS, P. B. C. (1951). *J. Physiol.* (Lond.) **113**, 475.
HENSEL, H. & BOMAN, K. K. A. (1960). *J. Neurophysiol.* **23**, 564.
HENSEL, H., IGGO, A. & WITT, INGRID (1960). *J. Physiol.* (Lond.) **153**, 113.
HENSEL, H. & ZOTTERMAN, Y. (1951). *J. Physiol.* (Lond.) **115**, 16.
HUNT, C. C. & MCINTYRE, A. K. (1960 a). *J. Physiol.* (Lond.) **153**, 88.

Hunt, C. C. & McIntyre, A. K. (1960 b). *J. Physiol.* (Lond.) **153**, 99.
Iggo, A. (1958). *J. Physiol.* (Lond.) **142**, 110.
Iggo, A. (1959). *Quart. J. exp. Physiol.* **44**, 362.
Iggo, A. (1960). *J. Physiol.* (Lond.) **152**, 337.
Iggo, A. (1963 a). *Acta Neuroveg.* **24**, 175.
Iggo, A. (1963 b). *Acta Neuroveg.* **24**, 225.
Iggo, A. (1964). *Nature* **204**, 481.
Iggo, A. & Muir, A. R. (1963). *J. Anat.* **97**, 151.
Iriuchijima, J. & Zotterman, Y. (1960). *Acta physiol. scand.* **49**, 267.
Lippold, O. C. J., Nicholls, J. G. & Redfearn, J. W. T. (1960). *J. Physiol.* (Lond.) **153**, 218.
Loewenstein, W. R. (1961 a). *Ann. N. Y. Acad. Sci.* **94**, 510.
Loewenstein, W. R. (1961 b). *J. Neurophysiol.* **24**, 150.
Pinkus, F. (1904). *Arch. mikr. Anat.* **65**, 121.
Ritchie, J. M. & Straub, R. W. (1956). *J. Physiol.* (Lond.) **136**, 80.
Straile, W. E. (1960). *Amer. J. Anat.* **106**, 133
Weddell, G. (1961). In: *Brain and Behaviour I*, Washington: Am. Inst. Biol. Sci.
Witt, Ingrid & Hensel, H. (1959). *Pflüg. Arch. ges. Physiol.* **268**, 582.

THE ORIGIN OF CEREBELLAR INHIBITION ON DEITERS' AND INTRACEREBELLAR NUCLEI

by M. Ito

The inhibitory function of the cerebellum was first demonstrated by the suppression of the Sherringtonian decerebrate rigidity that was caused by stimulation of the cerebellar cortex (Sherrington 1897; Lowenthal and Horsley 1897). Another type of rigidity, produced by the anaemic decerebration, has also been attributed to removal of the cerebellar inhibition which was acting tonically on the bulbar postural centres (Pollock and Davis 1927, 1930). However, since Moruzzi's finding (cf. Moruzzi 1950) that the effect of cerebellar stimulation could be reversed by altering the stimulus frequency, it appears to be a general assumption that the cerebellum includes two antagonistic systems, inhibitory and facilitatory.

In the present series of experiments upon cats the electrical events produced by cerebellar stimulation have been investigated in neurones impinged upon by cerebellar corticofugal projections. The cells of the nucleus of Deiters, as well as of the intracerebellar nuclei, were impaled with glass microelectrodes. The nucleus of Deiters contains giant neurones, and so these have been studied under the most favorable conditions (Ito, Hongo, Yoshida, Okada and Obata 1964 a, b). Since inhibitory postsynaptic potentials (IPSPs) were found in these cells to be characteristic of cerebellar stimulation, the investigation has been extended to correlate the time course of the IPSPs with the electrical activity of the cerebellar Purkinje cells under various conditions.

1. Latency of the cerebellar-induced IPSPs in Deiters' neurones

When the lobule III, IV or V of the cerebellum was stimulated at the vermis by applying brief pulses through concentric electrodes (Fig. 1 A), a marked

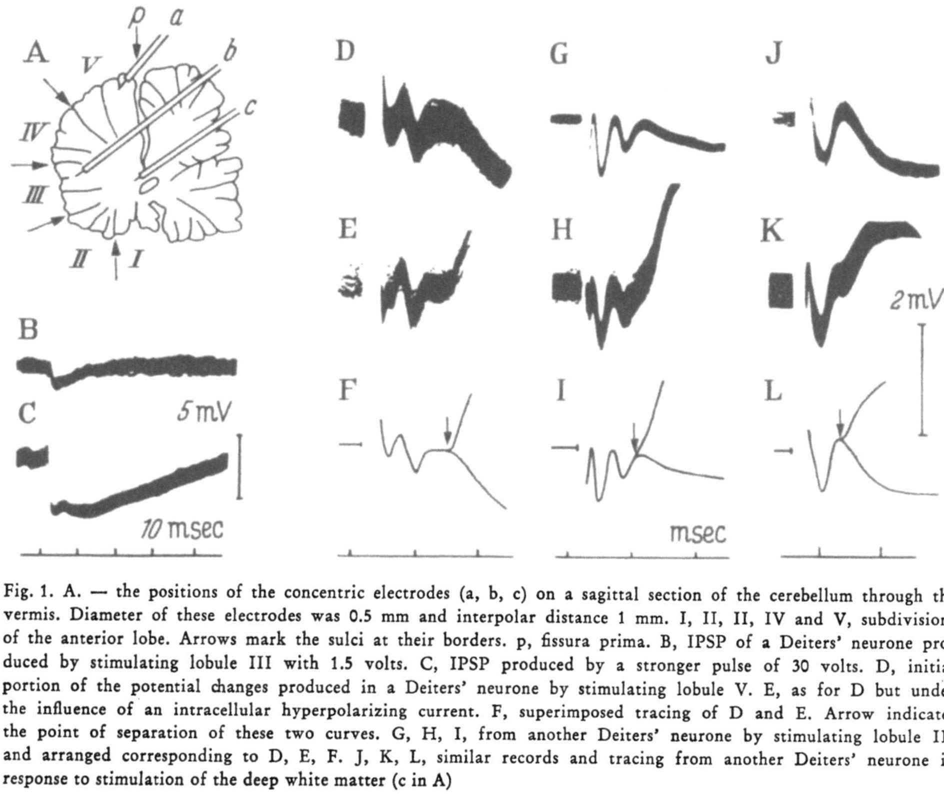

Fig. 1. A. — the positions of the concentric electrodes (a, b, c) on a sagittal section of the cerebellum through the vermis. Diameter of these electrodes was 0.5 mm and interpolar distance 1 mm. I, II, II, IV and V, subdivisions of the anterior lobe. Arrows mark the sulci at their borders. p, fissura prima. B, IPSP of a Deiters' neurone produced by stimulating lobule III with 1.5 volts. C, IPSP produced by a stronger pulse of 30 volts. D, initial portion of the potential changes produced in a Deiters' neurone by stimulating lobule V. E, as for D but under the influence of an intracellular hyperpolarizing current. F, superimposed tracing of D and E. Arrow indicates the point of separation of these two curves. G, H, I, from another Deiters' neurone by stimulating lobule III and arranged corresponding to D, E, F. J, K, L, similar records and tracing from another Deiters' neurone in response to stimulation of the deep white matter (c in A)

membrane hyperpolarization was observed in the majority of Deiters' neurones which immediately followed the stimulus and showed the properties of an IPSP (Ito and Yoshida 1964). This finding agrees with Pompeiano and Cotti's observation (1959) that the spontaneous firing of Deiters units was suppressed by stimulation of the vermis. With relatively weak stimulation near the threshold (usually around 1 volt), the IPSP was of a simple configuration with the peak time of about 1 msec, and a half time of decay of about 10 msec (Fig. 1 B). With increasing stimulus intensity, the membrane hyperpolarization was increased by delayed IPSPs, and its total duration often exceeded 100 msec (Fig. 1 C). The initial time course of these potential changes was complicated by spike field potentials (Fig. 1 D, G, J) from which the IPSPs could be distinguished by applying hyperpolarizing currents through the microelectrode. Such currents reversed the IPSPs in the depolarizing direction (Fig. 1 E, H, K). The time of the onset of the IPSPs was then determined accurately by superposing the original hyperpolarizing and the reversed depolarizing responses (Fig. 1 F, I, L). The latency of the IPSPs was thus revealed to be as short as 1.3 msec on the average when lobule V was stimulated (Fig. 1 D, E, F). It was shorter (1.1 msec) when lobule III or the rostral part of lobule IV was excited (Fig. 1 G, H, I). Further reduction was obtained, down to 0.8 msec on the average (Fig. 1 J, K, L), when the stimulating electrode was inserted deep into the white matter in the roof of the fourth ventricle (Fig. 1 A, c). The value of the latency as short as 0.8 msec indicated that the IPSPs were induced monosynaptically by impulses travelling down from the roof of the fourth ventricle to the nucleus of Deiters and impinging on Deiters' neurones directly. Prolongation of the latency according to the distance between the cerebellar electrodes and the nucleus of Deiters would be accounted for by assuming that the inbibitory path is continuous throughout the cerebellum with a conduction velocity of 20 m/sec or even less.

2. Localization of the IPSP-receptive field in the cerebellum for Deiters' neurones

When two or three concentric electrodes were combined to stimulate separate lobules in the anterior vermis, it was usually found that Deiters' neurones receive monosynaptic IPSPs from a wide homolateral region covering these lobules. In contrast, when the electrode tips were arranged on a right-left axis through a lobule, it was regularly observed that the homolateral vermis was almost exclusively effective in producing the monosynaptic IPSPs. For example, in Fig. 2 A, seven electrodes, associated in two separate sets, were inserted into lobule V, the spacing of their tips being indicated in Fig. 2 B on the surface of the culmen. Stimuli of three different intensities (duration, 0.2 msec) were applied in turn to these electrodes (Fig. 2 C). With 10 volts a monosynaptic IPSP of significant size occurred only from the spot No. 3 located in the homolateral vermis, and with 20 and 30 volts an IPSP was produced also from Nos. 2, 4, 5. The IPSPs induced from the other sites were delayed. The peak amplitudes of the monosynaptic IPSPs (indicated by interrupted lines) are plotted in Fig. 2 D as a function of the distance of the stimulating sites from the mid-sagittal line, for various values of the stimulating voltages. There is a clear dominance around the spot No. 3. The filled circles of Fig. 2 D plot the threshold intensity for inducing the monosynaptic IPSPs which, as would be expected, was lowest at No. 3 and increased sharply with movement to either side. Further investigation has been made with multipolar cerebellar electrodes which provided 16 to 20 separate spots for stimulation in the cerebellum at intervals of 2 or 3 mm (Ito and Kawai 1964). In seventy-eight Deiters' neurones thus examined, the spatial pattern of the IPSP-receptive field varied from case to case, but there was invariably a predominance of the homolateral vermis. It was further revealed that in about half of the cells IPSPs were induced monosynaptically also from the posterior lobe, exclusively from its homolateral vermian part. These distributions of the

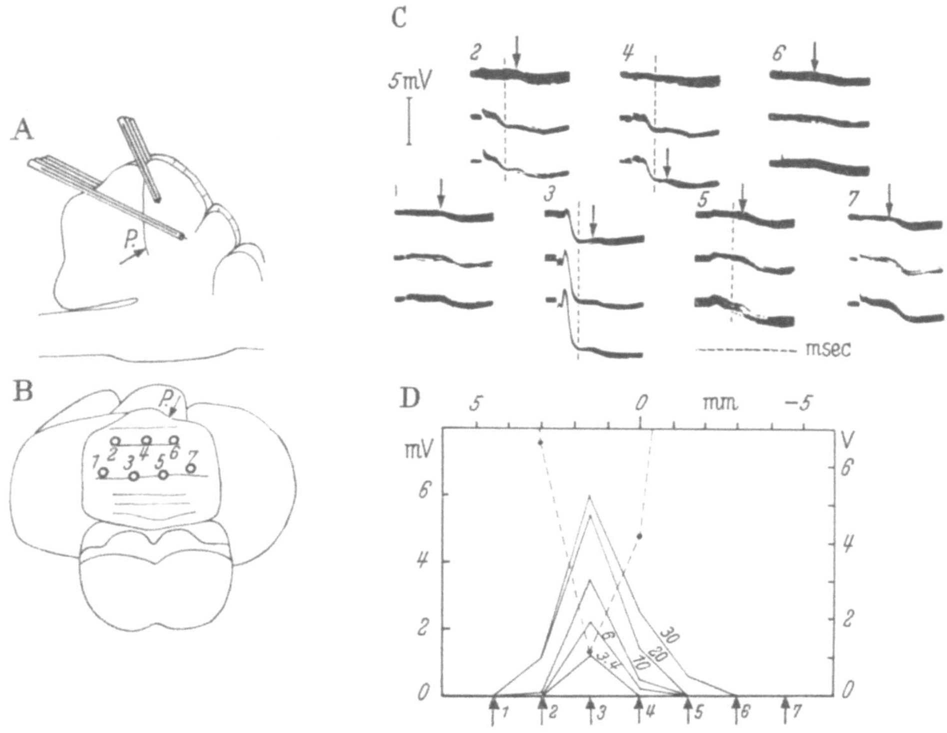

Fig. 2. A, side view of the cerebellum showing the positions of seven concentric electrodes. p, fissura prima. B, cranio-dorsal view of the culmen, indicating the spacing of the electrode tips 1 to 7. C, IPSPs recorded in a Deiters' neurone of the right nucleus by stimulating at each of the spots 1 to 7 in B, with pulses of 10 volts (uppermost traces), 20 volts (middle) and 30 volts (bottom). Vertical interrupted lines indicate the summits of the monosynaptic IPSPs and downward arrows the onsets of delayed IPSPs. D, Ordinates, amplitude (mV on left) of the monosynaptic IPSP (solid lines) and threshold (V on right) for inducing them (closed circles and broken lines). Abscissae, position of stimulated spots. Zero indicates the sagittal mid-line, figures with minus sign the contralateral side. Arrows indicate the position of the spots 1 to 7 in B. The figures on the solid lines show stimulus intensity

IPSP-receptive fields correspond well to the histologically determined cerebellar corticofugal projections onto Deiters' neurones (cf. Brodal, Pompeiano and Walburg 1962).

3. Monosynaptic inhibition of the intracerebellar nuclei

When neurones in the nuclei fastigii and interpositus were impaled with microelectrodes, stimulation of the homolateral anterior lobe regularly induced IPSPs with latencies as short as 0.8 msec. Likewise the dentate neurones were inhibited from the hemisphere on the homolateral side. The short latency for these IPSPs ensures that they occured monosynaptically (Ito, Yoshida and Obata 1964). Thus this cerebellar inhibition also conforms with the distribution of the short corticofugal projections (Jansen and Brodal 1954).

4. Correlation of the time course of the cerebellar-induced IPSP with the activity of cerebellar Purkinje cells

It was pointed out in connection with Fig. 2 C that, in addition to monosynaptic IPSP's, late IPSP's with latencies of 3—4 msec (marked with downward arrows) were recorded after sitmulation of the various cerebellar areas. It may be recalled that, when the deep white matter was stimulated, Purkinje cells were excited trans-synaptically after a latency of a few msec (Granit and Phillips

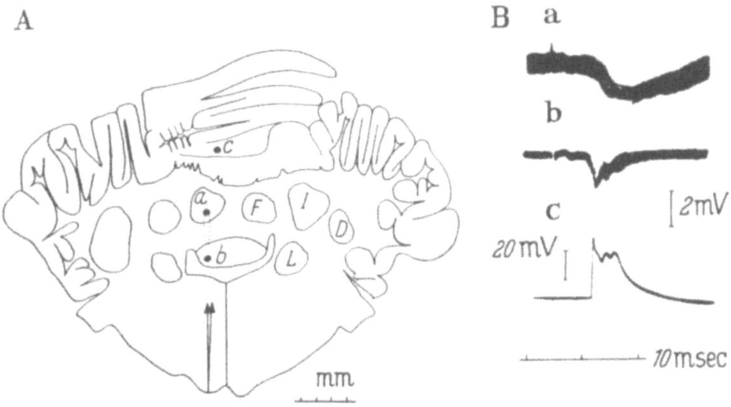

Fig. 3. A, Tracing of a transverse section of the cerebellum and the brain stem. Frozen section 40 microns thick. Arrows on the bottom indicate the directions of inserting the microelectrode. F, nucleus fastigii, I, nucleus interpositus, D, nucleus dentatus, L, nucleus of Deiters. Dotted line marks the microelectrode track found on this section. a, b, c, positions of taking the records in B. B, a, IPSP recorded from a fastigial neurone in response to C_3-stimulation. b, field potentials in lingula induced by the same stimulation as in a. C, activation of a Purkinje cell by the same means. Voltage calibration of 2 mV applies to a and b, and that of 20 mV to c

1956), there being a large depolarization which these authors called D-potentials. In the experiments of Ito, Obata and Ochi (1964) Purkinje cells were impaled near the tip of the concentric stimulating electrode which was inserted deep into the lobule III (see Fig. 1 A, a). These cells were found to be activated trans-synaptically with latency of 1.8 msec on the average, while the delayed IPSPs of Deiters' neurones had latencies of 3.1 msec. The discrepancy of 1.3 msec between these measurements would be accounted for chiefly by the conduction time and synaptic delay from Purkinje cells to Deiters' neurones (section 1).

IPSPs could also be recorded in Deiters' and fastigial neurones by stimulating the ventrolateral funiculi of the contralateral spinal cord at level C 3 (Ito et al.

1964 b), with latencies of 8 to 9 msec. For example IPSPs were recorded intracellularly (Fig. 3 B, a) from a fastigial neurone impaled at the position marked 'a' of Fig. 3 A. On the other hand, in synchronism with this IPSP, an appreciable negative field potential (Fig. 3 B, b), was detected in the lingula at position 'b', presumably due to the activity of Purkinje cells. Intracellular recording at position 'c' revealed that a Purkinje cell was indeed activated trans-synaptically a little earlier than the IPSPs (Fig. 3 B, c). The pathway for this spinal activation has been shown to be relayed by the inferior olive (ITO et al. 1964), the stimulation of which causes the Purkinje cells to fire and Deiters' neurones to be inhibited at latencies which are about half of those for cervical spinal cord stimulation. Thus there was a causal relationship between the activities of Purkinje cells and the inhibition of their target neurones.

Discussion

The present investigations have substantiated the inhibitory effects of cerebellar stimulation by revealing monosynaptic IPSPs of neurones in Deiters' and intracerebellar nuclei. By analysis of the spatial as well as temporal patterns of initiating these IPSPs it has further been suggested that the cerebellar Purkinje cells are responsible for this inhibition by being inhibitory in their nature. The cerebellar Purkinje cells would thus control, by *inhibition*, the reflex activities mediated through Deiters' as well as intracerebellar nuclei.

In addition to IPSPs, monosynaptic EPSPs were recorded in some Deiters' neurones, which, together with the polysynaptic EPSPs which were often superposed on IPSPs, would offer at least a partial explanation for the facilitatory effect of cerebellar stimulation. In contrast with the IPSPs, the receptive field for the monosynaptic EPSPs was wide, often covering the whole surface of the culmen and the posterior lobe, either homolateral or contralateral (ITO and KAWAI 1964). The cerebellar IPSPs have been attributed to the corticofugal projections as discussed above, and these EPSPs could be accounted for by another pathway. Thus, the possibility is pointed out that some cerebellar afferent fibres have the cerebellar cortex and the nucleus of Deiters as their common targets. Impulses in dorsal spinocerebellar fibres, primary vestibular afferents and the olivocerebellar fibres would eventually fire back from the cerebellar cortex to Deiters' neurones.

References

BRODAL, A., POMPEIANO, O. & WALBURG, F. (1962). In: *The vestibular nuclei and their connections, anatomy and functional correlations*, p. 64, Edinburgh: Oliver and Boyd.

GRANIT, R. & PHILLIPS, C. G. (1956). *J. Physiol.* (Lond.) **133**, 520.

ITO, M. & KAWAI, N. (1964). *Proc. Jap. Acad.* 40, 762.
ITO, M. & YOSHIDA, M. (1964). *Experientia* (Basel) 20, 515.
ITO, M., HONGO, T., YOSHIDA, M., OKADA, Y. & OBATA, K. (1964a). *Experientia* (Basel) 20, 295.
ITO, M., HONGO, T., YOSHIDA, M., OKADA, Y. & OBATA, K. (1964b). *Jap. J. Physiol.* 14, 638.
ITO, M., YOSHIDA, M. & OBATA, K. (1964). *Experientia* (Basel) 20, 575.
ITO, M., OBATA, K. & OCHI, R. (1964). *Proc. Jap. Acad.* 40, 765.
JANSEN, J. & BRODAL, A. (1954). *Aspects of Cerebellar Anatomy*, p. 189, Oslo: Johan Grundt Tanum.
LOWENTHAL, M. & HORSLEY, V. (1897). *Proc. Roy. Soc. Lond.*, B. 61, 20.
MORUZZI, G. (1950). *Problems in Cerebellar Physiology*, p. 24, Springfield: Charles G. Thomas.
POLLOCK, L. J. & DAVIS, L. (1927). *Brain* 50, 277.
POLLOCK, L. J. & DAVIS, L. (1930). *J. comp. Neurol.* 50, 377.
POMPEIANO, O. & COTTI, E. (1959). *Arch. sci. biol.* 43, 57 (quoted from BRODAL et al. [1962], p. 160).
SHERRINGTON, C. S. (1897). *Proc. Roy. Soc. Lond.* B. 61, 243.

DIFFUSION FROM CONSTRICTIONS

by J. C. JAEGER

I. Indroduction

Many cases appear to arise in physiology in which diffusion takes place from a restricted region into a large volume. Typical examples which will be considered later are:

(i) Diffusion from the ends of the micro-pipettes used in many experiments.
(ii) Diffusion from a tube with open ends immersed in liquid.
(iii) Diffusion from a thin slot such as the synaptic cleft.
(iv) Diffusion through a small hole in an impermeable membrane.

In many cases, simple approximate solutions are available and all that is needed is to determine possible inaccuracies in these or small corrections to them. In other cases, the actual behaviour near the constriction is of the greatest importance, for example in the related problem of heating by an electric current flowing into a half-space through a small contact. The electrical problem is discussed by HOLM (1946) and references to the thermal problems may be found in CARSLAW and JAEGER (1959).

Throughout, diffusion will be assumed to be linear (FICKIAN), D will be written for the diffusion coefficient and C for the concentration of the material diffusing. A case of diffusion combined with chemical reaction is treated in § VII. The notation

$$\operatorname{erf} z = \frac{2}{\pi^{1/2}} \int_0^z e^{-u^2}\, du, \quad \operatorname{erfc} z = 1 - \operatorname{erf} z, \tag{1}$$

$$\operatorname{ierfc} z = \int_z^\infty \operatorname{erfc} u\, du \tag{2}$$

will be used frequently. These functions are tabulated in most books of tables. It may be noted that erf $\infty = 1$. Also for large values of z

$$\pi^{1/2} z\, e^{z^2} \operatorname{erfc} z = 1 - {}^1/_2\, z^{-2}, \tag{3}$$

approximately, while for small values of z

$$\operatorname{erf} z = 2\,\pi^{-1/2} (z - {}^1/_3\, z^3 + \cdots). \tag{4}$$

II. Diffusion from instantaneous and continuous point sources

These are the fundamental solutions from which many others may be derived, they also provide useful approximations to many practical situations.

If a quantity M of diffusing substance is instantaneously liberated at the origin $r = 0$ at time $t = 0$ in an infinite medium in which its diffusion constant is D, the concentration C at distance r and time t is given by [CRANK (1956) § 3.2; CARSLAW and JAEGER (1959) § 10.2]

$$C = \frac{M}{8(\pi D t)^{3/2}}\, e^{-r^2/4Dt}. \tag{5}$$

For given r, this rises to a maximum value C_{max} and subsequently falls again. The maximum values C_{max} is given by

$$C_{max} = \frac{M}{8 r^3} \left(\frac{6}{\pi}\right)^{3/2} e^{-3/2} = 0.0736\ \frac{M}{r^3} \tag{6}$$

and is attained at a time t_{max} given by

$$t_{max} = r^2/6\,D. \tag{7}$$

These equations have been used by DEL CASTILLO and KATZ (1955) when dealing with the action of electrophoretically administered acetylcholine upon end-plate receptors.

In many experimental applications, the diffusing substance is supplied at a constant rate, m per unit time, at the origin for times $t > 0$. In this case [the "continuous point source", CRANK 1956, § 3.21, CARSLAW and JAEGER 1959, § 10.4] the concentration at distance r at time t is

$$C = \frac{m}{4\pi D r} \operatorname{erfc} \frac{r}{2(D t)^{1/2}} \tag{8}$$

The concentration under any conditions may be found immediately from tables of the error function. CURTIS, PERRIN and WATKINS (1960) have used this as an approximation for diffusion from micropipettes and show some typical concentration-distance curves.

An important practical case is that in which material is supplied at a constant rate m per unit time at the origin from zero time up to time t_1, after which time the supply ceases. In this case the concentration is given by (4) for $0 < t < t_1$, whilst for times $t > t_1$ it is

$$C = \frac{m}{4\pi D r}\left\{\operatorname{erfc}\frac{r}{2(Dt)^{1/2}} - \operatorname{erfc}\frac{r}{2[D(t-t_1)]^{1/2}}\right\} \tag{9}$$

Since (9) involves the three quantities r, t and t_1 it is difficult to display the results in a general fashion and it is most satisfactory to calculate special cases numerically. Fig. 1 shows typical curves of concentration against time.

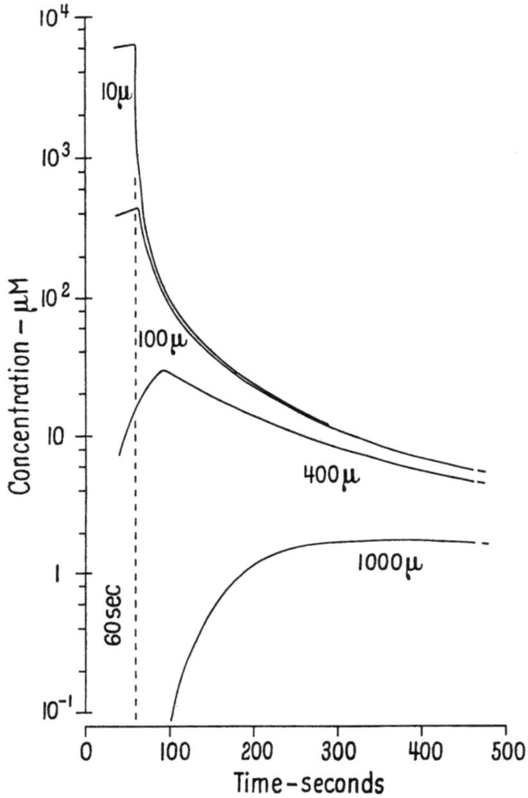

Fig. 1. Concentration C in micromoles at times 0—500 sec. and distances of 10, 100, 400, 1000 μ for the case of $D = 5 \times 10^{-6}$ cm²/sec., rate of supply $m = 4 \times 10^{-13}$ moles/sec., and time of supply $t_1 = 60$ sec. The cessation of supply is indicated by the broken vertical line

When discussing the variation of concentration at any point after the supply has been cut off it is convenient to take the time of supply t_1 as a basic unit of time and to express other times as multiples of it. The ratio of the concentration $C\,(n\,t_1)$ at any point at time $n\,t_1$ to its value $C\,(t_1)$ at time t_1 is given by

$$C(n\,t_1)/C(t_1) = \left\{\operatorname{erfc}\frac{r}{2(n\,D\,t_1)^{1/2}} - \operatorname{erfc}\frac{r}{2[(n-1)\,D\,t_1]^{1/2}}\right\} \Big/ \operatorname{erfc}\frac{r}{2(D\,t_1)^{1/2}}\,. \tag{10}$$

This is a function only of n and $r/2(D\,t_1)^{1/2}$ and so can be displayed as a family of curves. The ratio of the values of the concentration at time t with and without cut-off of supply at time t_1, i.e. the ratio of (9) and (8), has the same property.

As in the case of the instantaneous point source, it is interesting to calculate the maximum concentration $C_{\max}$ attainted at any point and the time $t_{\max}$ at which is attained. This is given by

$$t_{\max} = N\,t_1 \tag{11}$$

where N is given by

$$N(N-1)\ln[N/(N-1)] = r^2/6\,D\,t_1 \tag{12}$$

and when t_{max} has been found, C_{max} is given by (9). Values of these quantities are given in Table I. The case $N = \infty$ is that of the instantaneous point source for which results are given in (6) and (7). Table I suggests that for times greater than about 5 t_1 the effects of a pulse of length t_1 are approximated to reasonably well by an instantaneous point source.

Table I. *The time* $t_{max} = Nt_1$ *at which maximum concentration* C_{max} *is attained at distance* r *from a continuous point source at which material is supplied at rate* m *for time* t_1

N	1,5	2	3	4	5	10	∞
$r^2/6Dt_{max}$	0,549	0,693	0,811	0,863	0,892	0,948	1
r^3C_{max}/mt_1	0,0681	0,0714	0,0728	0,0732	0,0735	0,0736	0,0736

III. Diffusion across a small area into a large volume

It is difficult to discuss the conditions very close to the source of supply, partly because the physical conditions there are largely unknown, and partly because, even if they could be described adequately, the mathematics would be intractable. Most discussion has centred around the flow of electric current or heat from a circular contact into a half-space. In this case some mathematical solutions are available which allow the validity of simple approximations to be studied.

Stated in terms of diffusion, the problem is as follows. Material is to diffuse into the half-space $z > 0$ over the circle AB of radius a in the plane $z = 0$, Fig. 2 (a). There are of course two cases, the transient one in which the early stages of diffusion are considered, and the steady-state which is attained after an infinite time.

Considering first the steady-state case, the simplest assumption is that the concentration is maintained constant at C_0 over the circle of radius a, and that there is no diffusion across the plane outside this circle. In this case the total rate Q at which material crosses the circle per unit time can be calculated [CARSLAW and JAEGER 1959, § 8.2 (10)] and is

$$Q = 4\,D\,a\,C_0\,. \tag{13}$$

The quantity $R = C_0/Q$ which is analogous to contact resistance in the electrical or thermal case may be described as the resistance to diffusion through the contact and in this case is

$$R = C_0/Q = 1/4\,D\,a. \tag{14}$$

It is by no means clear that the assumption that the concentration is constant over the circle of radius a is the one most appropriate physically to the problem of diffusion from a circle into a half space. An alternative simple assumption is that the rate of transfer of material per unit time per unit area across the circle has a constant value. In this case the concentration C varies across the circle, but its mean value $\overline{C}$ can be calculated [Carslaw and Jaeger 1959 § 8.2 (15)] and the resistance R of the contact on this assumption is found to be

$$R = \overline{C}/Q = 8/3\,\pi^2 D\,a. \tag{15}$$

The fact that (14) and (15) differ numerically by a factor of 1.08 is a measure of the uncertainty introduced into the problem in its mathematical specification.

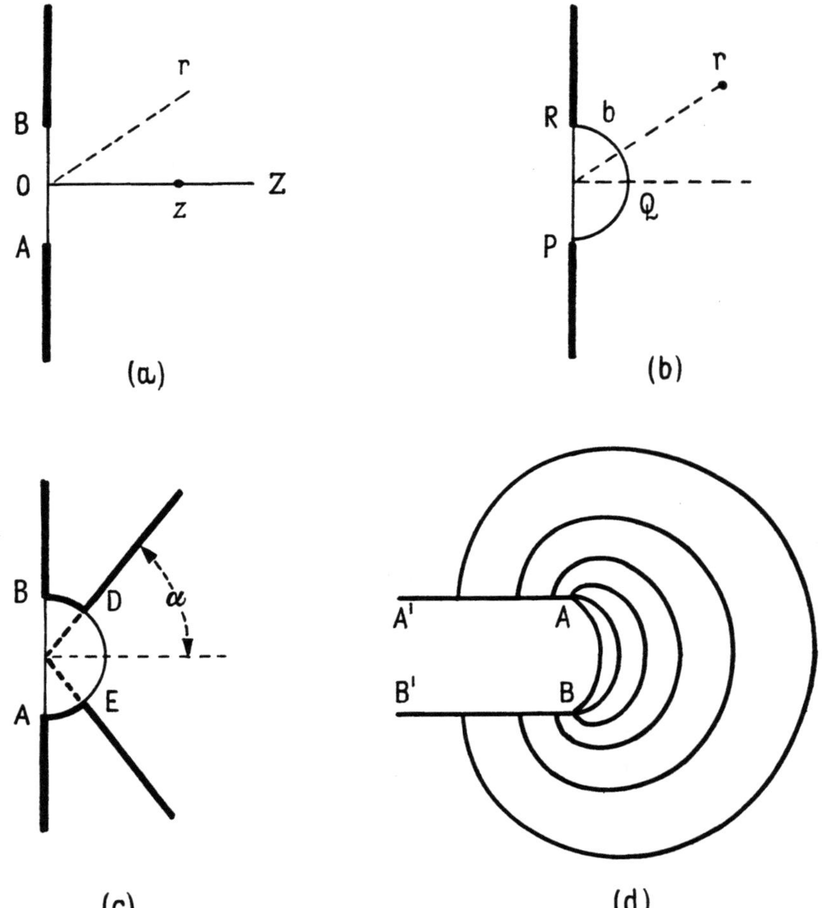

Fig. 2. Configurations for diffusion from a small source. (a) Disk source. (b) Hemispherical approximation. (c) Diffusion into a cone. (d) Diffusion from a slit in two dimensions

Because of this uncertainty and the rather complicated mathematics involved in dealing with the circular area AB, it has been usual to ignore effects in its immediate neighbourhood and to replace it by a hemisphere PQR, of radius b Fig. 2 (b), from which diffusion into the half-space would be radial. In this case if the rate of transfer of material per unit time per unit area across the hemisphere has the constant value $Q/2\pi b^2$, the concentration C at radius r in the steady-state is found to be $C = Q/2\pi D r$ and the resistance R of the contact is

$$R = 1/2\,\pi D\,b. \tag{16}$$

It appears that the resistance given by (16) on this assumption is the same as (14) for the circle if

$$b = 2\,a/\pi \tag{17}$$

Passing now to the variable-state case in which concentration C is zero at zero time $t=0$, the most important case, and the simplest for purposes of calculation, is that of supply of material at a constant rate. We shall consider a number of cases of this type and compare concentrations calculated from them.

(a) Material transfer at a constant rate $Q/\pi a^2$ per unit time per unit area for $t>0$ across the circle of radius a, Fig. 2 (a).

In this case the concentration at any point may be expressed as an infinite integral (JAEGER 1952) but on the axis of symmetry O Z takes the simple form

$$C = \frac{2\,Q\,t^{1/2}}{\pi\,a^2\,D^{1/2}} \left\{ \operatorname{ierfc} \frac{z}{2\,(D\,t)^{1/2}} - \operatorname{ierfc} \frac{(z^2+a^2)^{1/2}}{2\,(D\,t)^{1/2}} \right\} \tag{18}$$

where z is the distance along O Z.

(b) Material supply at the origin at the rate Q per unit time for $t>0$ diffusing into a half space.

This corresponds to replacing the disc-source of (a) by a point source emitting material at the same rate. This is the continuous point source previously discussed, except that material is diffusing into a half-space instead of an infinite region so that m in (4) has the value 2 Q and

$$C = \frac{Q}{2\,\pi\,D\,r} \operatorname{erfc} \frac{r}{2\,(D\,t)^{1/2}}, \tag{19}$$

where r is the radial distance from the origin 0.

(c) Material supplied at the rate $Q/2\,\pi\,b^2$ per unit time per unit area over a hemisphere of radius b and diffusing into a half-space outside it.

This model, Fig. 2 (b), has been used by HOLM (1946) and others in connection with flow of electric current near a contact. The concentration at radius r at time t is [CARSLAW and JAEGER (1959), § 9.10 (4)]

$$C = \frac{Q}{2\,\pi\,D\,r} \left\{ \operatorname{erfc} \frac{r-b}{2\,(D\,t)^{1/2}} - \exp\left[\frac{r-b}{b} + \frac{D\,t}{b^2}\right] \operatorname{erfc}\left[\frac{r-b}{2\,(D\,t)^{1/2}} + \frac{(D\,t)^{1/2}}{b}\right]\right\} \tag{20}$$

Numerical values calculated on these assumptions are compared in Fig. 3. In all cases the abscissa is $\log_{10}(D\,t/r^2)$ and the total rate of supply is Q per unit time. Curve I is the value (19) for the continuous point source. Curve II is calculated from (20) with $b=a$ and $r=2\,a$. Curve III is calculated from (20) with $b=2\,a/\pi$, as suggested by (17), and $r=2\,a$. Curve IV is calculated from (18) with $z=r=2\,a$. Since all these curves correspond to distances of the order of one diameter from the source, and the discrepancies between them are small, it appears that any of the approximations (a) to (c) may safely be used at greater distances. However if effects in the immediate neighbourhood, say $r<a$, of the source are under consideration, these approximations are unsatisfactory and also it becomes uncertain what actual boundary conditions should be used. This

question is of some importance since it is in this region that concentration-dependent effects are most likely to appear, and in the case of electrical contacts it is the region in which heating is greatest.

Clearly, if diffusion takes place into a cone of semi-vertical angle α, Fig. 2 (c), the approximations (b) and (c) may still be used. If the total rate of supply of material is Q, (19) and (20) have to be divided by $1-\cos\alpha$. The least satisfactory case will be that of diffusion from the surface AB into a whole space. This is the case of diffusion from the end of a micro-pipette $A'ABB'$, Fig 1 (d). From the geometry of Fig. 1 (c) and (d) it is clear that $\alpha=\pi$ and (8) are not likely to be good approximations for distances less than several times the diameter of the

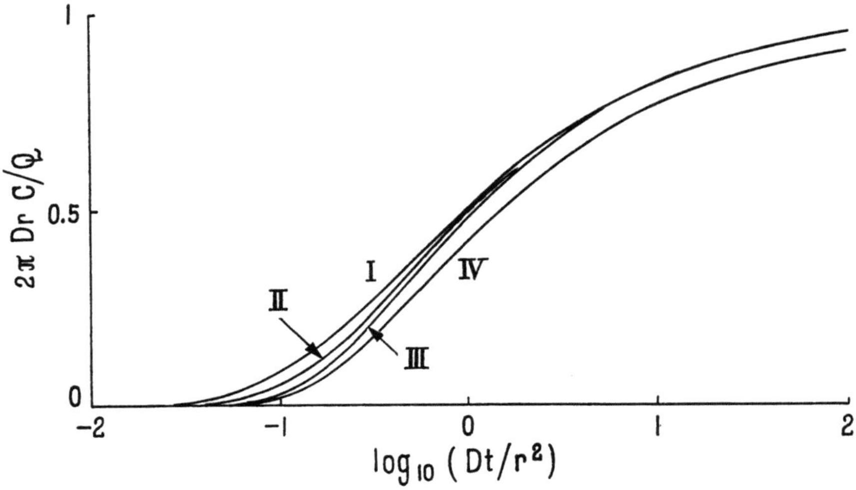

Fig. 3. Variation of concentration with Dt/r^2 for diffusion into a half-space at the rate Q per unit time from sources of various shapes. Curve I, a point source. Curve II, a hemisphere of radius $^1/_2\, r$. Curve III, a hemisphere of radius r/π. Curve IV, a disk of radius $^1/_2\, r$

tube. In general, concentrations will be less than those given by (8) in the "backward" direction and greater in the forward. The sort of effect to be expected is shown by Fig. 2 (d) which shows lines of equal concentration for two-dimensional diffusion in the steady state from a slit (JEANS 1925, p. 275).

IV. Diffusion from a tube into a half space

In the method of ANDERSON and SADDINGTON (1940) for measuring diffusion coefficients a tube of length $2\,l$ and diameter $2\,a$ is filled with liquid containing the material whose diffusion constant D is to be measured at constant concentration C_0, and at time $t=0$ the tube is immersed in a large volume of liquid. The average concentration C_{av} in the tube is measured at time t. In the theory it is assumed that the boundary condition at the ends of the tube is that of zero concentration and on this assumption [CARSLAW and JAEGER 1959, § 3.3 (10)]

$$C_{av}=\frac{8\,C_0}{\pi^2}\sum_{n=0}^{\infty}\frac{1}{(2\,n+1)^2}\exp\left\{-\frac{\pi^2\,D\,t\,(2\,n+1)^2}{4\,l^2}\right\} \tag{21}$$

This simple assumption ignores the fact that at the end O of the tube, Fig. 4 inset, there is in fact radial diffusion from the tube into the region around it.

To see the order of magnitude of this effect the approximations of the last section may be used. The case of diffusion into a half-space (tube with very thick walls) will be considered first. The problem is broken up into two parts, firstly the tube $0<x<l$ in which the concentration C satisfies

$$\frac{\partial^2 C}{\partial x^2} - \frac{1}{D}\frac{\partial C}{\partial t} = 0,\ 0<x<l \tag{22}$$

with $C = C_0$, when $t=0$, $\quad 0<x<l$, (23)

and $\frac{\partial C}{\partial x} = 0$, when $x=l$, $\quad t>0$. (24)

Secondly, following the discussion of the last Section, diffusion will be taken to be radial in the region $r>a$ outside the hemisphere H, Fig. 4 inset, so that the concentration C_1 in this region satisfies

$$\frac{\partial^2 C_1}{\partial r^2} + \frac{2}{r}\frac{\partial C_1}{\partial r} - \frac{1}{D}\frac{\partial C_1}{\partial t} = 0,\ r>a,$$

with

$$C_1 = 0,\ r>a, \quad t=0. \tag{25}$$

The assumption is now made that effects in the hemisphere can be neglected so that C and C_1 are connected by continuity of concentration and of rate of transfer of material so that

$$[C]_{x=0} = [C_1]_{r=a} \tag{26}$$

$$\pi a^2 \left[\frac{\partial C}{\partial x}\right]_{x=0} = -\mu \pi a^2 \left[\frac{\partial C_1}{\partial r}\right]_{r=a} \tag{27}$$

where $\mu = 2$ for the present case, but varying μ will take into account the other cases discussed later.

The solution of (22) to (27) follows easily by the LAPLACE transformation method (CARSLAW and JAEGER 1959, Chap. XII) and the average concentration C_{av} at time t is found to be

$$C_{av} = \frac{2\mu\beta^2 C_0}{\pi}\int_0^\infty \frac{1}{\Delta(u)}\, e^{-\alpha u^2}\sin^2 u\, du, \tag{28}$$

where

$$\Delta(u) = (\mu\cos u - \beta u \sin u)^2 + \mu^2\beta^2 u^2\cos^2 u \tag{29}$$

and

$$\alpha = D\,t/l^2,\ \beta = a/l. \tag{30}$$

Some values of C_{av}/C_0 as a function of $D\,t/l^2$ for various conditions are shown in Fig. 4. It can be shown that the integral (28) tends to the series (21) as $\beta \to 0$: values of this latter are shown in Curve I. The case $\beta = 0.1$, $\mu = 2$, corresponding to diffusion into a half-space is given in Curve II and shows the order of magnitude of the correction to be expected. For diffusion into a half-space through a hemisphere of radius $2a/\pi$, as suggested by (17), the value of μ is $8/\pi^2$ and Curves III and IV correspond to this case with $\beta = 0.1$ and $\beta = 0.05$, respectively. For diffusion into a whole space, $\mu = 4$ and the curve for $\beta = 0.1$ differs very little from Curve I.

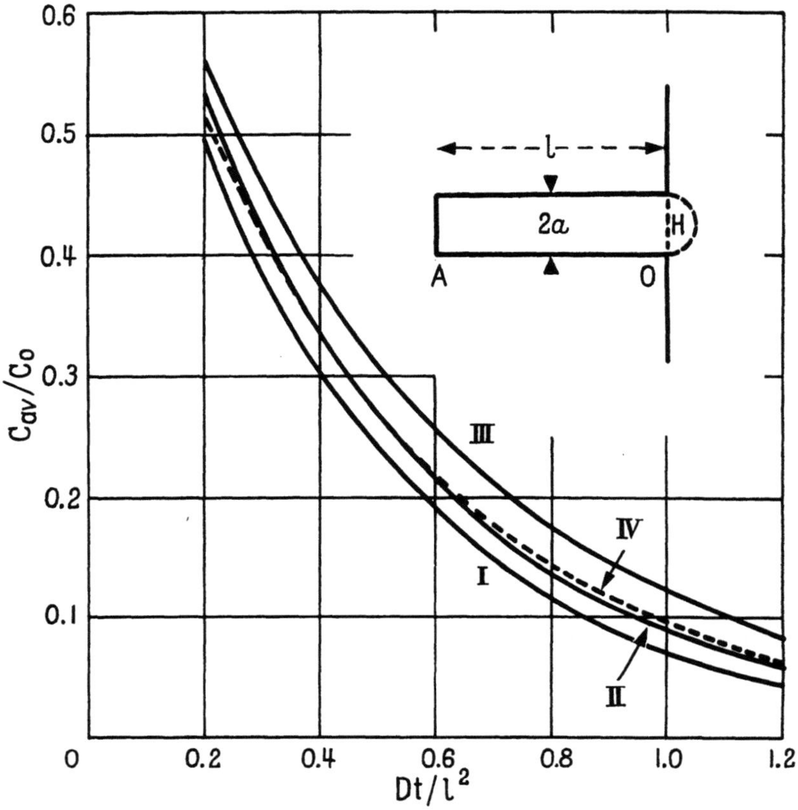

Fig. 4. Average concentration (C_{av}) as a function of Dt/l^2 for diffusion from a tube under various conditions. Curve I, $C = 0$ at the end or $\beta = 0$. Curve II, diffusion into a half space, $\beta = 0.1$, $\mu = 2$. Curve III, $\beta = 0.1$, $\mu = 8/\pi^2$. Curve IV, $\beta = 0.05$. $\mu = 8/\pi^2$.

V. Diffusion through a small hole in a thin impermeable membrane

Suppose that liquid to the left of an infinite plane impermeable membrane of negligible thickness contains liquid with an initial concentration C_0 of diffusing material, whilst the initial concentration to the right of the membrane is zero. Suppose that for $t > 0$ diffusion takes place through a small circular hole of radius a with its centre at the origin in the membrane. As before, effects in a sphere of radius a about the origin will be neglected and diffusion taken to be radial in the two regions. Then if C_1 and C_2 are the concentrations at radius r in the regions to the left and right of the membrane the conditions to be satisfied at the hole are

$$C_1 = C_2, \quad r = a, \tag{31}$$

$$\frac{\partial C_1}{\partial r} = \frac{\partial C_2}{\partial r}, \quad r = a. \tag{32}$$

The solution is found to be

$$C_1 = C_0 - \frac{a\,C_0}{2\,r}\ \operatorname{erfc} \frac{r-a}{2\,(D\,t)^{1/2}} \tag{33}$$

$$C_2 = \frac{a\,C_0}{2\,r}\ \operatorname{erfc} \frac{r-a}{2\,(D\,t)^{1/2}} \tag{34}$$

These results are readily extended to the case in which the diffusion constants in the two regions are different. Allowance can also be made for the thickness of the membrane by assuming a resistance to diffusion across the hole.

Problems in which there are sources in the media may also be treated approximately in the same way.

VI. Diffusion from a slit into a half space

This is the two-dimensional problem analogous to that treated in Section IV. It has been discussed by ECCLES and JAEGER (1957) in connection with diffusion of transmitter substance from the synaptic cleft. They give a family of curves similar to those of Fig. 4 but in this case the effect of the constriction appears at lower values of β. JAEGER (1960) has extended the present approximate method to study diffusion in clefts or tubes which divide into several branches.

VII. Radial diffusion from a point with chemical reaction

A further complication arises if molecules supplied at a point and diffusing away from it react with molecules present in the fluid around it.

The simplest problem of this type may be stated as follows: the spherical region $r>0$ initially contains material M_2 at uniform concentration C_0, its

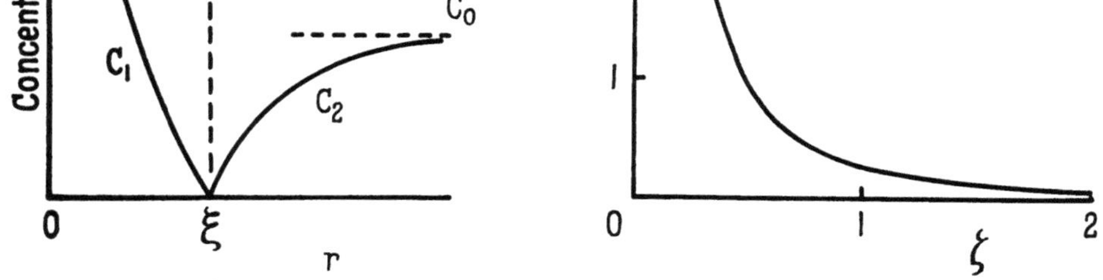

Fig. 5. (a) Variation of concentrations C_1 and C_2 with distance for the case of chemical reaction. (b) The function $z = \zeta^{-1} - \pi^{1/2}\ e^{\zeta^2} \operatorname{erfc} \zeta$

diffusion coefficient being D_2. For times $t>0$, material M_1 is supplied at the origin at a constant rate Q per unit time, its diffusion constant being D_1. The two materials M_1 and M_2 combine chemically in the proportion of μ molecules of M_1 to one of M_2. The effect of this is that there will be a moving spherical boundary of radius $r=\xi$, where ξ is a function of time, such that only M_1 is present for $r<\xi$ and only M_2 for $r>\xi$, whilst at $r=\xi$ the concentration of both M_1 and M_2 will be zero.

If C_1 and C_2 are the concentrations of M_1 and M_2, respectively, the distribution of concentration will be as shown in Fig. 5 (a). The equations to be satisfied are as follows:

$$\frac{D_1}{r^2}\frac{\partial}{\partial r}\left(\frac{(r^2\partial C_1)}{\partial r}\right)-\frac{\partial C_1}{\partial t}=0, \quad 0<r<\xi, \quad t>0 \tag{35}$$

$$-4\pi r^2 D_1\left[\frac{\partial C_1}{\partial r}\right]_{r=0}=Q, \tag{36}$$

$$\frac{D_2}{r^2}\frac{\partial}{\partial r}\left(r^2\frac{\partial C_2}{\partial r}\right)-\frac{\partial C_2}{\partial t}=0, \quad r>\xi, \quad t>0, \tag{37}$$

$$C_2\to C_0 \text{ as } r\to\infty, \tag{38}$$

$$C_1=C_2=0, \quad r=\xi, \tag{39}$$

$$-D_1\frac{\partial C_1}{\partial r}=D_2\frac{\partial C_2}{\partial r}, \quad r=\xi, \tag{40}$$

$$\xi=0, \quad t=0, \tag{41}$$

$$C_2=C_0, \quad t=0, \quad r>0. \tag{42}$$

It is known that there is no accurate elementary solution of these equations. An approximate solution may be found by a method which has proved to be valuable in the study of freezing. This consists of assuming that the processes are so slow that the concentration C_2 may be taken at all times to have its value for steady diffusion in the region $0<r<\xi$ so that $\partial C_1/\partial t$ in (35) may be neglected. On this assumption, (35), (36) and (39) are satisfied by

$$C_1=\frac{Q}{4\pi D_1}\left(\frac{1}{r}-\frac{1}{\xi}\right). \tag{43}$$

Also (37), (38) and (42) are satisfied by

$$C_2=C_0+B\left\{\frac{(D_2 t)^{1/2}}{r}e^{-r^2/4D_2t}-1/2\,\pi^{1/2}\operatorname{erfc}\frac{r}{2(D_2 t)^{1/2}}\right\} \tag{44}$$

where B is an unknown constant.

Substituting (43) and (44) in the remaining equations (39) and (40) gives

$$B\left\{\frac{(D_2 t)^{1/2}}{\xi}e^{-\xi^2/4D_2t}-1/2\,\pi^{1/2}\operatorname{erfc}\frac{\xi}{2(D_2 t)^{1/2}}\right\}=-C_0 \tag{45}$$

$$4\pi\mu D_2 B(D_2 t)^{1/2}e^{-\xi^2/4D_2t}=-Q. \tag{46}$$

Writing

$$\zeta=\frac{\xi}{2(D_2 t)^{1/2}}, \quad k=8\pi\mu C_0 D_2/Q, \tag{47}$$

(45) and (46) give

$$k(D_2 t)^{1/2}=\zeta^{-1}-\pi^{1/2}e^{\zeta^2}\operatorname{erfc}\zeta \tag{48}$$

from which ζ and hence ξ can be determined numerically as a function of t for given k. The right-hand side of (48) as a function of ζ is shown in Fig. 5 (b). It follows that for small values of the time ζ must be large, in fact $\zeta = (2k)^{-1/3}(D_2 t)^{-1/6}$ and $\xi = 2^{2/3} k^{-1/3} (D_2 t)^{1/3}$ so that $\xi \to 0$ as $t \to 0$ as required by (41). For large values of the time, ζ must be small and (48) gives approximately

$$\xi = \frac{2(D_2 t)^{1/2}}{\pi^{1/2} + k(D_2 t)^{1/2}} . \tag{49}$$

As $t \to \infty$, $\xi \to 2/k$ which is the radius at which the supply Q of M_1 is in equilibrium with the amount of M_2 diffusing from the region outside. (49) has been used by CURTIS, PERRIN and WATKINS (1960).

It was remarked that there was no simple and accurate solution of this problem. In fact, only three problems of this type have simple solutions and in all of them the distance ξ of the surface of separation is accurately proportional to $t^{1/2}$. These problems are (a) linear flow in the region $x > 0$ with prescribed concentration at $x = 0$, (b) radial diffusion in two dimensions from a slit or line source with a constant rate of supply of material, (c) radial diffusion in three dimensions from a point source at which material is supplied at a rate proportional to $t^{1/2}$. The solutions of the first two of these problems are given by CARSLAW and JAEGER (1959), §§ 11.2, 11.6, and that of the third may be obtained in the same way. Comparison of these accurate results with those obtained by the present approximate method allows its validity to be checked. The first of these problems, (a) above, and others of physiological interest are discussed in detail by HILL (1928) who uses approximations of the present type freely. The second problem mentioned, (b) above, is the exact analogue for radial diffusion of the one treated here so its solution may be of some practical interest.

Acknowledgements. I am greatly indebted to Dr. D. R. CURTIS for suggesting some of these problems to me and to Mrs. T. CHAMALAUN for the computation of the integral involved in Fig. 4.

References

CARSLAW, H. S. & JAEGER, J. C. (1959). *Conduction of Heat in Solids,* 2nd ed. Oxford: Oxford University Press.

CASTILLO, J. DEL & KATZ, B. (1955). *J. Physiol.* (Lond.) **128**, 157.

CRANK, J. (1956). *The Mathematics of Diffusion.* Oxford: Oxford University Press.

CURTIS, D. R., PERRIN, D. D. & WATKINS, J. C. (1960). *J. Neurochem.* **6**, 1.

ECCLES, J. C. & JAEGER, J. C. (1957). *Proc. roy. Soc.* B **148**, 38.

HILL, A. V. (1928). *Proc. roy. Soc.* B **104**, 39.

HOLM, R. (1946). *Electric contacts.* Stockholm: Hugo Gebers Förlag.

JAEGER, J. C. (1952). *Aust. J. Sci. Res.* A **5**, 1.

JAEGER, J. C. (1960). *Proc. Camb. Phil. Soc.* **56**, 55.

JEANS, J. H. (1925). *The mathematical theory of electricity and magnetism,* **5***th ed.* Cambridge: University Press.

THE QUANTAL RELEASE OF TRANSMITTER SUBSTANCES

by B. KATZ and R. MILEDI

We should like to take this opportunity not only of presenting our best wishes to our friend JACK ECCLES, but of doing something which is generally not done, namely to discuss a teleological question. The question we want to pose is: why should chemical transmitters be released in quantal amounts, that is, in large multi-molecular packets whose size is pre-set and independent of the event which causes their release?

Before doing so, it is advisable to survey briefly the existing evidence which shows that chemical transmitters are, in fact, released in quantal amounts, and not by random molecular leakage, from the nerve ending.

It started with the accidental discovery that the end-plate region of a muscle fibre is the site of spontaneous discharges of small subthreshold potential changes, the miniature end-plate potentials (FATT and KATZ 1950, 1952). In their reaction to various drugs, these miniature e.p.p.'s resemble very closely the local potentials which are produced by applying acetylcholine to the end-plate (FATT and KATZ 1952; DEL CASTILLO and KATZ 1955 a, 1957; MILEDI 1962). The view is therefore widely accepted that the miniature e.p.p.'s are due to a spontaneous impact of acetylcholine on the end-plate receptors of the muscle fibre. There is one important difference between miniature e.p.p.'s and artificially produced "ACh-potentials": ACh-potentials can be continuously reduced in size by lowering the applied amount, while the miniature e.p.p. has the character of a discrete all-or-none phenomenon. The fact that the ACh-potentials are "non-quantal" means simply that the effect of a single molecule of ACh cannot be detected. It also means that the appearance of discrete miniature e.p.p.'s must be due to the spontaneous delivery from some local source of large packets of acteylcholine molecules. That the source is in the motor nerve terminals has been shown in several ways: for instance by the fact that the frequency of the discharge can be controlled by passing a current through the terminal portion of the motor nerve (DEL CASTILLO and KATZ 1954 c; LILEY 1956 c; KATZ and MILEDI 1965 a), and because spontaneous discharge stops after nerve section at a time when the motor nerve endings degenerate (FATT and KATZ 1952; LILEY 1956 a).

At this point we must digress a little. Some years ago, we came across a curious observation, namely that in denervated frog muscle the stoppage of spontaneous end-plate activity is temporary. After a short period of complete "end-plate silence", spontaneous potentials begin to turn up again at the majority of the end-plates — though at a frequency much lower than in normal muscle. When doing the usual tests (external and intracellular microapplication of acetylcholine, application of curare and anti-cholinesterases), it became quite clear that a focal source had re-appeared on the outside of the muscle fibre, capable of delivering packets of acetylcholine to the end-plate. When these muscles were examined in the electron microscope, the only structures which were found in the appropriate positions were nucleated Schwann cells which had taken over the place of the disintegrated nerve endings and draped themselves over the end-plate surface of the muscle fibre forming a close "synaptic" contact with it.

The suggestion was therefore put forward (Birks, Katz and Miledi 1960) that Schwann cells, after disposing of the axon terminals, become capable of producing packets of acetylcholine and of releasing them, even though only on a rudimentary scale. This was a suggestion which some of our colleagues probably found difficult to accept, but to us at any rate it seemed reasonable and based on a good deal of converging, though not by any means conclusive, evidence. Since then, this suggestion has been further investigated. In the diaphragm of the rat, Miledi and Slater (1963) found that denervation causes the miniature e.p.p.'s to vanish permanently, except at very few scattered end-plates, say, one out of a hundred at which spontaneous activity was observed during the third week after nerve section. When the muscles were examined in the electron microscope, the great majority of the denervated end-plates were found to be covered with collagen: the Schwann cells had failed to establish close contact. One preparation, however, did show "synaptic" contacts between Schwann-type cells and muscle end-plate. This specimen came from a region in which resumption of miniature e.p.p.'s had occurred. Thus, the suggestion of Schwann cells being able to take over some of the biochemical functions of their lately associated neuron still stands and, in fact, is doing not too badly.

Going back to our main theme, the evidence that the miniature e.p.p.'s in normal resting muscle are due to a spontaneous multi-molecular release of acetylcholine from the motor nerve terminals, is strong enough to be taken as a basis from which to proceed. It is, at any rate, certain that the processes which follow the release, i. e., diffusion towards the receptors and the reaction leading to depolarization, lack any detectable quantal properties, as is shown by the continuously graded effect of ionophoretic micro-application of acetylcholine. It does not follow that the nerve endings *cannot* release acetylcholine in any other way: there may well be diffusive leakage quite apart from the quantal release, but only the latter produces an effect which can be detected with our recording methods. Suppose that the spontaneous quantal release, which produces miniature e.p.p.'s at a rate of about one per second at most end-plates, accounts for only 1/10 of the total efflux of acetylcholine from the resting nerve endings, and that 9/10 escape by diffusion, i. e., random molecular leakage at a steady average rate. Such constant leakage, spread out evenly in time, would cause only a minute steady depolarization, at the most some 50 μV, which is well below the limits of error in measuring resting membrane potentials. Thus, small diffusive leakage of acetylcholine — unlike the synchronous quantal release — would escape detection, and we have therefore no right to assert that it does not exist.

We can, however, be quite certain that the extra amount of transmitter released from the endings by a nerve impulse is made up of an integral (and normally large) number of quantal packets, and not of a mixture of quanta plus acceleration of any diffuse membrane leakage. This can be stated with assurance because the work of del Castillo and Katz (1954 b), Boyd and Martin (1956), Liley (1956 b) and Katz and Miledi (1965 b) has shown that the end-plate potential, i. e., the response to a nerve impulse, consists of an integral number of units which are identical with the spontaneous miniature e.p.p.'s. All these investigators found excellent agreement between the step-wise fluctuations of the e.p.p. and the statistical predictions of Poisson's theorem, with the spontaneously occurring miniature e.p.p. as the basic unit of response. The result of this work left no room for the assumption of any "non-quantal" transmitter release by the nerve impulse.

What was left quite open, however, was the question of the origin and the mechanism of the quantal release. The first attempts to explain it were based on the idea that there might be spontaneous excitation of small regions or "patches" within the nerve endings, and that the nerve impulse might occasionally fail to invade these regions and activate them only in a statistical random manner. There is no need nowadays to discuss this possibility: it has been ruled out definitely by a number of findings. Thus, the quantal spontaneous release still occurs when electric excitability has been totally suspended, in an isotonic solution of potassium sulphate. And recently, it has been shown (Katz and Miledi 1965 b, c) by direct recording from a minute patch of the junction, that the quantal fluctuations of acetylcholine release occur even though the nerve spike invades this region without fail and retains constant amplitude throughout.

The second way of explaining quantal release was by what one may call a membrane "shutter" mechanism (del Castillo and Katz 1954 a, c). Suppose there are specialized areas in the terminal axon membrane which act as acetylcholine "gates" and, though usually closed, can reach a degree of instability when they open for a brief period during which a flood of acetylcholine is released from the interior of the nerve ending. A process of this kind is feasible, but does not seem very attractive. For one would have to explain why such a membrane-controlled "flip-flop" process remains completely unaltered in size and duration even when the potential difference across the membrane changes from the resting level to the peak of spike activity, i. e., during variations of the electric field of the membrane of the order of 10^5 volt/cm. Any alteration in the "gating action" would necessarily be reflected in a change of the quantal size during the nerve impulse, but this is ruled out by a large body of experimental evidence.

The relative constancy of the quantal packet suggested that its size is controlled, not by the rapidly changing properties of the cell membrane, but by some cellular process which is not so easily disturbed from the outside. The idea arose that the transmitter quantum may be pre-formed inside the cell and "parcelled up" in submicroscopic particles from which it can be released at the cell surface in an all-or-none fashion (DEL CASTILLO and KATZ 1955 b). The synaptic vesicles which were discovered at about the same time by de ROBERTIS and BENNET (1954) and by PALADE and PALAY (1954) seemed just the right structural counterparts: they were found at the right place inside motor nerve endings (ROBERTSON 1956; BIRKS, HUXLEY and KATZ 1960), and no other structure could be seen nearby which would fit the case.

All this was apparent ten years ago. It was also recognized that the vesicular hypothesis would have a difficult time before it could finally be accepted or discarded. After 10 years, it is still a hypothesis; and while some would maintain that it has lost none of its attraction, others will no doubt feel that the appeal of novelty has gone, and that progress in providing experimental evidence has been disappointingly slow. Without wishing to dispute these sentiments, we think it is fair to say that the vesicular story remains a good working hypothesis and continues to provide the best explanation for the quantal release of acetylcholine.

In assessing the present state of this hypothesis, the following evidence should be taken into account:

(1) Synaptic vesicles in motor nerve endings appear to take up and retain particles like ferritin (PAGE, unpublished observations) from the surrounding medium. While other interpretations are possible, these observations suggest that at some stage the synaptic vesicles have direct communication with the extracellular medium (which is, of course, one of the postulates of the vesicular-release hypothesis).

(2) It has been known for many years (FELDBERG 1945) that most of the acetylcholine stored in the central nervous system, is protected or "bound" within subcellular particles to which it remains attached even during processes of blending and high-speed centrifugation. Some years ago, DE ROBERTIS, PELLEGRINO DE IRALDI, RODRIGUEZ and GOMEZ (1961) and GRAY and WHITTAKER (1962) found, by differential centrifugation, that a layer could be separated from the supernatant fluid which contained most of the acetylcholine in bound form. In the electron microscope, this layer was shown to consist largely of recognizable nerve endings which had evidently been torn off their parent axon without losing their characteristic accumulation of mitochondria and vesicles. This showed, at least, that a large part of the neuronal acetylcholine was located in the nerve endings. More recently, DE ROBERTIS, RODRIGUEZ DE LORES ARNAIZ, SALGANICOFF, PELLEGRINO DE IRALDI and ZIEHER (1963) and WHITTAKER, MICHAELSON and KIRKLAND (1964) succeeded in pushing the analysis one step further. They managed to disrupt the "nerve ending particles" and separate their contents in the centrifuge without causing all the acetylcholine to be immediately released into the supernatant fluid. Their recent reports indicate

that most of the residual acetylcholine is indeed contained within the vesicles themselves. It will be most exciting if this line of work can be developed further. The technical problems in producing pure and well-preserved vesicular fractions from various nerve tissues are clearly formidable, but a great deal of important quantitative information will have to await further technical progress along this line.

It is also obvious that the problem of the synaptic vesicles will not be solved to everybody's satisfaction by a single line of approach. Attempts to isolate vesicles in the centrifuge are certainly one important way of tackling it; but a direct localization of the ACh stores by histochemical or autoradiographic techniques will be needed to provide further proof.

At present the evidence is converging from various avenues, and one hopes the process will speed up as one gets closer to the target! The physiological part of the evidence, showing a quantal mechanism of transmitter release, is about as secure as one could wish. Moreover, it has in recent years achieved a wider generalization by being found to apply to a variety of synapses, cholinergic or otherwise, at which chemical transmission operates. Among these are smooth muscles (BURNSTOCK and HOLMAN 1962), arthropod muscles (DUDEL and KUFFLER 1961), sympathetic ganglia (NISHI and KOKETSU 1960; BLACKMAN, GINSBORG and RAY 1963) and various excitatory and inhibitory synapses in the spinal cord (KATZ and MILEDI 1963). Much evidence has also accumulated that chemical transmission and the presence of presynaptic vesicles are two closely correlated features. All these converging signs make the search for a more direct confirmation of the vesicular hypothesis an even more challenging task.

For the remainder of our discussion, let us assume that the vesicular explanation of the quantal transmitter release is correct and consider the question: what is the advantage of this mechanism? Why should chemical transmitters be released in packets, rather than by random molecular permeation through the surface membrane? It may not be strictly scientific to pose a question, the answer to which can be no more than a guess, but at least it is an interesting subject to think about.

The first thing that comes to one's mind is that the storage of a transmitter substance in small particles is merely another example of the structural organization of the cell components into membrane-bound subcellular compartments. Physiologists have been used for nearly a century to the idea that some structural separation must exist within the cell, between substrates and destructive enzymes, and one might imagine that, for this reason, a useful secretory product like a transmitter substance must be protected within a subcellular particle.

This is not likely to be the whole story. To protect ACh which has been synthesized inside the nerve ending against immediate destruction by presynaptic cholinesterase, it would after all be sufficient to screen the latter. There must be some other reason why the transmitter is released in large packets at relatively few points, rather than diffusing from a vast number of scattered points of the presynaptic membrane. It may be that a high local concentration of ACh is more effective in depolarizing the postsynaptic membrane, and also in surviving enzymic hydrolysis, than the same amount would be, if released at a lower concentration over a wider area.

But there may be a more fundamental reason. Perhaps the quantal release is a consequence of the way cell membranes are constructed. There is reason to believe that the critical permeability changes which occur in the axon membrane during the impulse, are restricted to a minute fraction of the membrane surface and probably take place at specialized molecular sites. Suppose that, similarly, the presynaptic membrane cannot release a transmitter substance except at relatively few, scattered points. If this is so, a quantal mechanism of release becomes almost a necessity. Furthermore, seen from this view, the synaptic vesicle takes on much greater significance than that of a subcellular storage particle; it becomes the means of releasing an effective quantity of transmitter from one point of the terminal surface. To do so, the synaptic vesicles must be able to build up a high concentration of transmitter molecules within them, either by active synthesis or by active accumulation from a low concentration in the surrounding cytoplasm. WHITTAKER's (1964) recent work indicates that (a) enzymic synthesis, by choline acetyltransferase, and (b) the storage of ACh are two processes which go on in separate parts of the terminal, the latter in the vesicles, the former in smaller cytoplasmic components. It would appear that one of the main functions of the vesicles is to concentrate ACh, from a low level built up in the cytoplasm to a high level inside the vesicle ready for "point-release".

This leads us to an interesting report by ELMQVIST, QUASTEL and THESLEFF (1963), that blockage of ACh synthesis by a hemicholinium compound leads to progressive diminution in the size of the quantal packets which are being delivered by a continued series of nerve impulses. That the size of the ACh-quantum would depend on the rate of intracellular synthesis, had been anticipated long ago (DEL CASTILLO and KATZ 1956), but ELMQVIST et al. (1963) were the first to produce experimental evidence for it. In conjunction with WHITTAKER's (1964) work, these findings suggest that the quantity of ACh accumulated within a vesicle varies with the (much lower) level of ACh concentration built up by the choline acetyltransferase in the surrounding cytoplasm. Such a relationship is not unusual, and analogous situations have been found in other cases of active transport: for example, the accumulation of potassium in muscle fibres depends within certain limits on the potassium level in the external medium. It seems from ELMQVIST et al.'s experiments that the intra-vesicular ACh-concentration rapidly readjusts itself and falls to a lower level when that in the surroundings is reduced. This may again be compared with the analogous case of muscle. The potassium concentration in a 100 μ muscle fibre takes hours to re-equilibrate when the external concentration is lowered; with similar flux rates, re-equilibration in a 500 Å vesicle would occur within seconds.

One of the missing links in our present information is the number of ACh molecules which are concentrated within one quantal unit. Estimates based on the amount of ACh-like substance collected in perfusion experiments are rather uncertain; indeed, the assayed quantities reported in the recent literature are very variable and their precise interpretation is still surrounded by some controversy (Krnjević and Mitchell 1961; Hayes and Riker 1963; Krnjević and Straughan 1964). It is in any case desirable to try and base one's estimate on measurements made on single end-plates. One way would be to apply ACh with a micropipette in close proximity to an end-plate, and determine the "ACh-equivalent" of a miniature end-plate potential. This has been done by Miledi (1961) with the result that even under the most effective conditions of ionophoretic application, approximately 10^5 ACh molecules had to be discharged from the pipette to produce a depolarization whose potential × time area was equal to that of a miniature e.p.p. 10^5 molecules seems too much for a 500 Å vesicle, but this figure must be regarded as an exaggerated value for two reasons. In the first place, even with the most effective micro-application, the ACh-potential takes at least 2—3 times longer to reach its peak than a miniature e.p.p. Evidently, the pipette cannot be positioned as closely nor as effectively to the receptors as is the presynaptic membrane, and so the release of ACh from the latter is bound to have a larger and quicker effect. Secondly, diffusion within the synaptic cleft is probably slower than in the bulk of the extracellular space: this would further favour the efficacy of ACh released within the cleft, relative to a dose applied just outside the cleft where much of the substance would leak away without reaching the synaptic receptors. If one takes these factors into account, it would not be surprising if the "true ACh-equivalent" of the miniature e.p.p. turns out to be no more than a few thousand molecules, which would be well within the capacity of a 500 Å vesicle.

In discussing the present state of the vesicular hypothesis, we hope it has become clear that as a "working proposition" it has lost none of its interest, in spite of the fact that 10 years after its conception, its legitimacy still remains to be proved. It provides, however, plenty of scope for renewed experimental attack from several directions, and one may hope that, with improved chemical and ultrastructural techniques, the basis of the quantal liberation of transmitter substances will be settled before another decade has gone by.

References

Birks, R., Huxley, H. E. & Katz, B. (1960). *J. Physiol.* (Lond.) **150**, 134.
Birks, R., Katz, B. & Miledi, R. (1960). *J. Physiol.* (Lond.) **150**, 145.
Blackman, J. G.,. Ginsborg, B. L. & Ray, C. (1963). *J. Physiol.* (Lond.) **167**, 389.
Boyd, I. A. & Martin, A. R. (1956). *J. Physiol.* (Lond.) **132**, 74.
Burnstock, G. & Holman, M. E. (1962). *J. Physiol.* (Lond.) **160**, 446.

DE ROBERTIS, E. D. P. & BENNETT, H. S. (1954). *Fed. Proc.* **13**, 35.
DE ROBERTIS, E., PELLEGRINO DE IRALDI, A., RODRIGUEZ, G. & GOMEZ, C. J. (1961). *J. biophys. biochem. Cytol.* **9**, 229.
DE ROBERTIS, E., RODRIGUEZ DE LORES ARNAIZ, G., SALGANICOFF, L., PELLEGRINO DE IRALDI, A. & ZIEHER, L. M. (1963). *J. Neurochem.* **10**, 225.
DEL CASTILLO, J. & KATZ, B. (1954 a). *J. Physiol.* (Lond.) **124**, 553.
DEL CASTILLO, J. & KATZ, B. (1954 b). *J. Physiol.* (Lond.) **124**, 560.
DEL CASTILLO, J. & KATZ, B. (1954 c). *J. Physiol.* (Lond.) **124**, 586.
DEL CASTILLO, J. & KATZ, B. (1955 a). *J. Physiol.* (Lond.) **128**, 157.
DEL CASTILLO, J. & KATZ, B. (1955 b). *J. Physiol.* (Lond.) **128**, 396.
DEL CASTILLO, J. & KATZ, B. (1956). *Progr. Biophys.* **6**, 121.
DEL CASTILLO, J. & KATZ, B. (1957). *Proc. Roy. Soc.* B. **146**, 357.
DUDEL, J. & KUFFLER, S. W. (1961). *J. Physiol.* (Lond.) **155**, 514.
ELMQVIST, D., QUASTEL, J. M. & THESLEFF, S. (1963). *J. Physiol.* (Lond.) **167**, 47 P.
FATT, P. & KATZ, B. (1950). *Nature,* (Lond.) **166**, 597.
FATT, P. & KATZ, B. (1952). *J. Physiol.* (Lond.) **117**, 109.
FELDBERG, W. (1945). *Physiol. Rev.* **25**, 596.
GRAY, E. G. & WHITTAKER, V. P. (1962). *J. Anat.* (Lond.) **96**, 79.
HAYES, JR. A. H. & RIKER, JR. W. F. (1963). *J. Pharmacol. exp. Therap.* **142**, 200.
KATZ, B. & MILEDI, R. (1963). *J. Physiol.* (Lond.) **168**, 389.
KATZ, B. & MILEDI, R. (1965 a). *Proc. Roy. Soc.* B., **161**, 439.
KATZ, B. & MILEDI, R. (1965 b). *Proc. Roy. Soc.* B., **161**, 469.
KATZ, B. & MILEDI, R. (1965 c). *Proc. Roy. Soc.* B., **161**, 482.
KRNJEVIĆ, K. & MITCHELL, J. F. (1961). *J. Physiol.* (Lond.) **155**, 246.
KRNJEVIĆ, K. & STRAUGHAN, D. W. (1964). *J. Physiol.* (Lond.) **170**, 371.
LILEY, A. W. (1956 a). *J. Physiol.* (Lond.) **132**, 650.
LILEY, A. W. (1956 b). *J. Physiol.* (Lond.) **133**, 571.
LILEY, A. W. (1956 c). *J. Physiol.* (Lond.) **134**, 427.
MILEDI, R. (1961). *Discovery* **22**, 444.
MILEDI, R. (1962). In *Ciba Foundation Symposium on Enzymes and Drug Action* (ed. by MONGAR, J. L. & DE REUCK, A. V. S.), London: Churchill Ltd.
MILEDI, R. & SLATER, C. R. (1963). *J. Physiol.* (Lond.) **169**, 23 P.
NISHI, S. & KOKETSU, K. (1960). *J. cell. comp. Physiol.* **55**, 15.
PALADE, G. E. & PALAY, S. L. (1954). *Anat. Rec.* **118**, 335.
ROBERTSON, J. D. (1956). *J. biophys. biochem. Cytol.* **2**, 381.
WHITTAKER, V. P. (1964). *Prog. Brain Res.* **8**, 90.
WHITTAKER, V. P., MICHAELSON J. A. & KIRKLAND, R. S. (1964). *Biochem. J.* **90**, 293.

MEMBRANE CALCIUM AND BIOELECTRIC POTENTIALS

by K. KOKETSU

The mechanism underlying the maintenance of the resting potential and the production of the action potential has been well interpreted by the ionic theory (HODGKIN 1951). According to this theory, the immediate energy source for establishing membrane potentials is the thermal energy of the ions in the extra- and intracellular solutions, and the potential is changed by the alteration of the membrane permeability to ions, such as Na, K and Cl. The crucial problem in this theory is how the membrane permeability to different ions is regulated or modified when the membrane is stimulated.

The most direct approach to this problem is obviously to study the physico-chemical characteristics of the membrane at the molecular level, and to clarify the physico-chemical reactions actually occurring in the membrane when the membrane potential, or the membrane permeability is altered. As the first step in this kind of analysis, it seems worthwhile to study the role of Ca ions, which may be bound to the membrane lipoprotein and may modify the physico-chemical characteristics of the membrane. Indeed, the importance of Ca ions for the maintenance of excitability of the membrane has been reviewed by several authors (HEILBRUNN 1952; BRINK 1954; SHANES 1958; CERF 1963), and it is speculated that the role of Ca ions is to regulate the membrane permeability to Na and K ions (e. g. FRANKENHAEUSER and HODGKIN 1957; TOBIAS 1958).

I. Binding of Ca ions to membrane

Although the role of Ca ions bound to the membrane has been frequently discussed, no direct experimental evidence showing that Ca ions actually bind to the membrane has been so far provided. Efforts to obtain decisive evidence for Ca binding have been attempted in our laboratory. Inasmuch as the membrane is composed of lipoproteins containing phospholipids, it is likely that Ca ions are bound to the negative phosphoric sites. The experiment carried out on surface films of phospholipids (KIMIZUKA and KOKETSU 1962) demonstrated that Ca ions were capable of binding to a monolayer film of lecithin within a physiological range of pH. Furthermore, results indicated that Ca ions might bind with two negative oxygen sites to form a bridge between two adjacent phospholipid molecules; as suggested by TOBIAS (1958). It is also of interest that the binding of Ca ions to a lecithin monolayer is saturated at a relatively small concentration which corresponds to the minimum Ca concentration necessary to maintain the physiological function of living excitable cells. To obtain more direct evidence that Ca ions are bound to living excitable membranes, the binding of Ca ions to muscle membrane has been studied. In these experiments, isolated membrane fragments of frog skeletal muscle fibers were used, and it was found that Ca ions were able to bind to these fragments or to the surface film of lipoproteins extracted from these fragments (KOKETSU, KITAMURA and TANAKA 1964). The physico-chemical properties of these membrane fragments would be essentially the same as those of the muscle fiber membrane *in vivo*, in view of the fact that the fine structure of the isolated membrane, as demonstrated by electron microscopy, was well preserved, and the preparation procedure was mild enough so that various enzyme activities were maintained (KONO 1963). The possibility exists that Ca ions may also bind to the membrane protein, and further investigation is necessary to confirm the exact site of binding.

II. Dissociation of membrane Ca

The membrane Ca (the Ca ions bound to the membrane) may be exchangeable and displaced by other ions, such as Na, K and H ions. It was found that these monovalent cations do impede the binding of Ca ions to a monolayer of phospholipids (KIMIZUKA and KOKETSU 1962; ROJAS 1964). Similar results were observed with isolated muscle membrane fragments and with surface films of lipoproteins extracted from these membrane ghosts (KOKETSU et al. 1964). It was also noted that the time course of the desorption of Ca ions from the phospholipid monolayer was much faster when the solution contained K ions than when it contained Na ions (KIMIZUKA and KOKETSU 1964 a), indicating that the rate of dissociation of Ca ions is larger in the presence of K ions than in the presence of Na ions. The selective permeability of the membrane to Na and K ions might be explained according to this phenomenon.

It has been reported that there are two kinds of surface bound Ca in frog skeletal muscle fibers; one is displaced only by self-exchange and the other can be displaced by other ions, such as Na, K and H ions (GILBERT and FENN 1957; SHANES and BIANCHI 1959). These studies have been re-examined and the results all confirmed. Furthermore, the changes in the efflux of radioactive Ca ions from isolated frog muscles or nerve preparations were studied when the membrane was depolarized under various experimental conditions, *viz.*, by removing the external Ca ions, adding external K ions, adding other depolarizing agents to the external solution or applying electrical stimulation. These experiments demonstrated that the rate constant of Ca^{45} efflux from the muscles increased when the membrane was depolarized by some means (cf. KOKETSU and MIYAMOTO 1961 a, b). The fact that similar results were obtained with nerve preparations (KOKETSU and NODA, unpublished) suggests that the increase of Ca^{45} efflux from muscle may be at least partly associated with the process of membrane depolarization (cf. SHANES and BIANCHI 1960). Since most of the intracellular Ca ions may be in bound form, and the Ca efflux is not simply proportional to the electro-chemical energy of the intracellular free Ca ions (cf. HODGKIN and KEYNES 1957; SHANES and BIANCHI 1959), the increase in the Ca efflux may be considered to be, at least in part, due to the increase in the dissociation of the membrane Ca.

The simplest hypothesis to account for the mechanism of dissociation of the membrane Ca would be expressed as follows (cf. FRANKENHAEUSER and HODGKIN 1957),

$$2\,PL^- + Ca_m^{++} \rightleftharpoons Ca{-}PL_2 \tag{1}$$

$$Ca_m^{++} = Ca_0^{++} \exp\left\{-\frac{2\,a\,FV}{RT}\right\} \tag{2}$$

$$\frac{(Ca{-}PL_2)}{(PL^-)^2} = K\,Ca_0^{++} \exp\left\{-\frac{2\,a\,FV}{RT}\right\} \tag{3}$$

where $Ca-PL_2$, PL, Ca_m, Ca_o, V, F, R, T, K and *a* represent the membrane Ca, membrane phospholipid, Ca ions in the membrane, Ca ions in the external solution, membrane potential, Faraday's constant, gas constant, absolute temperature, a constant and the correction factor of the membrane potential, respectively. These equations indicate that the fraction of the membrane Ca should be decreased either by reducing the external Ca concentration or by reducing the potential difference across the membrane. The actual reaction is not, of course, so simple, for the increase in the concentration of monovalent cations in the external solution would also decrease the fraction of the membrane Ca, because these ions presumably impede the binding of Ca ions to the membrane. It is also possible that the dissociation constant of the membrane Ca may be altered according to the changes in the physico-chemical characteristics of the membrane lipoproteins which are probably controlled by the concentration of membrane Ca. The removal of membrane Ca would cause depolarization by increasing the membrane permeability to Na ions (Section IV). Thus, the process of the removal of membrane Ca and membrane depolarization takes place in a regenerative manner.

III. Two states of the membrane

The membrane of excitable cells is capable of maintaining two different states, namely, either the resting state (resting potential) or the active state (action potential). If the active state of the membrane were maintained, the action potential would be prolonged and, in the extreme case, the membrane would stay in the depolarized state. In this case, hyperpolarizing responses may be produced by applying anodal pulses to the membrane, the membrane being shifted from the active to the resting state. The idea that the resting state of the membrane corresponds to the association of Ca ions with the membrane, and the active state, their dissociation from the membrane has been proposed (Koketsu, Nishi and Soeda 1963). According to this idea, the resting state of the membrane can be shifted to the active state when Ca ions bound to the membrane are removed by any means (cf. Section II), and the active state can be shifted to the resting state when Ca ions re-bind to the membrane. Such reaction must be dependent on the concentration of the Ca ions in the external solution, and it is expected that the membrane will tend to remain in the resting state when the external Ca concentration is high, and, on the other hand, tend to remain in the active state when the external Ca concentration is low or nullified. Indeed, it is well known that, in low or zero Ca Ringer's solution, the membrane is depolarized and often exhibits spontaneous activity. The increase in the excitability of the membrane in such solutions has been demonstrated in various excitable cells (cf. Frankenhaeuser and Hodgkin 1957; Koketsu and Noda

1962). Furthermore, the plateau of the action potential of heart muscle fibers was prolonged in Ca-free EDTA Ringer's solution (Abe and Goto 1964), and the action potential of frog nerve fibers (Spyropoulus 1961) and spinal ganglion cells (Koketsu and Koyama 1962) was prolonged in Ca-deficient Ringer's solution. It was also noted that the prolongation of the action potential of frog spinal ganglion cells observed under various experimental conditions was enhanced or depressed when the external Ca concentration was reduced or raised, respectively (Koketsu et al. 1963). Similar effects of the external Ca concentration on the two stable states observed in K-rich solutions were also observed.

IV. Membrane in the Ca-dissociated state

It is known that, in general, the membrane permeability to K ions is much larger than that to Na ions in the resting state of the membrane, and that the permeability to Na ions is selectively increased in the active state. If the active state is the Ca-dissociated state, the permeability to Na ions will be increased when the membrane Ca is removed experimentally by some means. The membrane Ca is sufficiently removed when a strong chelating agent is applied to the external solution. When frog skeletal muscles were soaked in Ca-free EDTA Ringer's solution, the membrane potential dropped, finally reaching a fairly steady potential of about -25 mV within 90—120 minutes (Kimizuka and Koketsu 1963); such depolarization was completely reversible. The membrane seemed to be in the active state when it was strongly depolarized in this solution. Indeed, in this solution, the membrane of frog spinal ganglion cells was found to be gradually depolarized to a critical potential level (-25 to -30 mV) from which the potential abruptly shifted (spontaneous firing) to a steady level of approximately -15 mV, which corresponded to the plateau level of the prolonged action potential produced in Ca-free EDTA Ringer's solution (cf. Koketsu and Koyama 1962). The membrane permeability to Na and K ions in Ca-free EDTA Ringer's solution was studied with frog skeletal muscle fibers, and it was found that the membrane permeability to Na ions was greatly increased while that to K ions was not increased (Kimizuka and Koketsu 1963).

V. The role of Ca in the external solution

The Ca in the external solution seems to not only regulate the concentration of the membrane Ca, but also to penetrate the membrane as free ions. The membrane permeability to free Ca ions appears to be increased particularly during the active state, i. e., during the Ca-dissociated state. Indeed, the possibility that the inward movement of Ca during the action potential is carrying part of the inward positive current has been mentioned in the case of squid giant axon (Hodgkin and Keynes 1957). The membrane permeability to Ca ions seems to vary widely depending on the particular membrane. In certain kinds of excitable membranes, the action potential may be due mainly to the inward movement of external Ca ions (Hagiwara, Naka and Chichibu 1964). It is well known that alkali-earth cations, such as Ba, can substitute for Na ions in the production of the action potential in crustacean muscle (Fatt and Ginsborg 1958), insect muscle (Werman, McCann and Grundfest 1961 a, b) and frog spinal ganglion cells (Tasaki 1959). Recent studies on a series of alkali-earth cations, *viz.* Mg, Ca, Ba and Sr, with frog spinal ganglion cells revealed that alkali-earth cations exert a dual action on the membrane (Nishi, Soeda and Koketsu, unpublished). One is the Ca-like action, namely, to bind to the membrane and alter the membrane potential by controlling the membrane permeability to monovalent cations. The other is the Na-like action, namely, to to penetrate the membrane, particularly during the action potential, as the charge carrier of the positive inward current. Ba ions apparently have such a dual action on these cells; Ba can substitute for Ca while it also acts as Na ions during the active state. Which of the dual actions is more predominant may depend both on the kind of membrane and the particular cation involved. Presumably, the affinity of the membrane lipoprotein to a given alkali-earth cation determines the mode of action of this particular ion.

VI. Theoretical consideration

It has been pointed out that the potential gradient may not be constant in the membrane where mobile ions and fixed charges of lipoprotein exist (Kimizuka and Koketsu 1964 b). Because little is known of the exact distribution of fixed charges in the membrane, it would be advantageous if a theory could be obtained without taking such factors as the potential field and the detailed structure of the membrane into consideration. From this standpoint, a new theory of ion transport through the cell membrane was presented on the basis of the thermodynamics of irreversible processes (Kimizuka and Koketsu 1964 b). In this theory, a constant-field in the membrane was not assumed, and the equation of ion flux and that of the membrane potential was in the same form as that presented by Parlin and Eyring (1954) and that presented by Goldman (1943) and Eyring (1960), respectively.

According to this theory, the net flux (j) of an ion (α) is expressed as follows,

$$j_\alpha = P_\alpha \left(C_{\alpha 0} \exp\left\{\frac{-Z_\alpha FV + \eta_{\alpha i}}{RT}\right\} - C_{\alpha i} \exp\left\{\frac{Z_\alpha FV + \eta_{\alpha i}}{RT}\right\}\right) \qquad (4)$$

where P_α denotes the membrane permeability to the ion α, V, the membrane potential, Z, valency, F, Faraday's constant, R, gas constant, T, absolute temperature, C_α, the concentration of the ion α, η_α, the excess free energy of the ion α, and i and o refer to the inside and the outside, respectively.

In the case where $|Z_\alpha|$ is unity for all ions, the total ionic current (I) carried by these ions can be expressed as follows,

$$I = -\frac{RT}{F} G\left(\exp \frac{F(V-Vr)}{2RT} - \exp\left\{\frac{-F(V-Vr)}{2RT}\right\}\right) \tag{5}$$

$$G = (G_+ G_-)^{\frac{1}{2}} \tag{6}$$

$$G_+ = \frac{F^2}{RT}\sum_{c,a}\left(P_c C_{ci} \exp\left\{\frac{\eta_{ci}}{2RT}\right\} + P_a C_{a0} \exp\left\{\frac{-\eta_{ai}}{2RT}\right\}\right) \tag{7}$$

$$G_- = \frac{F^2}{RT}\sum_{c,a}\left(P_c P_{c0} \exp\left\{\frac{\eta_{ci}}{2RT}\right\} + P_a C_{ai} \exp\left\{\frac{-\eta_{ai}}{2RT}\right\}\right) \tag{8}$$

$$Vr = \frac{RT}{F} \ln \frac{G_-}{G_+} \tag{9}$$

where r refers to the resting, and c and a the cations and anions, respectively.

If $\eta_{\alpha i}$ is zero (α represents a as well as c) and if the total ionic current is assumed to be carried by only Na, K and Cl ions, the membrane potential in the absence of electric current is expressed as follows,

$$Vr = \frac{RT}{F} \ln \frac{P_K C_{Ki} + P_{Na} C_{Nai} + P_{Cl} C_{Cl0}}{P_K C_{K0} + P_{Na} C_{Na0} + P_{Cl} P_{Cli}} \tag{10}$$

This equation is the same as that presented by GOLDMAN (1943). It must be kept in mind, however, that the membrane permeability to an ion was defined differently in the presently discussed theory. By assuming that the electrical potential changes only within the membrane, that the potential gradient is constant throughout the membrane and that the diffusion coefficient (D_α), η_α and j_α are constants, the membrane permeability (P_α) to an ion (α) can be expressed as follows,

$$P_\alpha = P_\alpha' \frac{\frac{Z_\alpha FV}{RT}}{\exp\left\{\frac{Z_\alpha FV}{2RT}\right\} - \exp\left\{\frac{-Z_\alpha FV}{2RT}\right\}} \tag{11}$$

$$P_\alpha' = \frac{D_\alpha}{\delta} \exp\left\{\frac{-\eta_{\alpha i}}{RT}\right\} \tag{12}$$

where δ is the thickness of the membrane. It may be seen from these equations that P_α' is the membrane permeability defined according to GOLDMAN's constant-field theory (1943); the term $\exp(-\eta_\alpha/RT)$ is equivalent to the partition coefficient which HODGKIN and KATZ (1949) used in explaining the selective membrane permeability.

There are experimental results which cannot be easily explained according to the ionic theory based on GOLDMAN's constant-field hypothesis (1943). It is not impossible to explain these results by computing the experimentally unknown factors involved in Goldman's equation. It may be worthwhile, however, to consider the membrane as a fixed charge system and the possibility that the flux of an ion may not be driven simply by its own electrochemical gradient. Furthermore, there seem to be several fundamentally different ways in which bioelectric potentials observed by electrophysiologists may arise. In other words, membrane potentials may not be simple diffusion potentials. One possible source of potential which may be involved is the interfacial potential set up at the boundary between the membrane and the extra- or intracellular phases. Such a potential might be set up by appropriate orientation of fixed charges or dipoles in the membrane, and might be altered by rearrangement of these charges or by preferential adsorption of ions to the membrane. It is unlikely that the interfacial potentials are involved in the membrane potential in the steady state because the changes in the interfacial potential would be rapidly compensated by redistribution of ions between the phases on each side of the interface (Davies and Rideal 1961). The possibility that the interfacial potential is involved when the potential is changing rapidly, however, may not be simply dismissed. It is worthwhile to clarify whether or not such a potential is actually involved during the action potential.

A number of lines of evidence indicating the importance of Ca ions in the functioning of excitable membranes has been summarized but there are as yet no precise details of the mechanism involved. We may conclude that the elucidation of this mechanism at the molecular level constitutes one of the most promising approaches to the fundamental problem of bioelectric potentials.

Acknowledgement

This research was supported by the Institute of Neurological Diseases and Blindness, National Institute of Health Grant NB 01 650.

References

Abe, Y. & Goto, M. (1964). *Jap. J. Physiol.* **14**, 123.

Brink, F. (1954). *Pharmacol. Rev.* **6**, 243.

Cerf, J. A. (1963). *Handbuch der experimentellen Pharmakologie, Ergänzungswerk*, vol. 17, (ed. by Bacq, A. M.) p. 164. Berlin: Springer-Verlag.

Davies, J. T. & Rideal, E. K. (1961). *Interfacial Phenomena*, p. 56. New York and London: Academic Press.

Eyring, H. (1960). *Molecular Biology*, (ed. by Nachmansohn, D.) p. 77. New York and London: Academic Press.

Fatt, P. & Ginsborg, B. L. (1958). *J. Physiol.* (Lond.) **142**, 516.

Frankenhaeuser, B. & Hodgkin, A. L. (1957). *J. Physiol.* (Lond.) **137**, 218.

Gilbert, D. L. & Fenn, W. O. (1957). *J. gen. Physiol.* **40**, 393.

Goldman, D. E. (1943). *J. gen. Physiol.* **27**, 37.

Hagiwara, S., Naka, K. & Chichibu, S. (1964). *Science* **143**, 1446.

Heilbrunn, L. V. (1952). *An Outline of General Physiology*, Philadelphia: Saunders.

Hodgkin, A. L. (1951). *Biol. Rev.* **26**, 339.

Hodgkin, A. L. & Katz, B. (1949). *J. Physiol.* (Lond.) **108**, 37.
Hodgkin, A. L. & Keynes, R. D. (1957). *J. Physiol.* (Lond.) **138**, 253.
Kimizuka, H. & Koketsu, K. (1962). *Nature* (Lond.) **196**, 995.
Kimizuka, H. & Koketsu, K. (1963). *J. gen. Physiol.* **47**, 379.
Kimizuka, H. & Koketsu, K. (1964 a). *Biological Membranes and Artificial Membranes*, p. 19. Tokyo: Tokyo University Press.
Kimizuka, H. & Koketsu, K. (1964 b). *J. Theoret. Biol.* **6**, 290.
Koketsu, K. & Koyama, I. (1962). *J. Physiol.* (Lond.) **163**, 1.
Koketsu, K. & Miyamoto, S. (1961 a). *Nature* (Lond.) **189**, 402.
Koketsu, K. & Miyamoto, S. (1961 b). *Nature* (Lond.) **189**, 403.
Koketsu, K., Nishi, S. & Soeda, H. (1963). *Nature* (Lond.) **200**, 786.
Koketsu, K. & Noda, K. (1962). *J. cell. comp. Physiol.* **59**, 323.
Koketsu, K., Kitamura, R. & Tanaka, R. (1964). *Am. J. Physiol.* **207**, 509.
Kono, R. (1963). *J. Jap. Biochem. Soc.* **35**, 237.
Nishi, S., Soeda, H. & Koketsu, K. (1965). *J. Neurophysiol.*, in press.
Parlin, R. B. & Eyring, H. (1954). In *Ion Transport across Membrane*, (ed. by Clark, H. T.) p. 103. New York: Academic Press.
Rojas, E. (1964). Abstracts of the 9th Annual Meeting of the Biophysical Soc., U.S.A.
Shanes, A. M. (1958). *Pharmacol. Rev.* **10**, 59.
Shanes, A. M. & Bianchi, C. P. (1959). *J. gen. Physiol.* **42**, 1123.
Shanes, A. M. & Bianchi, C. P. (1960). *J. gen. Physiol.* **43**, 481.
Spyropoulos, C. S. (1961). *Am. J. Physiol.* **200**, 203.
Tasaki, I. (1959). *Nature* (Lond.) **184**, 1574.
Tobias, J. M. (1958). *J. cell. comp. Physiol.* **52**, 89.
Werman, R., McCann, F. V. & Grundfest, H. (1961 a). *J. gen. Physiol.* **44**, 979.
Werman, R., McCann, F. V. & Grundfest, H. (1961 b). *J. gen. Physiol.* **44**, 997.

TRANSMITTER LIBERATION AND MOBILIZATION IN DEGENERATING CENTRAL SYNAPTIC ENDINGS

by P. G. Kostyuk

If central synaptic endings are excited by repetitive nerve volleys their trans-synaptic function depends on the impulse frequency in a rather complicated manner. The relation between frequency of presynaptic impulses and the effectiveness of trans-synaptic action can be regarded as a result of an interplay of two different processes — the liberation of the immediately available quantity of transmitter and transmitter mobilization inside the synaptic endings (Curtis and Eccles 1960). The analysis of the nature of these two processes is of great importance for the understanding of the function of synaptic terminals. As the activity and submicroscopic structure of the synaptic endings are rapidly changed after trans-section of the corresponding afferent fibre (Vera and Luco 1958; Kostyuk and Savoskina 1959, 1962; Birks, Katz and Miledi 1960), an investigation of changes in trans-synaptic function of the ending in the course of its degeneration after fibres section could be of great interest for this problem.

Methods

Experiments were made on cats. About one half of the afferent fibres in one lumbar dorsal root was cut aseptically under ether narcosis. The same animal was used 24—36 hours later for intracellular recording from motoneurones in the corresponding lumbar spinal cord segment. Usual microelectrode technique and nembutal anaesthesia were employed. The postsynaptic potentials produced in the motoneurone by stimulation of the central part of the previously sectioned dorsal root were compared with postsynaptic potentials produced in the same cell by stimulation of the remaining ("normal") part of the dorsal root fibres.

A great disadvantage of this technique is the impossibility of separate stimulation of group I a muscle afferent fibres which produce monosynaptic excitation of motoneurones. The monosynaptic EPSPs were always complicated by some polysynaptic action from low-threshold cutaneous fibres. This difficulty could be minimized by careful adjustment of stimulus strength; it was also much less prominent during repetitive stimulation which readily depresses the polysynaptic activity.

Results

Monosynaptic EPSPs produced in a motoneurone by volleys from normal dorsal root fibres show the usual pattern of dependence upon the frequency of rhythmic stimulation (see Curtis and Eccles 1960). After a decline at the beginning of stimulation of the rate of rise and in the amplitude, they become stabilized at a particular plateau level. The presence of such a level can be used for superposition of several successive EPSPs and for plotting the relation between the average EPSP amplitude (or rate of rise) and the frequency of dorsal root stimulation (Fig. 1 — open circles). The shape of the curve resembles closely that for the central terminals of group I a muscle afferent fibres.

The monosynaptic excitation of the same motoneurone through afferent fibres cut 24—36 hours before shows a characteristic change in effectiveness of trans-synaptic action during stimulation with different frequencies. EPSPs are not maintained with a stable amplitude but rapidly decline with increasing frequency of stimulation. The difference is quite clear by comparing the curves of Fig. 1.

If it is supposed that the relation between the effectiveness of trans-synaptic action and stimulus frequency represents the level of balance between processes of transmitter mobilization and transmitter liberation in synaptic endings, then the degenerating ending can be described as lacking in transmitter mobilization. The trans-synaptic action of the ending is rapidly exhausted during even relatively low frequency stimulation, and the transmitter mobilization never produces a sufficient compensation for this exhaustion.

There is no significant difference at the beginning of this period after dorsal root section in the amplitude or rate of rise of EPSPs evoked by single nerve volleys. But the latency of the monosynaptic EPSP generated by degenerating terminals is already clearly increased (1.66 ± 0.02 msec compared with normal

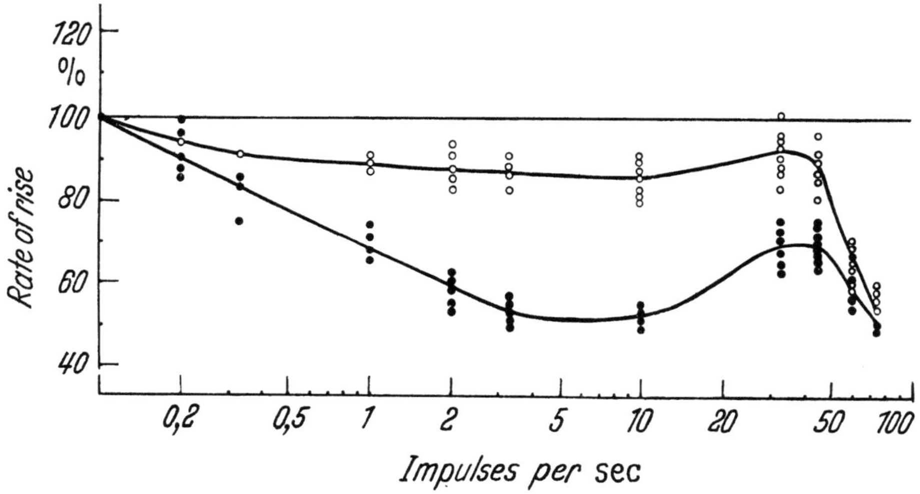

Fig. 1. Relation between the rate of rise of a monosynaptic EPSP (ordinate) during the steady state and frequency of afferent impulses (abscissa). The rate of rise was measured as a percentage of its value during the lowest frequency of stimulation. Open circles-stimulation of normal afferent fibres, filled circles-stimulation of fibres that were cut 35 hours before the measurement

value of 1.38 ± 0.02 msec; $P < 0.001$). This increase is probably due to slowing in conduction velocity in degenerating nerve fibres. With increased time after dorsal root section the EPSPs generated by single volleys become smaller. After 48 hours the monosynaptic transmission to motoneurones practically ceases.

The termination of repetitive stimulation also produces long-lasting changes in the effectiveness of trans-synaptic action of synaptic endings well known as "post-tetanic depression" and "post-tetanic potentiation" (PTP) (Eccles and Rall 1951; Lloyd 1949). Post-tetanic changes can be equally considered as an effect of changes in transmitter liberation and transmitter mobilization processes, although changes in membrane polarization of afferent terminals can also play an important role (Lloyd 1949; Eccles and Krnjević 1959).

A comparison of monosynaptic EPSPs generated in motoneurones by volleys from degenerating and normal terminals shows marked differences in their post-tetanic changes (see Fig. 2). A tetanic stimulation of previously trans-sected afferent fibres did not produce any PTP of the monosynaptic EPSP (Fig. 2B;

C: filled circles). On the contrary, there was a long-lasting depression in effectiveness of synaptic transmission to motoneurones. Tetanic stimulation of normal afferent fibres converging to the same motoneurone resulted in the usual PTP

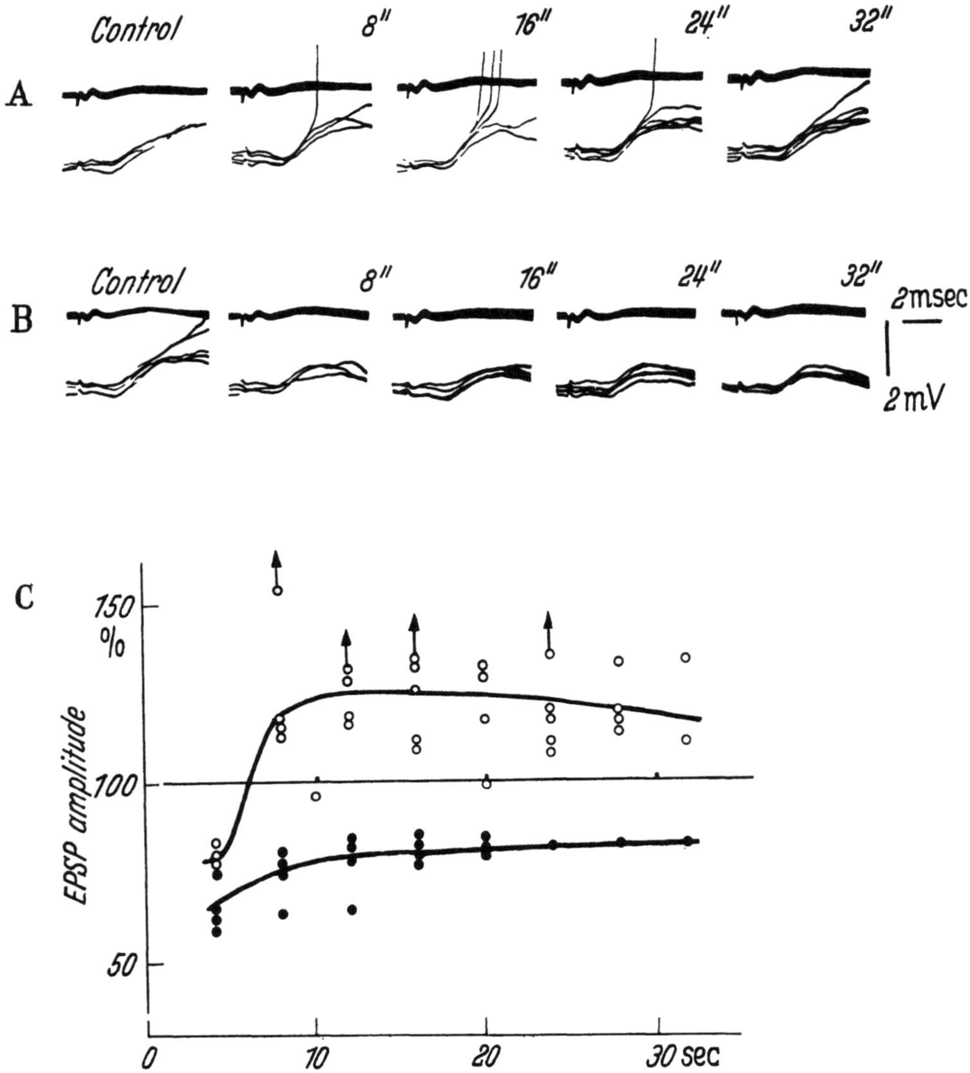

Fig. 2. Monosynaptic EPSP of a motoneurone before (CON) and at different time intervals after the end of tetanic stimulation of the dorsal root (60 sec with frequency 180 per sec). Recording was made by superimposing sweeps once every sec. A. — stimulation of the normal afferent fibres; B — stimulation of fibres cut 30 hours before the experiment. C — these results are plotted as a relation between EPSP amplitude (as a percentage of its value during low frequency stimulation) and time interval after the end of tetanization. Arrows indicate spike generation: Open circles-stimulation of normal afferent fibres; filled circles-stimulation of fibres cut 30 hours previously

of the EPSP (Fig. 2A; C: open circles). A great reduction or even complete disappearance of PTP of synaptic transmission from degenerating afferent terminals fits well with the assumption that the increased effectiveness of synaptic transmission after tetanic stimulation reflects mainly transmitter accumulation in a form of a reserve that can be easily used by the nerve impulse (transmitter mobilization).

Discussion

Present results show that initial changes in the function of synaptic terminals during the process of degeneration are not due to changes in transmitter synthesis. Such disturbances in the synthesis processes must be important during later stages of degeneration when the transmission of even single nerve volleys is reduced. Probably the process of transmitter mobilization is connected with redistribution of synaptic vesicles inside the ending and accumulation of them at strategic sites near the inner surface of the presynaptic membrane. An example of possible movements of synaptic vesicles can be seen in the slow increase in effectiveness of trans-synaptic action of the nerve ending during hyperpolarization of its membrane by direct current (Hubbard and Willis 1962; Vladimirova 1963). The movement of synaptic vesicles in an electric field must be connected with the presence of fixed electrical charges at their surface. But it is difficult to say what kind of electric field could produce the same electrophoretic redistribution of synaptic vesicles during natural repetitive stimulation of the nerve fibre. Therefore only speculative suggestions can be made about the nature of early changes in transmitter mobilization during synaptic degeneration. For instance, there can be a change in electric charges of the synaptic vesicles or a change in properties of the cytoplasm in the ending. It is important to notice that an early submicroscopic change in degenerating synaptic endings is an aggregation of synaptic vesicles in a single mass, which is quite consistent with the suggestion about loss of their electric charges.

Summary

Excitatory postsynaptic potentials were recorded from spinal cord motoneurones monosynaptically activated by volleys from normal dorsal root afferent fibers and fibers cut 24—36 hours before the experiment. Essential differences in trans-synaptic action of such fibers during rhythmic stimulation were shown. They can be considered to be a result of rapid exhaustion of the amount of transmitter immediately available for the nerve impulse in the synaptic ending.

References

Birks, R., Katz, B. & Miledi, R. (1960). *J. Physiol.* (Lond.) **150**, 134.
Curtis, D. R. & Eccles, J. C. (1960). *J. Physiol.* (Lond.) **150**, 374.
Eccles, J. C. & Krnjević, K. (1959). *J. Physiol.* (Lond.) **149**, 274.

Eccles, J. C. & Rall, W. (1951). *J. Neurophysiol.* **14**, 353.
Hubbard, J. I. & Willis, W. D. (1962). *J. Physiol.* (Lond.) **163**, 115.
Kostyuk, P. G. & Savoskina, L. A. (1959). *Fiziol. zh. AN Ukrain. SSR.* **66**, 719.
Kostyuk, P. G. & Savoskina, L. A. (1962). *Fiziol. zh. AN Ukrain. SSR.* **9**, 581.
Lloyd, D. P. C. (1949). *J. gen. Physiol.* **33**, 147.
Vera, C. & Luco, J. (1958). *J. Neurophysiol.* **21**, 319.
Vladimirova, I. A. (1963). *Bull. exper. Biol. Med. USSR.* **56**, 11.

POST-ACTIVATION CHANGES IN THE SALIVARY GLANDS OF THE DOG

by Wlodzimierz Kozak

Despite the fact that secretion of salivary glands has been extensively studied over the past century (for references see Burgen and Emmelin 1961), certain fundamental questions still remain unsettled. The methods of recording secretion which are generally used are inaccurate and this, we feel, has limited our knowledge of the secretory processes. The secretory effect of a single volley of nerve impulses is little known (Gayda 1925; Beznák and Farkas 1936—37). It is not clear how the previous function of the gland affects the subsequent secretion. The augmented secretion elicited by a parasympathetic stimulation preceded by another stimulation has been seldom studied or mentioned (Goldenberg 1923—24). Yet, salivation has been long used as an indicator of brain processes in the study of classical conditioning in dogs (Pavlov 1940; Konorski 1948). In our studies, we concentrated on improving the methods of recording salivation in acute and chronic experiments on dogs. We were interested in the changes of secretory capacity of the submaxillary (S) and parotid (P) glands following single and repetitive volleys of nerve impulses. The changes of gland function were also studied in unanaesthetized dogs.

I. Methods of recording salivation

Numerous volumetric methods have been used for recording salivation in the nerve-gland preparation and in freely moving animals. In some of them, pneumatic transmission has been used for remote recording of the volume of saliva (Podkopaev 1952). However, the latencies and volume of salivation are thus recorded with a considerable error due to the compressibility of air in the system. The drop-counting devices used either alone or in combination with the pneumatic transmission (PODKOPAEV 1952) introduce discontinuity of recording. Besides, the size of drops of saliva may change considerably during secretion due to changes of surface tension (KOZAK, unpublished). Nevertheless, both types of recording are widely used (BURGEN and EMMELIN 1961). In order to

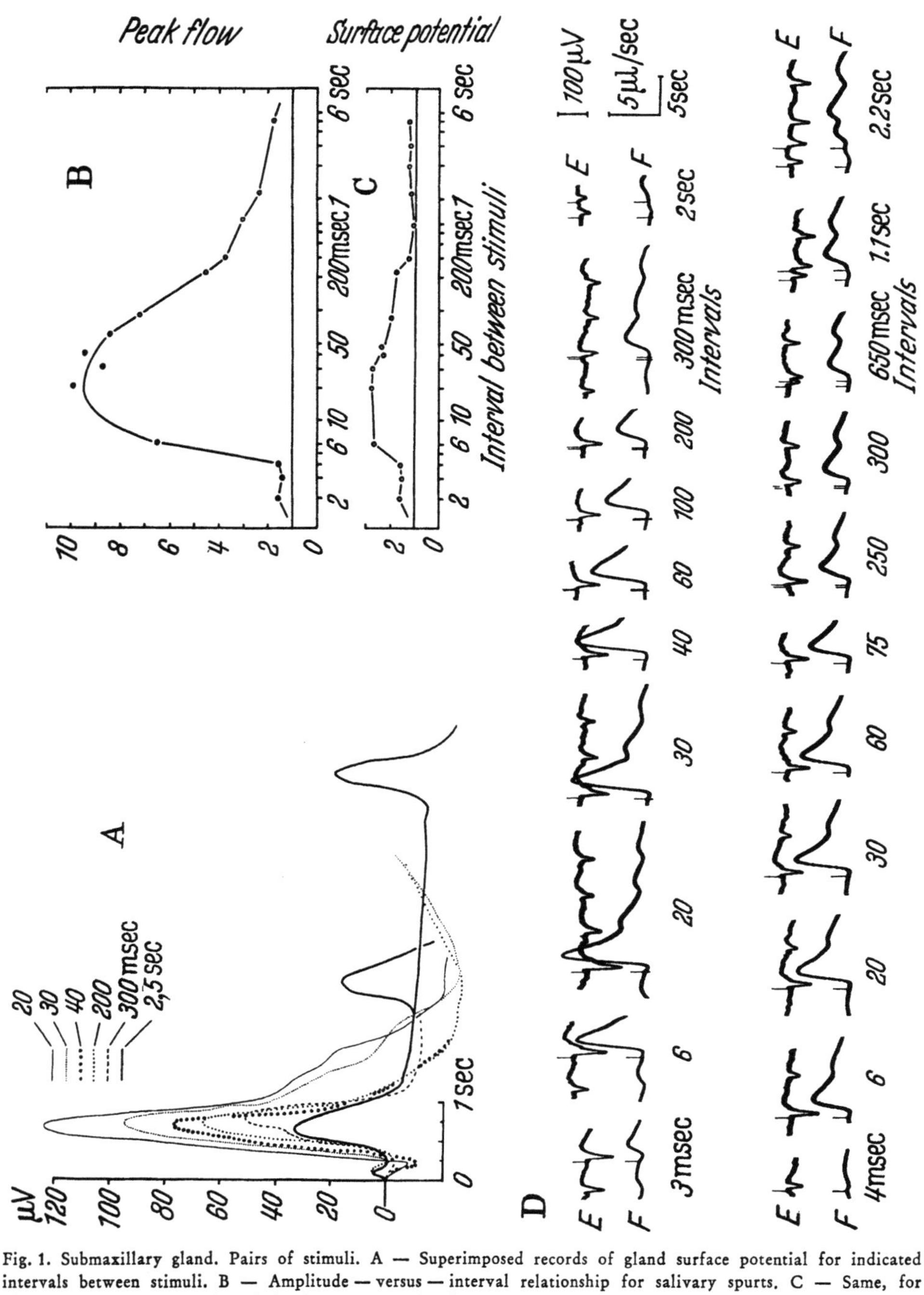

Fig. 1. Submaxillary gland. Pairs of stimuli. A — Superimposed records of gland surface potential for indicated intervals between stimuli. B — Amplitude — versus — interval relationship for salivary spurts. C — Same, for electrograms. D — examples of simultaneous records of potential (E) and flow (F) at different intervals between shocks

measure the varying rate of flow during secretion and the undistorted latencies, sensitive flow-meters have been constructed (Kozak 1954, 1957). A cathode-ray oscillograph recorded simultaneously the salivary flow rate and the gland surface potential (Fig. 1 D, 3 J, K). A steady current of water used for calibration is recorded as in Fig. 2 H and 3 A. Salivary flow rates between 0.3 and 100 μl/sec could be recorded with the instrument set for an appropriate sensitivity, without obstructing the flow. The device recorded the exact latency and form of the salivary "spurts" following single shocks, the superposition and fusion of these spurts, the changes of latency and rate of salivation.

II. Changes of secretory capacity of the salivary glands

1. *Short-lasting effects. Pairs of stimuli*

A supramaximal electric stimulus producing a synchronous volley in the secretory nerve evokes a spurt of salivation after latency of 0.25—1 sec (S) or 0.5—1 sec (P). After a rise-time of 1—1.5 sec (P) or 1.3—3 sec (S), the salivary flow reaches an amplitude of 0.3 to 8 μl/sec. The flow diminishes afterwards with a time-constant of 0.8—3.5 sec (P) or 1.2—3 sec (S) and lasts about ten seconds. The stimulation of the gland by a single volley produces a trace excitation lasting for about 10 sec, as revealed by both the technique of pairs: conditioning-testing stimuli and by repetitive stimulation. At the optimal interval between stimuli (20—50 msec), the resulting spurt of secretion exceeded the control ones by the factor of ten (S, Fig. 1 B), or five (P), while the gland surface potentials exceeded the controls three times (S, Fig. 1 A, C). The relationship: amplitude-versus-preceding interval was similar for both the electrograms and the sialograms (records of flow, Fig. 1 B, C). For intervals longer than 100 msec, a partial superposition of both electrograms and sialograms is shown (Fig. 1 A, D). The latencies of the compound spurts decrease whenever their amplitudes increase (Fig. 1 D).

2. *Effects of medium duration. Post-tetanic potentiation of unitary salivations*

By applying electric shocks to the secretory nerve at 0.1/sec rate, a series of distinct unitary salivations (spurts) is produced. For frequencies of 0.2—1/sec, single spurts become superimposed more and more closely and the rate of salivary flow increases (Fig. 2 A—G). At the frequency of 2/sec, a smooth flow is produced (cf. Bruner and Kozak 1957). With further increase of frequency of stimulation, the rate of flow increases to a maximum, attained at about 35/sec (Fig. 2 K), and decreases for still higher frequencies. The single spurts of parotid secretion to supramaximal shocks at 0.2/sec rate, are very small (Fig. 2 A). After the optimal (35/sec) stimulation of the auriculo-temporal nerve for 20 sec, the same stimulation at 0.2/sec elicits spurts up to 25 times higher (Fig. 2 E). The amplitude of the spurts diminishes afterwards. It falls to 1/e of the maximal value within one minute after the tetanus (P, Fig. 2 E, L). The post-tetanic potentiation in the submaxillary gland is less intense, but it lasts

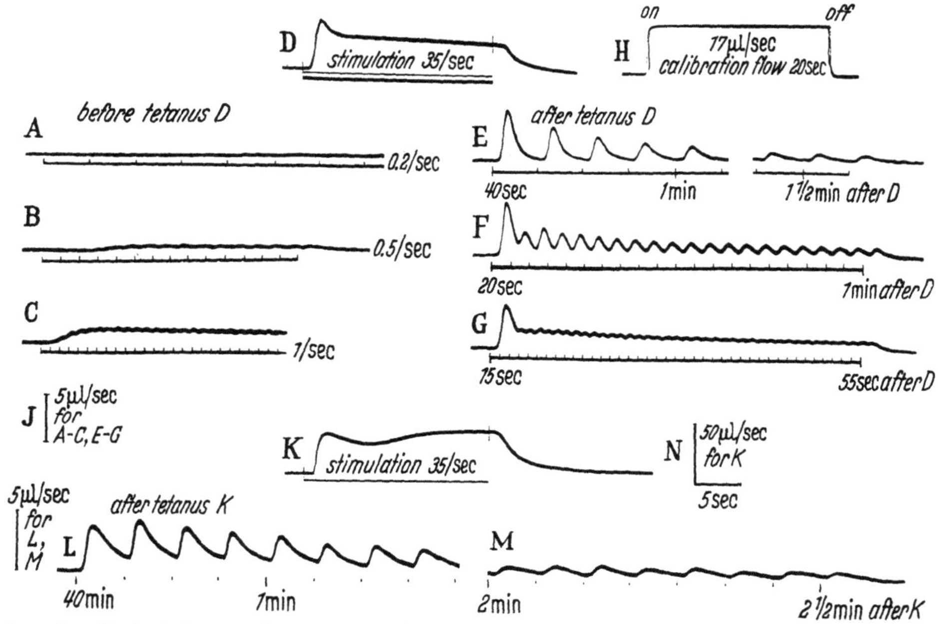

Fig. 2. Parotid gland. Post-tetanic potentiation of unitary salivations. A, B, C — before; E, F, G — after tetanic stimulation D at indicated delays. Calibration J applies to A—C, E—G. K, L, M — another gland; single spurts at indicated delays after tetanus K. Time scale uniform for A—M

longer. The reduction of spurt amplitude to 1/e takes about nine minutes. When frequencies of 0.5—1/sec are applied before and after tetanic stimulation, the potentiation is less marked than for 0.2/sec (Fig. 2 B—F, C—G).

3. *Long-lasting effects. Augmented secretion*

By applying supramaximal tetanic stimulation with the optimal frequency (20—50/sec, 1—2 msec pulses) and 20 secs duration, we elicit a flow of saliva in which two phases can be distinguished (Fig. 2 D, K; 3 B—H), each one lasting about ten seconds. The peak flow of the first phase is shown by arrows in Fig. 3 B—H. The length of the period of rest preceding each tetanic stimulation is indicated on Fig. 3 B—H. The relationship between the period of rest and the amplitude of the following peak flow is shown on Fig. 3 L, and the latency of salivation after different periods of rest in 3 M. Each dot in Fig. 3 L, M represent a mean of several data obtained from the one gland. Fig. 3 J shows a surface electrogram (E) and sialogram (F) taken from a submaxillary gland tetanically stimulated 1 min 40 sec after a preceding tetanus. Fig. 3 K — the same, but after 1½ hour of rest. The initial phase of salivation is reduced to 40—50 %

of the optimum in both glands after 1/2—1 hour of rest (cf. Bruner and Kozak 1954 b). The depressing effect of a long rest is abolished by a small dose of pilocarpine. Long after the secretion has stopped, the gland responds optimally to a tetanus (Bruner and Kozak 1954 d).

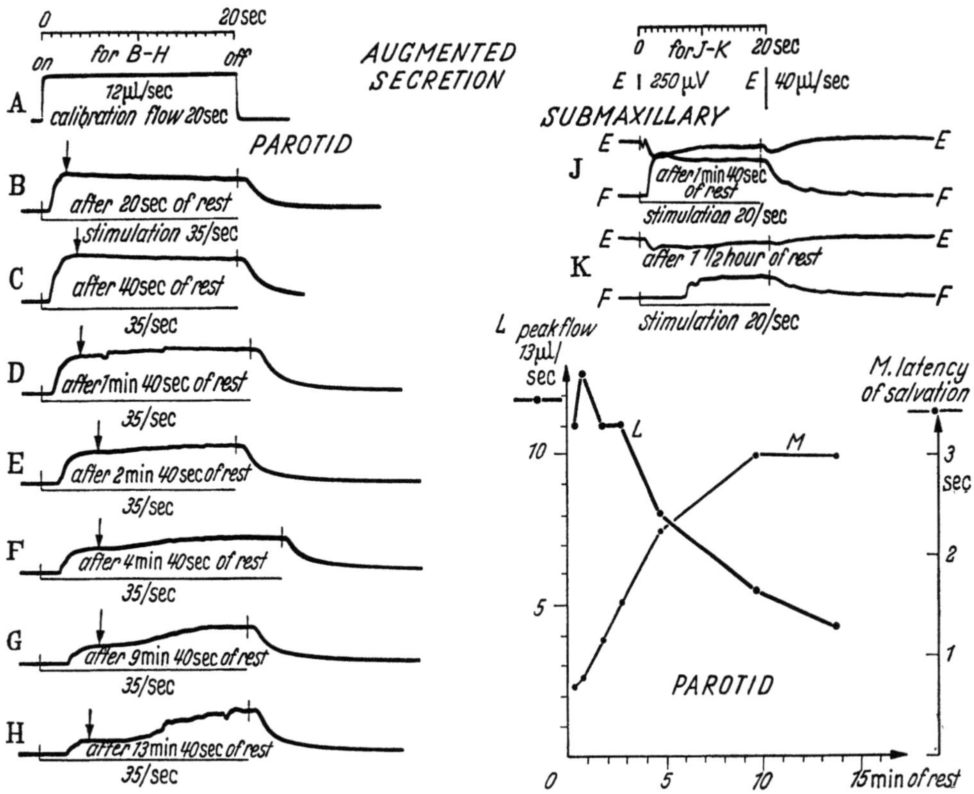

Fig. 3. Salivary flow to optimal stimulation after varying periods of rest. B—H — Parotid. Examples of sialograms taken after indicated periods of rest. J, K — Submaxillary. Records of surface potential (E) and flow (F) after indicated periods of rest. L, M — parotid; L — mean peak flow (arrowed at B—H) and M — mean latency-versus-preceding rest

III. Chronic experiments on unanaesthetized dogs

A similar relationship between the salivary flow and the preceding period of rest has been found in unanaesthetized, semi-freely moving dogs. Chronic fistulae of the shortened parotid ducts were made and healed several weeks beforehand. Salivation was then recorded remotely using a voluminograph (Kozak 1950). By giving the dog a standard amount of food (Bruner and Kozak 1954 a), or by introducing standard amounts of lactic acid solution into the mouth (Czarnecka and Soltysik 1962), and by varying the intertrial intervals, it was found that the flow rate during the first ten seconds of salivation was reduced to 55—62% after 15—20 min of rest. The character of the function was similar to that found in acute experiments. The latencies of salivation increased with the length of preceding rest (cf. Fig. 3 M).

The acute extinction of a conditioned salivary reflex has been long known to depress the subsequent salivation and this depression was interpreted as an irradiation of cortical inhibition (Pavlov 1940). However, an identical depressing effect could be produced by merely leaving an interval between salivations similar to that connected with the acute extinction procedure (Bruner and Kozak 1954 c). It seems that this depression can be explained by the post-activation changes in the salivary gland.

IV. General remarks and conclusions

The post-activation potentiation of salivary flow is very pronounced. It occurs in preparations where either pre-ganglionic fibres (tympanicolingual nerve), post-ganglionic fibres (auriculo-temporal nerve) or the gland cells (pilocarpine) are stimulated. It occurs in unanaesthetized dogs to a comparable degree. The trace excitation lasts for about ten seconds after a single conditioning volley, with a maximum at about 30 msec interval. After a 20 secs conditioning tetanus of physiological frequency (20—50/sec), the trace excitation lasts for about half an hour, with a maximum at about 40 sec after tetanus. The salivation to single shocks can be potentiated up to 25 times the pre-tetanic value. The latencies of salivation are variable; during the potentiation they become shorter, approximating the latencies of the electrograms — 250 msec. The potentiation of salivary flow is accompanied by an increase of amplitude of the electrogram. The post-activation changes in the salivary glands should be taken into consideration when interpreting the salivary conditioning in dogs. Numerous kinds of inhibition described by Pavlov were invariably associated with lengthening of interval between successive salivations. This lengthening of interval depressed the secretory capacity of the gland, and the reduced salivation has been misinterpreted as an effect of central inhibition.

References

Beznák, M. & Farkas, E. (1936—37). *Quart. J. Exper. Physiol.* **26**, 265.
Bruner, J. & Kozak, W. (1954 a). *Acta Physiol. Polon.* **5**, 107.
Bruner, J. & Kozak, W. (1954 b). *Acta Physiol. Polon.* **5**, 229.
Bruner, J. & Kozak, W. (1954 c). *Acta Physiol. Polon.* **5**, 507.
Bruner, J. & Kozak, W. (1954 d). *Acta Physiol. Polon.* **5**, 512.
Bruner, J. & Kozak, W. (1957). *Acta Physiol. Polon.* **8**, 291.
Burgen, A. S. V. & Emmelin, N. G. (1961). *Physiology of the salivary glands*, p. 56. London: Edward Arnold.
Czarnecka, M. & Soltysik, S. (1962). *Acta Biol. Exper.* **22**, 15.

Gayda, T. (1925). *Arch. di Scienze Biol.* **7**, 178.
Goldenberg, E. E. (1923—1924). *J. Physiol.* (Lond.) **58**, 267.
Konorski, J. (1948). *Conditioned reflexes and neuron organization.* pp. 7, 251. Cambridge: University Press.
Kozak, W. (1950). *Acta Biol. Exper.* **15**, 185.
Kozak, W. (1954). *Acta Physiol. Polon.* **5**, 559.
Kozak, W. (1957). *Acta Physiol. Polon.* **5**, 395.
Pavlov, I. P. (1940). *Conditioned Reflexes,* pp. 156, 164, 168. Oxford: University Press.
Podkopaev, N. A. (1952). In *Metodika Izucheniya Uslovnykh Refleksov,* pp. 90, 106. Moscow: Izdatelstvo Akad. Med. Nauk.

CHOLINERGIC INNERVATION OF THE CEREBRAL CORTEX

by K. Krnjević

As soon as it became clear that acetylcholine (ACh) may transmit nerve signals across certain peripheral synapses, Dale (1934) suggested that cholinergic transmission may also play a significant role in the central nervous system. However, in spite of much indirect evidence, which has been reviewed extensively by Feldberg (1945, 1957), it is still not at all certain that any central fibres are cholinergic, except for the recurrent branches of motor axons which excite Renshaw cells in the spinal cord, as shown very convincingly by Eccles and his associates (Eccles, Fatt and Koketsu 1954; Eccles, Eccles and Fatt 1956; Curtis and Eccles 1958).

Recent experiments on the mammalian cerebral cortex have shown directly that some cortical neurones can be excited with ACh (Krnjević and Phillis 1961, 1963 a; Spehlmann and Kapp 1961; Spehlmann 1963). Is this sensitivity to ACh an isolated phenomenon or are there good reasons for believing that cholinergic transmission is a significant element in the complex mechanism of cortical function? The available evidence concerning the cerebral cortex will be reviewed briefly, and a description given of some features of a likely cholinergic pathway.

I. ACh metabolism

All the components of the ACh metabolic system are found in the cortex: ACh is present (MacIntosh 1941; Elliott, Swank and Henderson 1950), being particularly concentrated in nerve endings and synaptic vesicles (Whittaker and Sheridan 1965); as well as cholineacetylase (Feldberg and Vogt 1948;

HEBB and SILVER 1956) and acetylcholinesterase (BURGEN and CHIPMAN 1951; POPE 1952). ACh is released in the cortex and can be collected from the surface *in vivo* (ELLIOTT et al. 1950; MACINTOSH and OBORIN 1953; MITCHELL 1963).

II. ACh excites cortical neurones

1. Some suggestive observations made by early investigators (BONNET and BREMER 1937; MILLER, STAVRAKY and WOONTON 1939) have been confirmed recently by very localised microiontophoretic applications of small amounts of ACh (KRNJEVIĆ and PHILLIS 1963 a; SPEHLMANN 1963). However, when cells were tested at random, only a certain proportion were obviously excited by ACh. The exact percentage is difficult to assess since it varies a great deal between different animals, being affected by various factors such as anaesthesia. In anaesthetized cats the proportion has varied between 4 and 45%; 13 cats gave a mean of 15%. In two unanaesthetized *cerveaux isolés,* 27% of cells were excited by ACh.

An example of the effect produced by ACh on a cortical cell is shown in Fig. 1 (C). The most important features are the relatively slow onset and the

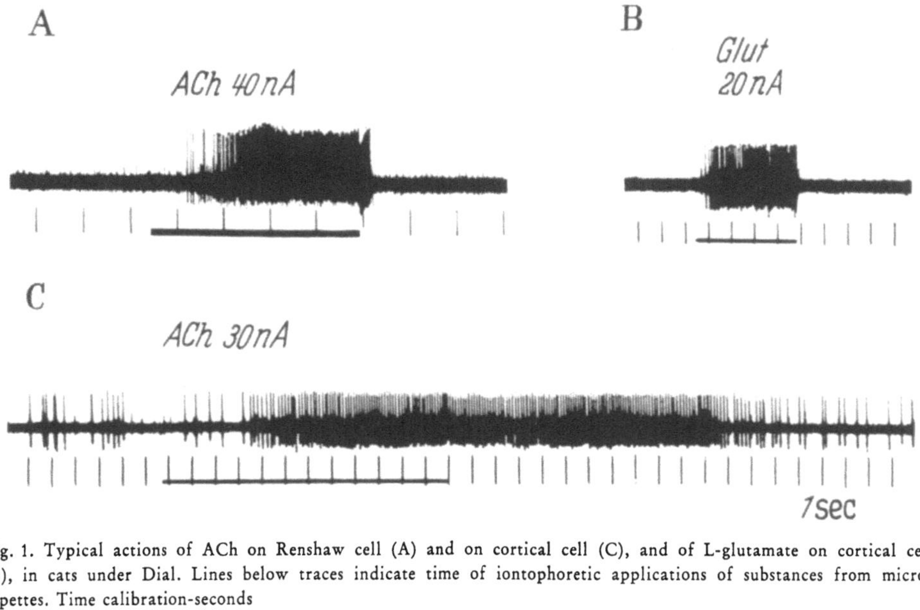

Fig. 1. Typical actions of ACh on Renshaw cell (A) and on cortical cell (C), and of L-glutamate on cortical cell (B), in cats under Dial. Lines below traces indicate time of iontophoretic applications of substances from micropipettes. Time calibration-seconds

prolonged after-discharge. These are particularly striking when compared with the quicker effects produced by ACh on the Renshaw cells (Fig. 1 [A]), or by L-glutamate on cortical cells (Fig. 1 B), both of which are clearly compatible with a detonator type of transmitter action. Although some possible depressant effects of ACh on cortical neurones have also been described (RANDIĆ, SIMINOFF and STRAUGHAN 1964) their significance is not certain; the strong inhibition, elicited by surface and by indirect cortical stimulation does not appear to depend on a cholinergic mechanism (KRNJEVIĆ, RANDIĆ and STRAUGHAN 1964).

2. *Pharmacology of ACh action.* The actions of ACh in the cortex are predominantly due to its muscarine-like properties (KRNJEVIĆ and PHILLIS 1963 b); they are therefore blocked very effectively by atropine and hyoscine, but largely unaffected by curariform antagonists, including tubocurarine, dihyro-β-erythroidine and excess nicotine. Various muscarinic agents reproduce the actions of ACh, but nicotine and other nicotinic substances usually do not. In this respect cortical cells show a further difference from Renshaw cells which are excited particularly by nicotinic agents (ECCLES et al. 1954; ECCLES et al. 1956).

ACh sensitive cells in the thalamus (CURTIS and ANDERSEN 1962; CURTIS and DAVIS 1963) have receptors with both nicotinic and muscarinic properties, perhaps because they receive a different kind of innervation, as suggested by the presence of relatively large amounts of pseudo-cholinesterase (SHUTE and LEWIS 1963). In the cortex practically all the cholinesterase activity can be ascribed to acetylcholinesterase (KOELLE 1950; BURGEN and CHIPMAN 1951; KRNJEVIĆ and SILVER 1963 b, 1965 a).

III. Identity of ACh sensitive cells

1. From a detailed study of cortical cells excited by ACh (KRNJEVIĆ and PHILLIS 1963 a) it appeared that these cells may be found in all regions of the neocortex, though particularly near the main afferent areas. Very few occurred in the superficial layers and they were particularly concentrated at depths between 0.8 mm and 1.3 mm from the surface. Since most of the Betz cells identified by antidromic excitation from the pyramids were sensitive to ACh, it was concluded that cholinoceptive cells in general are likely to be deep pyramidal cells.

2. For further identification, studies have been made of the distribution of acetylcholinesterase (AChE) in the cerebral cortex, using a modification of Koelle's technique (KRNJEVIĆ and SILVER 1963 a, b; 1965). Only relatively weak cellular staining was observed; too weak to be strong evidence for the presence in the cortex of definite cholinergic neurones. But many large pyramidal cells in layer V consistently showed moderate AChE activity, and they often formed a relatively clear layer at the depth where cholinoceptive cells are mainly found. A direct comparison gave excellent agreement between the distribution of ACh sensitive cells and that of stained cells in the same area. It is therefore highly probable that these are in fact the cholinoceptive neurones.

These deep pyramidal cells seem to acquire their staining mainly because they are closely invested by a network of fibres and fibrils which contain much AChE and with which they may have synaptic contacts. These fibres probably have their origin from distant cell bodies because cortical isolation by undercutting causes a large reduction in their staining, as well as in the level of cortical choline acetylase activity (HEBB, KRNJEVIĆ and SILVER 1963).

IV. Identity of postulated cholinergic afferent pathways

1. Electrophysiological evidence

There was no obvious association between cortical cholinoceptive neurones and the short latency afferent responses, evoked either from the periphery or from the relay nuclei in the thalamus (KRNJEVIĆ and PHILLIS 1963 a). Similarly, although some of these neurones receive transcallosal signals, or respond to non-specific thalamic activation (recruiting responses), other, non-cholinoceptive, cells were just as likely to be excited by these pathways.

The main types of activity that seemed to bear a significant relation to the cholinoceptive cells were irregular spontaneous discharges ("projection activity"), and prolonged repetitive responses to single pulse stimulation of thalamic relay nuclei, similar to those first described by ADRIAN (1941) and MORISON and DEMPSEY (1943). The exact route by which the relevant impulses reach the cortex is not known, but it is likely to be relatively indirect.

Both types of activity could be enhanced by local applications of anticholinesterases and they were depressed or abolished by systemic doses of atropine or chloralose; the latter is known to reduce the liberation of ACh in the cortex (MACINTOSH and OBORIN 1953; MITCHELL 1963), though probably not by a general inactivation of central cholinergic endings, since the excitation of Renshaw cells by motor axon collaterals is not affected (BISCOE and KRNJEVIĆ 1963). Even local iontophoretic applications of atropine can depress the spontaneous and evoked (repetitive) activity of cholinoceptive cells; but this effect is more variable and usually less prolonged than the block of responses to local applications of ACh, as is commonly the case with several types of cholinergic synapses (DALE and GADDUM 1930; DALE 1938; AMBACHE 1955).

2. Histochemical evidence

a) In the adult cat (KRNJEVIĆ and SILVER 1963 b; 1965 a) the AChE-containing fibres form a predominantly tangential system. It is seen best under the sulci, where bundles of stained fibres are in the well-known U-formations. In the gyrus these fibres ascend on either side of the relatively unstained central core, which contains the main afferent, efferent and commissural pathways; they give off branches which form a complex network in the deeper half of the cortex, with rather diffuse endings, especially in relation to pyramidal cells of layer V.

A continuous system of AChE-containing fibres is thus present over the whole hemisphere, especially in the outer part of the white matter and deeper layers of the cortex. Their main subcortical connexions can be traced to the corpus striatum and the septal region. The lateral and frontal parts of the cortex receive fibres from the putamen via the external capsule, while the medial and superior aspects are innervated by septal fibres which travel mostly around the corpus callosum in the cingulum. A substantial number of isolated fibres also come from

Fig. 2. AChE-containing fibres from lenticular nucleus invading lateral wall of primitive forebrain. Note absence of staining in cortical lamina. Cat embryo at about 40 days: line = 0.5 mm. (K. Krnjević and A. Silver, unpublished observations)

the septum via the subcallosal striatum which is mainly a lateral extension of the dorsal fornix, though its outermost portion is probably made up of striatal fibres from the caudate nucleus.

It seems that at least some AChE-containing U-fibres from links between adjacent cortical areas. Such fibres presumably have their origin in cell bodies in the cortex. Although no cortical cells show very strong AChE activity, comparable with that of known cholinergic neurones, some spindle cells in the deepest layer of the cortex often have relatively large amounts of intracellular AChE; as they are also particularly closely related to U-fibres, they may be cholinergic cells whose axons enter the U-formation (cf. MEYNERT 1867), but the evidence is at present inconclusive.

Many very strongly stained cells are to be seen in the septal region and in the corpus striatum, especially in the pallidum. Whether these are sufficient to account for the cortical projections described above or whether some fibres may have their origin even further back is not certain. SHUTE and LEWIS (1963) consider these AChE containing pathways as forebrain extensions of the mid-brain and hypothalamic reticular formation.

b) Further studies of the development of AChE staining in embryos (KRNJEVIĆ and SILVER 1964, 1965 b) have confirmed the essentially non-cholinergic nature of the primary cortical elements. In the primitive forebrain there is no AChE except in the developing lenticular nucleus and in the septum, both of which probably arise from the striatal ependyma. AChE containing fibres later spread out from these regions and invade the rest of the hemisphere (Fig. 2) but they do not begin to penetrate neocortical areas much before birth. This relatively late development of the postulated cholinergic innervation may account for the late appearance of definite sensitivity to ACh in the cortex of young kittens (KRNJEVIĆ, RANDIĆ and STRAUGHAN 1964 b).

V. Possible significance of cholinergic pathways

In view of the evidence presented above there is a strong likelihood that an ascending cholinergic system innervates some deep cortical cells. SHUTE and LEWIS (1963) have suggested that the AChE-containing fibres in the forebrain mediate by a cholinergic mechanism the cortical arousal known to be produced by activation of the mid-brain reticular formation (MORUZZI and MAGOUN 1949). These fibres would thus provide the final corticipetal link which has been missing in previous analyses of this pathway (NAUTA and KUYPERS 1958; DROOGLEEVER-FORTUYN 1960). This suggestion is in agreement with previous evidence that cortical arousal is prevented by atropine, probably by a direct action on the cortex (LOEB, MAGNI and ROSSI 1960; MONNIER and ROMANOWSKI 1962; LING 1963).

Chloralose, which reduces the liberation of ACh in the cortex, is also known to interfere with cortical arousal (MORUZZI and MAGOUN 1949; BREMER and STOUPEL 1959). From the observations of the last authors, chloralose seems to depress the excitability of the reticular formation. Unlike atropine, chloralose prevents behavioural arousal as well as cortical desynchronization, presumably because of its central action on the reticular formation. The dissociation between the action of atropine on cortical activity and on behavioural arousal (WIKLER

1952; Bradley and Elkes 1957) may be due to the fact that the mechanisms controlling sleep are likely to be in the brain stem (Cairns 1952), where cholinoceptive elements do not show clear muscarinic properties and therefore would not be fully blocked by atropine. The situation in the brain stem is further complicated by evidence that activation of certain ACh-sensitive systems may induce sleep (Hernandez-Péon, Chavez-Ibarra, Morgane and Timo-Iaria 1963) and that the "limbic circuit", apparently concerned with emotions, probably contains many cholinergic elements (MacLean 1957).

The level of cortical 'arousal' may thus be determined by the rate of release of ACh from the ascending cholinergic fibres, which would control the excitability of deep pyramidal cells particularly; the relatively slow action of ACh is clearly ideally suited for this kind of function. It may be relevant that young kittens have little fast activity in their electrocorticograms and minimal spontaneous unit discharges in the first two weeks (Scheibel 1962; Marley and Key 1963), when the cortical cells are insensitive to ACh (Krnjević et al. 1964 b).

Recent observations show that reticular stimulation, which causes a particularly marked increase in the release of ACh in the cortex (Kanai and Szerb 1965) also activates cortical cholinoceptive units (K. Krnjević and M. Randić, unpublished observations). However, this latter effect, though suggestive, is not conclusive because the responses have a relatively long latency; this is not necessarily negative evidence, since a slow effect may well be due to slow conduction along fine fibres, multisynaptic pathways or a slow action of the transmitter.

Conclusions

There is adequate evidence that ACh can be released in the cerebral cortex, where it excites many deep pyramidal cells. On the other hand, AChE-containing fibres from the corpus striatum and the septal region project diffusely over the cortex. These fibres are likely to provide a cholinergic innervation of cortical cholinoceptive cells which controls their excitability and mediates the characteristic slow and prolonged activation seen during spontaneous discharges and evoked repetitive activity; this may be the ascending reticular pathway responsible for cortical arousal.

References

Adrian, E. D. (1941). *J. Physiol.* (Lond.) **100**, 159.
Ambache, N. (1955). *Pharmacol. Rev.* 7, 467.
Biscoe, T. J. & Krnjević, K. (1963). *Exp. Neurol.* **8**, 395.
Bonnet, V. & Bremer, F. (1937). *C. R. Soc. Biol.* (Paris) **126**, 1271.
Bradley, P. B. & Elkes, J. (1957). *Brain* **80**, 77.
Bremer, F. & Stoupel, N. (1959). *Arch. int. Pharmcodyn.* **122**, 234.

BURGEN, A. S. V. & CHIPMAN, L. M. (1951). *J. Physiol.* (Lond.) **114**, 296.
CAIRNS, H. (1952). *Brain* **75**, 109.
CURTIS, D. R. & ANDERSEN, P. (1962). *Nature* (Lond.) **195**, 1105.
CURTIS, D. R. & DAVIS, R. (1963). *J. Physiol.* (Lond.) **165**, 62.
CURTIS, D. R. & ECCLES, R. M. (1958). *J. Physiol.* (Lond.) **141**, 435.
DALE, H. H. (1934). *Brit. med. J.* p. 1.
DALE, H. H. (1938). *J. Mt. Sinai Hosp.* **4**, 401.
DALE, H. H. & GADDUM, J. H. (1930). *J. Physiol.* (Lond.) **70**, 109.
DROOGLEEVER FORTUYN, J. (1960). In: *Structure and function of the cerebral cortex* (ed. by TOWER, D. D. & SCHADÉ, J. P.) p. 88. Amsterdam: Elsevier.
ECCLES, J. C., ECCLES, R. M. & FATT, P. (1956). *J. Physiol.* (Lond.) **131**, 154.
ECCLES, J. C., FATT, P. & KOKETSU, K. (1954). *J. Physiol.* (Lond.) **126**, 524.
ELLIOTT, K. A. C., SWANK, R. L. & HENDERSON, N. (1950). *Amer. J. Physiol.* **162**, 469.
FELDBERG, W. (1945). *Physiol. Rev.* **25**, 596.
FELDBERG, W. (1957). In: *Metabolism of the Nervous System* (ed. by W. RICHTER), p. 493. London: Pergamon Press.
FELDBERG, W. & VOGT, M. (1948). *J. Physiol.* (Lond.) **107**, 372.
HEBB, C. O., KRNJEVIĆ, K. & SILVER, A. (1963). *Nature* (Lond.) **198**, 692.
HEBB, C. O. & SILVER, A. (1956). *J. Physiol.* (Lond.) **134**, 718.
HERNANDEZ-PÉON, R., CHAVEZ-IBARRA, G., MORGANE, P. J. & TIMO-IARIA, C. (1963). *Exp. Neurol.* **8**, 93.
KANAI, T. & SZERB, J. C. (1965). *Nature* (Lond.) **205**, 81.
KOELLE, G. B. (1950). *J. Pharmacol.* **100**, 158.
KRNJEVIĆ, K. & PHILLIS, J. W. (1961). *Experientia* **17**, 469.
KRNJEVIĆ, K. & PHILLIS, J. W. (1963 a). *J. Physiol.* (Lond.) **166**, 296.
KRNJEVIĆ, K. & PHILLIS, J. W. (1963 b). *J. Physiol.* (Lond.) **166**, 328.
KRNJEVIĆ, K., RANDIĆ, M. & STRAUGHAN, D. W. (1964 a). *Nature* (Lond.) **201**, 1294.
KRNJEVIĆ, K., RANDIĆ, M. & STRAUGHAN, D. W. (1964 b). *J. Physiol.* (Lond.) **175**, 21 P.
KRNJEVIĆ, K. & SILVER, A. (1963 a). *J. Physiol.* (Lond.) **165**, 3 P.
KRNJEVIĆ, K. & SILVER, A. (1963 b). *J. Physiol.* (Lond.) **168**, 39 P.
KRNJEVIĆ, K. & SILVER, A. (1964). *J. Physiol.* (Lond.) **175**, 22 P.
KRNJEVIĆ, K. & SILVER, A. (1965 a). *J. Anat.* (Lond.) (in press).
KRNJEVIĆ, K. & SILVER, A. (1965 b). *J. Anat.* (Lond.) (in press).
LING, G. M. (1963). *J. Neuropsychiat.* **4**, 255.
LOEB, C., MAGNI, F. & ROSSI, G. F. (1960). *Arch. ital. Biol.* **98**, 293.
MACINTOSH, F. C. (1941). *J. Physiol.* (Lond.) **99**, 436.
MACINTOSH, F. C. & OBORIN, P. E. (1953). *Abstr. XIX int. physiol. Congr.* p. 580.
MACLEAN, P. (1957). *Arch. Neurol. Psychiat.* (Chicago) **78**, 113.
MARLEY, E. & KEY, B. J. (1963). *Electroenceph. clin. Neurophysiol.* **15**, 620.
MEYNERT, T. (1867). *Vierteljahresschrift Psychiat.* p. 198.
MILLER, F. R., STAVRAKY, G. W. & WOONTON, G. A. (1940). *J. Neurophysiol.* **3**, 131.
MITCHELL, J. F. (1963). *J. Physiol.* (Lond.) **165**, 98.
MONNIER, M. & ROMANOWSKI, W. (1962). *Electroenceph. clin. Neurophysiol.* **14**, 486.
MORISON, R. S. & DEMPSEY, E. W. (1943). *Amer. J. Physiol.* **138**, 297.
MORUZZI, G. & MAGOUN, H. W. (1949). *Electroenceph. clin. Neurophysiol.* **1**, 455.
NAUTA, W. J. H. & KUYPERS, H. J. G. M. (1958). In *Reticular Formation of the Brain* (ed. by H. H. JASPER et al.) p. 3. London: Churchill.
POPE, A. (1952). *J. Neurophysiol.* **15**, 115.
RANDIĆ, M., SIMINOFF, R. & STRAUGHAN, D. W. (1964). *Exp. Neurol.* **9**, 236.
SCHEIBEL, A. B. (1962). *Recent Adv. biol. Psychiat.* **4**, 313.
SHUTE, C. C. D. & LEWIS, P. R. (1963). *Nature* (Lond.) **199**, 1160.
SPEHLMANN, R. (1963). *J. Neurophysiol.* **26**, 127.
SPEHLMANN, R. & KAPP, H. (1961). *Pflüg. Arch. ges. Physiol.* **274**, 37.
WHITTAKER, V. P. & SHERIDAN, M. N. (1965). *J. Neurochem.* (in press).
WIKLER, A. (1952). *Proc. Soc. exp. Biol. Med.* **79**, 261.

AN APPROACH TO THE STUDY OF NEUROGLIA AND OF EXTRA-CELLULAR SPACE BASED ON RECENT WORK ON THE NERVOUS SYSTEM OF THE LEECH

by STEPHEN W. KUFFLER, JOHN G. NICHOLLS and DAVID D. POTTER

In 1938 two of the present authors (J. G. N. and D. D. P.) had recently started elementary school while the other (S. W. K.) had just arrived in Sydney, Australia, embarking on a career in pathology. The job in pathology was one week old when good fortune arranged a chance encounter with JOHN ECCLES on a Sunday afternoon on a tennis court. The result was an enthusiastic introduction to neurophysiology, an immediate conversion to the field of the nervous system and a sudden change in jobs. The conversion was enduring and so was the influence and friendship of JOHN ECCLES. Over the years J. C. E. has worked with eminent success in most areas of neurophysiology. He has, however, not yet turned his interest to the problems of the present essay which we therefore hastily and with pleasure dedicate to him. We will briefly and informally look at a large problem, that of neuroglia, from the restricted angle of the leech. For details a series of recent papers should be consulted (KUFFLER and POTTER 1964; KUFFLER and NICHOLLS 1964; NICHOLLS and KUFFLER 1964; NICHOLLS and KUFFLER 1965).

Many physiologists have long recognized the need for studying neuroglia which makes up a large portion, if not the greater part, of our nervous system (PENFIELD 1932). Work on glia had been planned for at least ten years but other problems of current research always took precedence. The deficiency of our knowledge became more obvious and painful when, in the process of planning a neurophysiology teaching program, it was felt that neuroglia should have its place. Textbooks of physiology disposed of the physiology of neuroglia, if they treated it at all, with a general statement such as "neuroglia is assumed to serve a supportive-nutritive function in the nervous system". The basis for this idea was established, like so many other physiological interpretations, from considerations of structure. For instance, in 1870 GOLGI (see GOLGI 1903) studied the various glial cells within the central nervous system and showed that astrocytes linked capillaries and neurons. A bridge or channel apparently existed and expressive terms like 'end feet' and 'sucker processes', strengthened the reasonable idea of glia (especially astrocytes) being a supply channel. Of course, the light microscopist did have alternatives for the passage of materials from the blood to the neurons, because all sorts of perivascular and intercellular spaces were also available (HELD 1909; STUDNIČKA 1899). The relatively large estimates of extracellular space from chemical measurements or ion composition were in accord with this concept (e. g., DAVSON and SPAZIANI 1959).

The old idea for a nutritive role of glia received a great boost from the revolution that was created by electron microscopy of the nervous system. The spaces seen by the light microscopist suddenly disappeared; they now resembled those seen in epithelia and became quite unlike those of skeletal muscle (HORSTMANN and MEVES 1959). All was very tightly packed; and wherever one looked, one saw only cells, except for extremely narrow clefts that were usually 100 to 200 Å wide. It was natural that a series of proposals, based on the new anatomy, should be put forward. Since extracellular space was apparently no longer available, materials from the blood had to pass through glia in order to reach neurons. This idea was combined with the observation that during cerebral swelling the fluid entered glial cells selectively. As a result it was inferred that the physiological "extracellular space" must be located in glial cells (DE ROBERTIS 1962; SJÖSTRAND 1961). It was also known that neurons contain little Na^+ but need it in their environment. Therefore, when a large quantity of Na^+ was measured in the nervous system, it was assumed to be located in the glial cells (KATZMAN 1961; KOCH, RANCK and NEWMAN 1962). The logic of all these deductions led to a widespread acceptance of the proposal that glia constitutes the extracellular space. This in turn favored the assumption that neurons and glia had a joint membrane across which metabolic pumps and enzyme systems operated (see HYDÉN and PIGON 1960; CUMMING and HYDÉN 1962; MITARAI, SVAETICHIN, VALLECALLE, FATECHAND, VILLEGAS and LANGER 1961). The numerous review articles, by repeatedly presenting views that had frequently started as tentative suggestions, reinforced the more recent notions. There were no good grounds to reject the concepts current early in 1961, except that one would have liked more direct evidence on almost each of the experimental results that were crucial for the conclusions.

The first research effort was directed to finding a suitable preparation for the more rigorous study of neuroglia. Studies on the intact brain of mammals seemed difficult because there the glial cells, although much more numerous than neurons, are small and cannot be readily impaled and identified by microelectrodes. A search was therefore made for a nervous system that would fulfill some of the following criteria:

1. The anatomical relationship between neurons and glia should be as intimate as in the vertebrate nervous system.

2. It should survive well in isolation and preferably be without blood vessels that penetrate between the cells.

3. It should be transparent enough for its component cells to be identified under direct visual observation.

4. Above all, it should have large glial cells, so that hopefully it would be suitable for chemical studies and also for experiments with microelectrodes, such as measuring the electrical properties of the cell membrane.

Two ingredients, necessary for a successful search, were available and they were, in order of importance: a) the excellent library of the Marine Biological Laboratory at Woods Hole, and b) the previous discovery and description of the desired nervous systems. To our pleasant surprise we found that some of the most capable anatomists had worked on nervous systems that fulfilled many of the above criteria. The leech was decided on when it turned out that all of the requirements were met and in addition some further unexpected advantages turned up. The roster of biologists who contributed to the study of the leech nervous system structure 50 or more years ago makes an illustrious list. A few of the names are: Valentin, Leydig, Hermann, Rohde, Biedermann, Retzius, Apáthy, Holmgren, Cajal, Held, Sanchez, Gaskell and Hortega (for many of the historical references see Ito 1936). The structure lies in a blood sinus and a feature of special interest is that a single large glial cell is associated with many nerve cells or axons.

The advantages of the leech for a detailed acquisition of general information were obvious, but it was also clear that our curiosity relating to glia and extra-cellular space in the vertebrate would not be satisfied directly. Numerous structural analogies, however, made us confident that we were working in the direction of our original interest in exploring the properties of glial cells and the characteristic neuron-glia relationship in the leech.

First of all, it is known that the histological staining properties of glial cells in leech and vertebrates are similar. This may be one of the reasons that during the last century the identification and intensive study of glial cell structure was carried out so extensively in the leech. The recent electron microscopic view of many portions of the leech nervous system brings out additional similarities. At an early stage we had the good fortune that Drs. Fawcett and Coggeshall, working on the next floor, became interested in the leech. They made electron microscopic studies that were essential and basic for the physiological experiments. At the same time Drs. Gray and Guillery independently began to work on the leech at University College, London. Here we wish to single out one outstanding feature of their studies; the neuron-glia relationship shows that the two cell types are interwoven and in intimate contact.

The electrical properties of neurons and glia were the first questions to be studied and subsequent investigations dealt with those problems which may be linked with the close and characteristic apposition of the cells. The first centered around the question whether glial cells possess a signalling mechanism that resembles the well-known neuronal impulse. The problem of the functional relationship between the two types of cell turned out to have numerous intriguing implications which required prolonged study with a number of different techniques. Relatively simple questions, such as the role of intercellular channels, the nature of the connections between glial cells, and the electrolyte content of various compartments of the nervous system have been answered with reasonable success. They in turn have led to the more complicated problems of the transport and uptake system of cells. There also beckons at the end the mystique of the blood-brain barrier(s).

We will not attempt here, for reasons of space, to give details of the initial steps that we have taken to understand the function of neuroglia and of extracellular channels in their relationship to neurons. On the whole, we have learned a good deal about properties of neuroglia in the leech. We have found that these glial cells have high resting potentials, high internal K^+, and a high membrane resistance. They are not involved in the exchange of NaCl, K^+, sucrose and choline between blood and neurons. They do not give action potentials or interact electrically with the neurons although they interact electrically with one another. Glial cells can take up large molecules from the bathing fluid and can rapidly make glycogen from glucose. These results lead us to hope that the leech nervous system will be suitable for more direct experiments to test the old hypotheses of exchange and transfer of substances between neurons and glia.

Acknowledgements

This investigation was supported in part by research grants NB-02253-05 and NB-03813-04 from the United States Public Health Service.

Dr. NICHOLLS is a Fellow of the National Multiple Sclerosis Society and in receipt of a Wellcome Research Trust travel grant.

References

COGGESHALL, R. E. & FAWCETT, D. W. (1964). *J. Neurophysiol.* **27**, 229.
CUMMING, J. & HYDÉN, H. (1962). *Biochem. biophys. Acta.* **60**, 271.
DAVSON, H. & SPAZIANI, E. (1959). *J. Physiol.* (Lond.) **149**, 135.
DE ROBERTIS, E. (1962). *World Neurol.* **3**, 98.
GOLGI, C. (1903). *Opera Omnia* **1**, 40.
GRAY, E. G. & GUILLERY, R. W. (1963). *Z. Zellforsch.* **60**, 826.
HELD, H. (1909). *Mschr. Psychiat. Neurol.* **26**, 360.
HORSTMANN, E. & MEVES, H. (1959). *Z. Zellforsch.* **49**, 569.
HYDÉN, H. & PIGON, A. (1960). *J. Neurochem.* **6**, 57.
ITO, T. (1936). *Okajimas Fol. Anat. Jap.* **14**, 111.
KATZMAN, R. (1961). *Neurology.* **11**, 27.
KOCH, A., RANCK, J. B., JR. & NEWMAN, B. L. (1962). *Exp. Neurol.* **6**, 186.
KUFFLER, S. W. & NICHOLLS, J. G. (1964). *Arch. exp. Path. Pharmak.* **248**, 216.
KUFFLER, S. W. & POTTER, D. D. (1964). *J. Neurophysiol.* **27**, 290.
MITARAI, G., SVAETICHIN, G., VALLECALLE, E., FATECHAND, R., VILLEGAS, J. & LANGER, M. (1961). In *The Visual System,* (ed. by JUNG, R. & KORNHUBER, H.). Berlin: Springer.
NICHOLLS, J. G. & KUFFLER, S. W. (1964). *J. Neurophysiol.* **27**, 645.
NICHOLLS, J. G. & KUFFLER, S. W. *J. Neurophysiol.* (in the Press).
PENFIELD, W., ed. (1932). *Cytology and Cellular Pathology of the Nervous System,* vol. 2. New York: PAUL B. HOEBER.
SJÖSTRAND, F. (1961). In *The Visual System* (ed. by JUNG, R. & KORNHUBER, H.), p. 13. Berlin: Springer.
STUDNIČKA, F. K. (1899). *Anat. Anz.* **16**, 397.

CENTRAL CONNECTIONS OF THE SENSORY PATHS FROM THE CAT'S TONGUE

by Sven Landgren

Histological studies using degeneration techniques have provided fundamental information concerning the central connections of the trigeminal system. This method has, however, certain limitations. It can, for instance, not differentiate between neuronal systems activated by different types of stimuli. Such a differentiation is significant when studying tongue afferents because they convey information from receptors subserving at least three different modalities, i. e. touch, thermal sensations and taste. Single unit analysis with microelectrode recording and with histological localization of the recording sites has provided a new method for studying functional aspects of central nervous pathways, and since this method during recent years has been used to a considerable extent in the study of the trigeminal system, a short review of the information gained may be of interest.

I. The tactile path

The central connections of the tactile path from the tongue have been extensively investigated with the single unit analysis method. The primary afferents of this path consist of fast conducting (50—80 m/sec) myelinated fibres travelling in the lingual nerve and the chorda tympani. They are discharged by sensitive mechanoceptors in the mucous membrane of the tongue (cf. Zotterman 1936). The primary afferents are monosynaptically relayed in the trigeminal nucleus and the secondary axons proceed to the *nucleus ventralis posteromedialis* (VPM) of the thalamus, mainly via the ipsilateral dorsal ascending trigeminal tract but also after crossing to the other side via the contralateral trigeminal lemniscus. The conduction velocity of the secondary axons ranges from 7 to 25 m/sec (cf. Eisenman et al. 1964). After relaying in VPM the path projects to the sensorimotor cortex. The projection area is located in the rostrobasal part of gyrus coronalis rostralto sulcus orbitalis and includes parts of areas 3 a, 3 b and 43 of Hassler and Muhs-Clement (1964). In the cat the total length of the tactile pathway from the receptors of the tongue to the cerebral cortex is approximately 150 mm. Latencies of action potentials recorded from neurones located in *nucleus trigemini sensibilis principalis* (NVsnpr), VPM and in the cortex are given in table 1. The stimulus was electrical stimulation of the tongue surface. Minimum latencies for the population studied and modal values are given.

A small and constant peripheral receptive fields is a feature common to the neurones at all levels of the tactile path. Another characteristic is the precise somatotopic organization of the central projections. The neurones of this system are sensitive to mechanical stimuli but do not respond to thermal or gustatory stimuli.

The secondary trigeminal neurones of the tactile path from the tongue form a column of cells occupying a dorso-medial position within NVsnpr and *nucleus tractus spinalis trigemini oralis* γ (NVspo γ), which is the rostral subdivision of

Table 1. *Latencies of action potentials of neurones in the tactile path*

	NVsnpr	VPM	Cortex
Minimum latency, msec	1.5	2.5	5.0
Mode, msec	2.0	4.5	6.5

the nucleus of the spinal trigeminal tract (cf. MEESEN and OLSZEWSKI 1949). The tactile path also projects to a similar dorso-medial column of neurones in *nucleus tractus spinalis trigemini caudalis* (NVspc). The oral connections within this subnucleus are more numerous at the level of the obex. The axons of the secondary neurones in NVsnpr travel to VPM via the ipsilateral dorsal ascending trigeminal tract and those of the neurones in NVspc via the contralateral trigeminal tract. The rostral and the caudal trigeminal projection areas are separated by a system of neurones with large receptive fields which do not project to the thalamus (cf. DARIAN-SMITH et al. 1963 b). The above described trigeminal relays in the tactile path from the tongue form a part of the somatotopically organized trigeminal system studied by GORDON et al. (1961), WALL and TAUB (1962), KRUGER and MICHEL (1962 a and b), DARIAN-SMITH et al. (1963 a and b) and by EISENMAN et al. (1963 and 1964).

The thalamic projection area of the tactile path from the tongue forms part of the somatotopically organized representation of the body surface in the ventro-basal nuclear complex (cf. MOUNTCASTLE and HENNEMAN 1949, ROSE and MOUNTCASTLE 1954, POGGIO and MOUNTCASTLE 1960 and 1963). Details of the localization of the tongue region are given by APPELBERG and LANDGREN (1958), LANDGREN (1960 a) and DARIAN-SMITH (1964). It is located medially in VPM but does not occupy the most medial part of this nucleus *(Nucleus ventralis posterior medialis parvocellularis)* (VPMpc) of OLSZEWSKI (1952). Dorso-laterally it is bordered by neurones discharged from the maxillary region and ventrolaterally by mandibular neurones. The majority of the thalamic tongue relay neurones have ipsilateral receptive fields but contra-lateral connections were also observed (cf. LANDGREN 1960 a).

The cortical projection area of the tactile path from the tongue was localized by recording cortical action potentials evoked by electrical stimulation of low threshold afferents in the trigeminal component of the lingual nerve and in the chorda tympani (cf. COHEN et al. 1957). The responsive regions overlapped but the trigeminal responses covered a larger area. This is not surprising as the low threshold components in both nerves are "touch fibres". The tongue area is located in a region where the cortical projection areas SI and SII approach and it is difficult to subdivide the responsive region in SI and SII fields. It overlaps with the per-ioral projection area found near the coronal sulcus and with glossopharyngeal responses in the suprasylvian gyrus. Records from single units show that a large group of the cortical neurones in the tongue projection area have features similar to those seen at the lower levels of the tactile path.

The specificity to tactile stimuli, the high conduction velocity, the small receptive fields and precise somatotopic organization speak in favour of the assumption that the tactile path subserves discriminative tactile perception.

II. The thermoceptive path

Sensitive and specific thermoceptive afferents from the cat's tongue were investigated by HENSEL and ZOTTERMAN (1951) and by DODT and ZOTTERMAN (1952). The trigeminal relay of this path has so far not been localized. Thalamic units responding specifically to cooling or warming of the tongue were, however, described by LANDGREN (1960 b) in the cat and by BENJAMIN (1963) in the rat and the squirrel monkey. They were highly sensitive to cooling of the tongue and the time course of their response was similar to that of the lingual thermoceptors. These thalamic neurones were located in the lateral part of VPMpc in a region overlapping with the nucleus of the tactile path. Ipsilateral as well as contralateral receptive tongue fields were observed.

Neurones specifically sensitive to cooling of the tongue were also found within the cortical projection area of the tactile path (cf. LANDGREN 1957). Some units responding to cooling were observed near "touch cells" along the same microelectrode track. At the thalamic as well as at the cortical level the thermoceptive units responded to electrical stimulation of the receptive field with a latency about three times longer than that of the tactile units. Minimum latencies of "cold cells" observed in the thalamus ranged from 3.5 to 16.4 msec (mode 7.5 msec). The shortest latency observed in a cortical "cold cell" was 14 msec.

III. The gustatory path

The central connections of the gustatory path are not sufficiently studied in the cat. Responses to gustatory stimuli were, however, recorded by MAKOUS et al. (1963) in the cat's *nucleus tractus solitarius,* and single units responding only to taste stimuli were sampled within the cortical tongue projection area by COHEN et al. (1957). More information may, however, be obtained from studies on the rat. ERICKSON (1958) and PFAFFMAN et al. (1961) applied taste solutions to the tongue and recorded responses from neurones located in the rostral part of *nucleus tractus solitarius* of this animal. The localization is in agreement with the conclusions drawn by ALLEN (1923 a and b) from anatomical studies on the guinea pig.

At the thalamic level the gustatory path is relayed in VPMpc as demonstrated by BENJAMIN (1963). The cortical projections were found ventrally in the somatic sensory face area of the rat and ablation of this area produced impairment of taste discrimination (BENJAMIN and PFAFFMAN 1955).

IV. The nociceptive systems

According to size of the receptive field the central neurones of the nociceptive systems can be divided into two groups i. e. the small and the wide receptive nociceptive cells. A group of trigeminal neurones responding to nociceptive mechanical stimuli from small receptive fields on the ipsilateral half of the tongue was thus described by EISENMAN et al. (1963). These units were not discharged from any other part of the face or the body. They did not respond to gustatory stimuli and were not sensitive to thermal stimuli. One of them did, however, respond to nociceptive cooling of the tongue.

Tongue afferents also contribute to the complex convergence seen in the group of wide receptive nociceptive cells. These neurones had large receptive fields; a part of which was often more sensitive and responded to tactile stimulation. Responses to nociceptive cooling or warming within the receptive field were common. The frequency of the response was slowly increasing and slowly adapting. The discharge often outlasted the application of the stimulus (cf. the B-cells of GORDON et al. 1961).

The wide receptive nociceptive trigeminal neurones were located centrally and medially in NVsnpr and NVspo γ. At the level of NVspc they were found in a border region between this nucleus and *nucleus reticularis ventralis*. In the population of neurones studied by EISENMAN et al. (1964) 25% of the wide receptive nociceptive neurones sent ascending axons in the contralateral trigeminal lemniscus.

Thalamic neurones with large receptive fields on the body surface were described by POGGIO and MOUNTCASTLE (1960). These neurones were located in the posterior group of nuclei at the caudal end of the ventrobasal complex. Tongue afferents also contribute to the convergence onto these neurones as shown by DARIAN-SMITH (1964). A nucleus of similar responses located to the *zona incerta* of the subthalamus was also described by this author.

The functional significance of the nociceptive systems described above is at present unknown. Some of these cells are obviously local interneurones but some of the trigeminal ones have ascending lemniscal projections and could possibly be related to the similar neurones found at the thalamic level. The wide receptive neurones of the interpolar subnucleus of the trigeminal complex were assumed to be trigemino-cerebellar relays (cf. DARIAN-SMITH et al. 1963 a).

References

ALLEN, W. F. (1923 a). *J. comp. Neurol.* **35**, 171.
ALEN, W. F. (1923 b). *J. comp. Neurol.* **35**, 275.
APPELBERG, B. & LANDGREN, S. (1958). *Acta physiol. scand.* **42**, 342.
BENJAMIN, R. M. (1963). In *Olfaction and Taste* (ed. by ZOTTERMAN, Y.) p. 309. Oxford: Pergamon Press.
BENJAMIN, R. M & PFAFFMAN, C. (1955). *J. Neurophysiol.* **18**, 56.
COHEN, M. F., LANDGREN, S., STRÖM, L. & ZOTTERMAN, Y. (1957). *Acta pnysiol. scand.* **40**, Suppl. 135.
DARIAN-SMITH, I. (1964). *J. Physiol.* (Lond.) **171**, 339.
DARIAN-SMITH, I., PHILLIPS, G. & RYAN, R. D. (1963 a). *J. Physiol.* (Lond.) **168**, 129.
DARIAN-SMITH, I., PROCTOR, R. & RYAN, R. D. (1963 b). *J. Physiol.* (Lond.) **168**, 147.
DODT, E. & ZOTTERMAN, Y. (1952). *Acta physiol. scand.* **26**, 345.
EISENMAN, J., LANDGREN, S. & NOVIN, D. (1963). *Acta physiol. scand.* **59**, suppl. 214.
EISENMAN, J., FROMM, G., LANDGREN, S. & NOVIN, D. (1964). *Acta physiol. scand.* **60**, 337.
ERICKSON, R. P. (1958). Unpublished PhD Thesis, Brown University, USA.
GORDON, G., LANDGREN, S. & SEED, W. A. (1961). *J. Physiol.* (Lond.) **158**, 544.
HASSLER, R. & MUHS-CLEMENT, K. (1964). *J. für Hirnforschung* **6**, 377.
HENSEL, H. & ZOTTERMAN, Y. (1951). *Acta physiol. scand.* **23**, 291.
KRUGER, L. & MICHEL, F. (1962 a). *Exp. Neurol.* **5**, 139.
KRUGER, L. & MICHEL, F. (1962 b). *Arch. Oral. Biol.* **7**, 491.
LANDGREN, S. (1957). *Acta physiol. scand.* **40**, 202.
LANDGREN, S. (1960 a). *Acta physiol. scand.* **48**, 238.
LANDGREN, S. (1960 b). *Acta physiol. scand.* **48**, 255.
MAKOUS, W., NORD, S., OAKLEY, B. & PFAFFMAN, C. (1963). In: *Olfaction and Taste* (ed. by ZOTTERMAN, Y.), p. 381. Oxford: Pergamon Press.
MEESEN, H. & OLSZEWSKI, J. (1949). *A cytoarchitectonic atlas of the rhombencephalon of the rabbit.* Basel: S. Karger.
MOUNTCASTLE, V. B. & HENNEMAN, E. (1949). *J. Neurophysiol.* **12**, 85.
OLSZEWSKI, J. (1952). *The thalamus of the Macaca Mulatta.* Basel: S. Karger.
PFAFFMAN, C., ERICKSON, R. P., FROMMER, G. P. & HALPERN, B. P. (1961): In: *Sensory Communication* (ed. by ROSENBLITH, W.), p. 455. New York: The M. I. T. Press and J. Wiley and Sons.
POGGIO, G. F. & MOUNTCASTLE, V. B. (1960). *Johns Hopk. Hosp. Bull.* **106**, 266.
POGGIO, G. F. & MOUNTCASTLE, V. B. (1963). *J. Neurophysiol.* **26**, 775.
ROSE, J. E. & MOUNTCASTLE, V. B. (1954). *Johns Hopk. Hosp. Bull.* **94**, 238.
WALL, P. D. & TAUB, A. (1962). *J. Neurophysiol.* **25**, 110.
ZOTTERMAN, Y. (1936). *Skand. Arch. Physiol.* **75**, 105.

SLOW SYNAPTIC RESPONSES IN AUTONOMIC GANGLIA

by BENJAMIN LIBET

The known synaptic potentials present a considerable range of durations, but at least in the vertebrates the maximum durations are limited to an order of a hundred milliseconds or so (ECCLES 1964). It has become clear, however, that postsynaptic responses with durations in the range of about 1 to 25 seconds can be elicited by preganglionic impulses in sympathetic ganglia. This presents a range up to 100 fold greater than the longest postsynaptic potentials (PSP) durations otherwise known. One of the slow ganglionic responses is a surface-negative, facilitatory postsynaptic potential which may be termed a "slow EPSP". The other is a surface-positive, presumably hyperpolarizing postsynaptic potential, for which it has not been practicable thus far to obtain direct evidence that it is inhibitory in nature. Although the slow excitatory postsynaptic potential (EPSP) appears to be generated in the same ganglion cells as the initial or fast EPSP, it has an extraordinary latency of about 250 msec and it is mediated by a second type of postsynaptic receptor for acetylcholine (ACh) which is atropine-sensitive or muscarinic in nature rather than by the well known nicotinic, curare-sensitive type. Since the long duration and latency of the slow EPSP in ganglia are evidently a function of a single level of synaptic terminations (i. e. of a "monosynaptic" type), these slow PSP's assume an especial significance for neurophysiological processes in general. Suitable intracellular studies have not yet been achieved, partly because of the fibrous impenetrability of mammalian ganglia. But the evidence obtained from surface recordings, especially with isolated, well-cleaned ganglia, seems decisive on many points.

A slow surface-positive postsynaptic potential, lasting some seconds was apparently first demonstrated by ECCLES (1943) in the curarized stellate ganglion of the cat, and LAPORTE and LORENTE DE NÓ (1959) in curarized sympathetic ganglia of the turtle. In curarized ganglia of the mammal R. M. ECCLES (1952) found that the surface-positive (P) potential was followed by a large, even longer lasting surface-negative potential (the late negative or LN wave). Only a small late negative component is visible in curarized turtle ganglia (LAPORTE and LORENTE DE NÓ 1950) and it is usually absent in the sympathetic ganglia of the frog (LIBET and CHICHIBU, unpublished). Large slow surface-positive responses could, however, be elicited by suitable preganglionic stimulation of frog ganglia (LIBET and CHICHIBU, unpublished), as in the other two classes of vertebrates. (The absence of slow PSP's in intracellular recordings made thus far by R. M. ECCLES 1963, and BLACKMAN, GINSBORG and RAY 1963, is explainable on certain technical grounds.)

The slow ganglionic potentials recorded in curarized mammalian ganglia are clearly postsynaptic potentials in the accepted sense. Both the slow surface-positive and negative potentials decrement rapidly along the postganglionic axons in the same way as does the EPSP (LIBET, unpublished; LAPORTE and LORENTE DE NÓ 1950). No such potentials are detected at the preganglionic nerve with an electrode within 1 mm or less of the ganglion.

Generation of the slow PSP's appears to involve presynaptic release of chemical transmitters; the responses are blocked by lowering Ca^{++}/Mg^{++} ratios (LIBET, unpublished) and by botulinum toxin (ECCLES and LIBET 1961). The slow PSP's are not, however, some sort of "synaptic after-potentials" which automatically follow the EPSP; rather each is apparently generated independently at a different type of receptor site. In contrast to the EPSP, the slow PSP's are relatively insensitive to curariform drugs (Eccles 1952). On the other hand, low concentrations of atropine which have no effect on the initial EPSP strongly depress the slow EPSP and partially the slow positive PSP. Dibenamine strongly depresses the slow positive PSP, and partially the slow EPSP (Eccles and Libet 1961).

Nor are the slow PSP's some kind of abnormal response which is unique to curarized ganglia. The slow EPSP, at least, can be demonstrated clearly in uncurarized ganglia. In these it can be seen as a negative component that is superimposed on the ganglionic recording of the true, large positive after-potential of the ganglion cells, following orthodromic activation of postganglionic discharge (Libet 1964). Atropine selectively removes this late negative component and reveals a simpler shape for the positive after-potential (see Figs. 2—4; Libet 1964).

With a slow EPSP available, post-tetanic potentiation (PTP) in sympathetic ganglia is not due simply to changes in the presynaptic terminals. Even with heterosynaptic testing, following a conditioning train, a large PTP can usually be seen in the stellate ganglion of the cat (Libet 1964). Atropine abolishes virtually all of such heterosynaptic PTP, as well as a portion of the homosynaptic PTP in the stellate ganglion; in the superior cervical ganglion even the homosynaptic PTP is virtually all atropine-sensitive. There are, then, two forms of PTP in sympathetic ganglia: 1) the well known presynaptic PTP, which is not atropine-sensitive and is best elicited by higher frequency trains of long durations; 2) a PTP due to the slow EPSP, which is blocked by atropine, is relatively insensitive to frequency of the conditioning volleys and is almost maximal with trains of shorter durations (about 1 sec) at low frequencies (5—10 per sec).

The evidence that the same ganglion cell generates both the initial and the slow EPSP is difficult to dispute, although, in the absence of suitable studies of responses of single cells, it is at present indirect in nature (Libet 1964). Since both EPSP's are mediated by ACh, it would follow that the atropine-sensitive "muscarinic" receptors which are responsible for the slow EPSP are located on the same cell as are the curare-sensitive "nicotinic" ones which mediate the initial EPSP. Curtis and Ryall (1964) have recently demonstrated more directly that both types of receptors for ACh exist in the same Renshaw cells in the spinal cord.

The question of how the slow PSP's are generated poses some intriguing problems in general neurophysiology, in view of the extraordinarily long latencies and durations and of the special sensitivities to blocking agents. It is accepted that all synaptic responses in sympathetic ganglia are developed without the mediation or support of interneuronal discharges. Thus the long durations cannot be ascribed to repetitive interneuronal activation of the ganglion cells. Nor can discharge in recurrent axon collaterals of the ganglion cells be invoked; the slow PSP's can be elicited by preganglionic volleys in fully curarized ganglia, in which no postganglionic discharge can be detected. A theoretical schema has been proposed (ECCLES and LIBET 1961; Fig. 7) to explain the origins and natures of the transmitters that mediate the slow PSP's in relation to their pharmacological sensitivities and other properties.

The long latency (about 250 msec) of the slow EPSP is similar with a single preganglionic volley or with a tetanus (LIBET, unpublished), although the duration and amplitude are less with the former. The long latency and duration is reminiscent of those found by BROWN and ECCLES (1934) for the inhibitory action of a single vagal nerve volley on heart rate. Time for diffusion of ACh, perhaps through barriers, from the site of its liberation to the appropriate postsynaptic receptors could account for such long latencies. In ganglia this presumably means that some fraction of the ACh released focally at presynaptic terminations must diffuse from the synaptic clefts to muscarinic receptors located in areas of the postsynaptic membrane outside these subsynaptic regions (ECCLES and LIBET 1961; ECCLES, ECCLES and FATT 1956). MILEDI (1960) has produced direct evidence showing that ACh receptors do exist normally in a region of membrane outside the neuromuscular junction. He also showed that if the point of application of ACh is moved 35 μ away from the junction, there appears a latency of over 300 msec, from the start of electrophoretic discharge of ACh to the beginning of depolarization of the end-plate. One should not lose sight of the alternative possibility, however, that release of transmitter might occur along regions of the presynaptic fibers other than those presently designated as synaptic contacts in electron microphotographs, i. e. the specialized contacts between pre- and postsynaptic membranes which have a space or "synaptic cleft" between them of only 100—200 Å. Perhaps there are, in addition, "loose synapses" in which the contact is not so close and in which neither the sites of presynaptic release of transmitter nor the locations of the postsynaptic receptors are so focal as in the accepted close contact synapse. There are indeed numerous examples of the location of so-called synaptic vesicles within regions of the presynaptic axon not engaged in close synaptic contacts. In any case, it is of considerable interest that close synaptic contacts are almost completely confined to the dendritic processes of the mammalian sympathetic ganglion cell (ELFVIN 1963). Perhaps the slow EPSP is generated in the cell body region.

The amplitude of the slow EPSP is small after a single orthodromic volley, but it rapidly builds up with repetition, even with short train durations at low frequencies, to values as much as a fourth or more of the ganglionic spike voltage. One function of the slow EPSP may be to provide a kind of long-lasting amplification of preganglionic excitatory effects. This could be especially significant in the recruitment and maintainence of postganglionic discharge in the face of depressant after-hyperpolarizations (LIBET 1964) at the rather low frequencies of preganglionic impulses found to occur during natural activities in the autonomic nervous system (e. g. PITTS, LARRABEE and BRONK 1941). It should be remembered that even small additional depolarizations, by providing the margins required for raising cells to firing levels, can have a large influence on total discharge patterns.

If the slow, positive PSP is indeed an inhibitory postsynaptic potential (IPSP), as seems likely, then it re-opens the possibility of the existence of specific inhibitory preganglionic fibres. Such fibres could be activated in certain reflexes that are specifically inhibitory to autonomic functions, or as part of regulatory patterns elaborated by the central nervous system. If the slow positive PSP is mediated instead by collaterals of non-specific preganglionic fibres then its requirement of a higher frequency preganglionic input (optimum at 40—60 per sec, ECCLES and LIBET 1961) may be important. It could mean that a significant slow IPSP was elicited only during unusually intense activity in the autonomic system; such an arrangement could serve as an automatic braking device, a "negative feed-forward".

Finally, there is the question of the possible further import that the slow PSP's may have for synaptic mechanisms other than those in autonomic ganglia. Neurophysiologists have continually sought for or devised theoretical models of neural functions at the cellular level which could mediate the comparatively slow processes which are so prominent in cerebral activities. We now have the relatively simple possibility of a chemically-mediated postsynaptic response which has a duration of many seconds, without the intervention of multiple interneuronal discharge patterns. Additionally, the long latency of the "monosynaptic" slow EPSP in ganglia shows that latency values cannot always be taken as simple indices of the complexity and length of synaptic chains interposed before the measured response.

The ability to generate the slow EPSP heterosynaptically and with low frequencies would make it much more suitable than presynaptic PTP for participation in slow central phenomena. This would be especially true where there is a requirement for interaction of separate inputs, as in the conditioning process (although it is possible to devise explanations of enduring heterosynaptic interactions for this purpose using only homosynaptic PTP as a facilitatory mechanism, as was done by FESSARD and SZABO 1961). The slow EPSP would also be more likely than presynaptic PTP to result in recordable slow potentials at the cerebral cortex. It should therefore be considered a possible source of the relatively steady potentials in the brain and of the slow shifts in these.

Direct evidence of a slow EPSP in cerebral tissues is not available but should be sought. The presence and nature of muscarinic properties of cells in the central nervous system (CURTIS and RYALL 1964; KRNJEVIĆ and PHILLIS 1963; ANDERSEN and CURTIS 1964), is suggestive of the possibility that a slow EPSP type of mechanism may not be restricted to autonomic ganglia.

Acknowledgement

The unpublished work cited in this article was carried out with the support of U. S. Public Health Service Grant NB-884.

References

ANDERSEN, P. & CURTIS, D. R. (1964). *Acta physiol. scand.* **61**, 85.
BLACKMAN, J. G., GINSBORG, B. L. & RAY, C. (1963). *J. Physiol.* (Lond.) **167**, 355.
BROWN, G. L. & ECCLES, J. C. (1934). *J. Physiol.* (Lond.) **82**, 211.
CURTIS, D. R. & RYALL, R. W. (1964). *Nature* (Lond.) **203**, 652.
ECCLES, J. C. (1943). *J. Physiol.* (Lond.) **101**, 465.
ECCLES, J. C. (1964). *The Physiology of Synapses,* p. 161. Berlin: Springer-Verlag.
ECCLES, J. C., ECCLES, R. M. & FATT, P. (1956). *J. Physiol.* (Lond.) **131**, 154.
ECCLES, R. M. (1952). *J. Physiol.* (Lond.) **117**, 196.
ECCLES, R. M. (1963). *J. Physiol.* (Lond.) **165**, 387.
ECCLES, R. M. & LIBET, B. (1961). *J. Physiol.* (Lond.) **157**, 484.
ELFVIN, L.-G. (1963). *J. Ultrastruct. Res.* **8**, 441.
FESSARD, A. & SZABO, T. (1961). In: *Brain Mechanisms and Learning* (ed. by DELAFRESNAYE, J. F.), p. 353. Springfield, Illinois: Charles C. Thomas.
KRNJEVIĆ, K. & PHILLIS, J. W. (1963). *J. Physiol.* (Lond.) **166**, 328.
LAPORTE, Y. & LORENTE DE NÓ, R. (1950). *J. cell. comp. Physiol.* **35**, Suppl. 2, 61.
LIBET, B. (1964). *J. Physiol.* (Lond.) **174**, 1.
MILEDI, R. (1960). *J. Physiol.* (Lond.) **151**, 24.
PITTS, R. F., LARRABEE, M. G. & BRONK, D. W. (1941). *Amer. J. Physiol.* **134**, 359.

PHYSIOLOGICAL OBSERVATIONS IN FOETAL TRANSFUSION

by A. W. LILEY

It has been shown (LILEY 1963 c; MCCROSTIE 1964) that intrauterine transfusion by the intraperitoneal route may ensure survival to a more viable maturity for the foetus afflicted by severe haemolytic disease very early in the third trimester. This procedure involves initially the timely discovery of the foetal disease by early amniocentesis (LILEY 1960) and spectrophotometric examination of the amniotic fluid (MACKAY 1961; LILEY 1961, 1963 b). By amniography hydrops and foetal malformation are excluded, and the placenta located in order that it may be avoided if possible. Swallowed contrast medium concentrated in the foetal gut provides a target for peritoneal puncture by a Tuohy needle which is withdrawn as soon as a nylon catheter with generous slack has been introduced. Radiological check of the catheter position (Fig. 1) is followed by the transfusion of blood packed to a haematocrit of about 75 %. This procedure permits some interesting if restricted physiological observations of the human foetus which would be difficult to justify ethically in other circumstances.

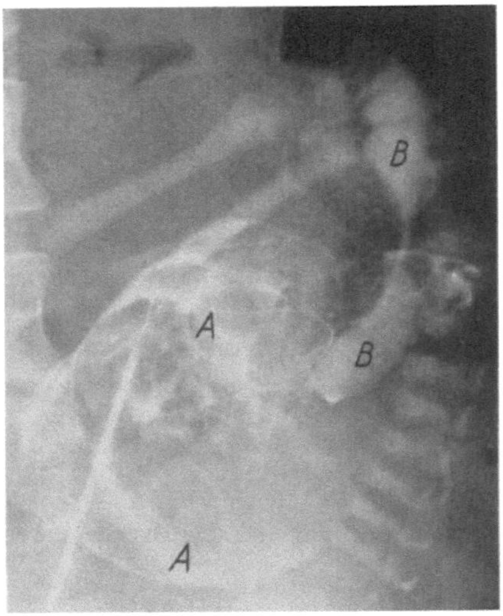

Fig. 1

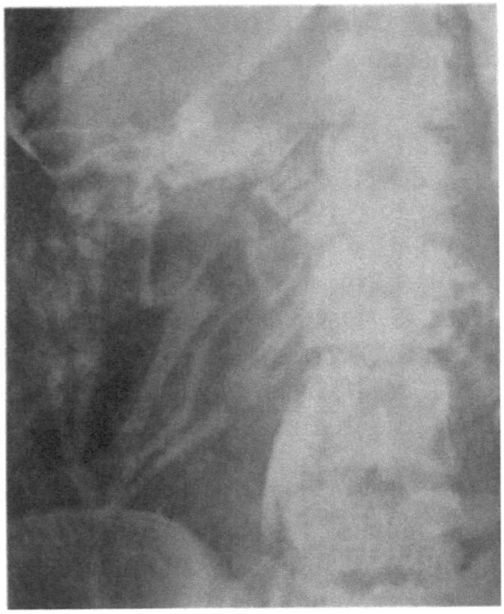

Fig. 2

Fig. 1. The coiled catheter and injected contrast medium A, A in the foetal peritoneum, and swallowed contrast medium B, B in the foetal sigmoid colon and rectum

Fig. 2. Foetal thumb sucking

I. The fate of intrauterine contrast medium

The contrast medium used is Urografin (Schering AG, Berlin), a pyelographic medium consisting of sodium and methylglucamine salts of diatrizoic acid in 76 % solution. When 20 ml. of this medium is injected into the amniotic cavity most of the dose is removed in 15—20 minutes. Most of the dye is removed directly into the maternal circulation with the appearance of a maternal pyelogram. However a significant proportion is swallowed by the foetus (LILEY 1963 a; McLAIN 1963). This swallowed dye is concentrated and travels the length of the foetal gut to persist indefinitely in the large bowel until passed with the meconium. Although during swallowing there may be a dense accumulation of dye in the pharynx and oesophagus in no circumstance has dye been seen in the tracheo-bronchial tree. This observations is consistent with the claims of ADAMS, FUJIWARA and RAWSON (1963) that foetal lungs have a kidney-like function with a nett outflow of fluid. Dye injected into foetal tissue planes or peritoneum is cleared rapidly by absorption and placental excretion. In view of the known foetal diuresis the absence of any foetal pyelogram even with an intraperitoneal dose of 16 ml. is remarkable since 20 ml. is the normal adult pyelographic dose, and in the neonate 2 ml. intramuscularly produces an excellent pyelogram. If the large foetal bladder surmounted by the urachus is needled, drained of perhaps 60 ml. of dilute urine and outlined with 3 or 4 ml. of dye, the rapid restoration of the original outline illustrates the high water output and the low solute load of the foetal kidneys.

II. The absorption of intraperitoneal blood

In the infant and toddler, the absorption of intact erythrocytes from the peritoneal cavity has been demonstrated by the use of isotope tagged or morphologically recognisable cells. Absorption occurs through the lymphatics and is enhanced by hyperventilation. The volume administered is limited by abdominal distension and respiratory restriction. Physiological cirumstances are different in the foetus. The peritoneal cavity is relatively more capacious since the bowel is empty of gas and the diaphragm immobile at the approximate level of the 6th rib. However there is no peritoneo-pleural pressure gradient and no respiratory excursion to aid absorption or venous or lymphatic return. Despite these differences intraperitoneal transfusion appears at least as effective in the foetus as in the neonate and possibly more rapid in effect. The quantities of packed cells given, 75—150 ml. at maturities from 28—32 weeks are very large in comparison with amounts given to neonates. Uptake may be checked by two independent tests. In comparison with untreated severely affected babies of comparable maturity, the neonates transfused antenatally show consistent high fractions of HbA in cord blood tested by alkali denaturation (Singer, Chernoff and Singer 1951). Even evidence of increased fractions of HbA in haemolytic disease (Oppé and Fraser 1961) can hardly compromise HbA fractions of 90% or greater encountered after foetal transfusion. Even less disputable is the second test, differential agglutination on the grand scale, whereby in 3 babies with near total destruction of their own circulating erythrocytes, the cord blood findings were those of donor blood-group O, Rh negative and negative Coombs reaction. Normally the rate of uptake of intraperitoneal cells cannot be checked except for the broad observation that clean peritoneal cavities have been encountered at repeat transfusions at intervals from 10—14 days, and that infants delivered at this interval after intrauterine transfusion have shown no sign of free peritoneal fluid. However in one instance, a foetus who had received a total of 62.5 g. of haemoglobin in three transfusions, unwelcome premature delivery occurred 64 hours after the 3rd transfusion which involved 125 ml. of blood containing 25.8 g. of haemoglobin. Death from extreme prematurity and kernicterus occurred at 32 hours or 96 hours from the last transfusion. At postmortem examination less than 10 ml. of blood remained in the peritoneum. There was no trauma, adhesions, red cell debris or fibrin deposition and only a minimal serosal reaction.

III. Foeto-placental blood volume

Intrauterine transfusion implies transfusion not only of the foetus but also the placenta. The effect of a given quantity of haemoglobin in relieving anaemia is therefore determined by the unknown combined foeto-placental blood volume. Foetal transfusion itself admits an estimate of at least an order of magnitude of this volume especially in those neonates where donor cells are present in sufficient quantity to compromise grouping and the Coombs test. It has been found that, in practice, 5 g. of intraperitoneal haemoglobin is required to raise the cord haemoglobin concentration 1 g %. This suggests that the foeto-placental blood volume is of the order of 500 ml. early in the third trimester. However a reduction must be allowed for the natural breakdown of donor cells, ideally 1 % per day, and if absorption of intact cells falls far short of 100 % then further reduction is necessary. As a check on this startling order of magnitude direct measurement of the foeto-placental blood volume has been attempted at human caesarean section by dye dilution and elution technique using Azovan Blue (T 1824). Preliminary experiments in foetal lambs with chronically indwelling aortic and cotyledonary venous catheters have shown that mixing to within 10 % of 10 minute value may be achieved in 45—75 seconds. At caesarean section, where foetal condition and operative circumstances permit, undisturbed mixing, times of 90—240 seconds have been obtained after injection of 1 ml. of Azovan Blue into an umbilical vessel. While there is no method by which the precision of any single result can be checked, 13 estimates at 35—40 weeks have produced a mean volume of 532 ml. and range 318—640 ml. Naturally estimation of total blood volume from plasma volume is compromised by ignorance of the ratio of cord venous haematocrit to whole body or, more accurately, "whole conceptus" haematocrit. Despite this handicap, the foeto-placental red cell mass — mean 19.3 g. Hb/kg., coefficient of variation 15 %, shows a greater constancy than the foeto-placental blood volume mean 128 ml./kg., coefficient of variation 23 %. This result suggests that the foetus and placenta, as would be expected, share with the mother the extracellular dehydration by sodium depletion so commonly seen in prolonged labour, and that the cord haemoglobin concentration is not an ideal parameter of foetal haemopoiesis. The large size of the foeto-placental blood volume indicates that in utero the placenta commonly contains more than half the total volume.

IV. Foetal behaviour

The foetus is not indifferent to his surroundings. Foetal position is largely determined by foetal comfort and external stimuli may lead to purposeful foetal movement if they produce foetal discomfort. The foetus attempts to evade the pressure of the microphone of a cardiotachometer and sustained local pressure on prominences will produce foetal movements if minor modifications in foetal position are necessary at foetal transfusion. Changes in maternal position, vigorous or repeated palpation, and the pressure of a phonendoscope produce a restless foetus and must be avoided prior to foetal puncture. The foetus responds promptly to painful stimuli, needle puncture, concentrated contrast medium or cold blood. As judged by abrupt movement and transient variation in heart rate with sudden loud noise, human voices included, the foetus can hear but with tympanic membranes damped by fluid there must be a relative high tone deafness. In the newborn, the so-called "seeking" reflex is elicited by stroking or touching the cheek or face. The baby turns his head to the stimulus and opens his mouth. This is clearly the reflex by which the newborn "homes" on the nipple, and indeed his mouth is the only part of a newborn baby which he can locate in space reliably. Since the foetal position is normally one of flexion with hands and feet often in proximity to the face, it is not surprising that the foetus should elicit a "seeking" reflex himself, and it is not uncommon in foetal radiology to see the foetus sucking thumb, fingers or toes (Fig. 2).

References

Adams, F. H., Fujiwara, T. & Rawson, G. (1963). *J. Pediat.* **63**, 881.
Liley, A. W. (1960). *New Zealand med. J.* **59**, 581.
Liley, A. W. (1961). *Amer. J. Obst. Gynec.* **82**, 1359.
Liley, A. W. (1963 a). In: *Modern Trends in Human Reproductive Physiology*, Vol. 1 (ed. by Carey, H. M.), p. 227. London: Butterworth.
Liley, A. W. (1963 b). *Amer. J. Obst. Gynec.* **86**, 485.
Liley, A. W. (1963 c). *Brit. med. J.* **2**, 1107.
Mackay, E. V. (1961). *Austr. N. Z. J. Obst. Gynaec.* **1**, 78.
McCrostie, H. H. (1964). *Brit. med. J.* **1**, 885
McLain, C. R. (1963). *Amer. J. Obst. Gynec.* **86**, 1079.
Oppé, T. E. & Fraser, I. D. (1961). *Arch. Dis. Child.*, Vol. **56**, p. 507.
Singer, K., Chernoff, A. I. & Singer, L. (1951). *Blood* **6**, 413.

CHOLINERGY AND ADRENERGY IN THE NEURAL CONTROL OF SWEAT GLANDS

by David P. C. Lloyd

Neural control of the sweat glands of the cat foot pad was established by Goltz (1875) who observed the appearance of sweat droplets on stimulating the sciatic nerve. Shortly thereafter Coyne (1878) and Ranvier (1879) described the terminal innervation of the sweat glands. Although Hermann and Luchsinger

(1878) discovered electromotive phenomena in association with secretory activity provoked by stimulation of the sciatic nerve, it remained for Waller (1901, 1904) to record in similar circumstances the potential changes of the foot pad.

More recently Wang and Lu (1930), Richter and Whelan (1943), Patton (1948) and Lloyd (1961) have described the relatively brief negative potential deflection now known as the action potential of the sweat glands, or galvanic skin response. A prolonged positive deflection evokable by nerve stimulation (Lloyd 1961) or by chemical stimulation (Hara 1932; Lloyd 1963), known as the secretory potential, waxes and wanes in relation to secretion and reabsorption in the sweat glands. There is, in association with the secretory potential and exactly parallel with it, a change in impedance across the foot pad. This too, is related to secretion and reabsorption, which is to say filling and emptying of the sweet ducts (Lloyd 1960, 1961). A general account of these findings has been presented (Lloyd 1962).

I. Cholinergic control of sweat glands

The first observation on chemical activation of sweat glands appears to be that of Cutinho (1874) who noted the sudorific result of injecting an extract of pilocarpus. Then Luchsinger (1877) noted the action of pilocarpine and antagonism by atropine. A year later (Ott and Wood Field 1878), the action of muscarine and its antagonism by atropine were demonstrated. Instances of chemical action upon autonomic effectors multiplied, and within about a quarter century the beginning concept of neurohumoral transmission was laid down both with respect to adrenergic systems (Elliott 1905) and cholinergic systems (Dixon 1906). Association between the action of acetylcholine and that of the parasympathetic nerves began with the synthesis of acetylcholine (Nothnagel 1894) and study by Reid Hunt and Taveau (1906) of its actions. But it was Dale (1914) who drew the clear distinction between nicotinic and muscarinic actions of acetylcholine, noted that acetylcholine, injected intravenously or into the foot pad, caused sweating, and emphasized the ready response of sweat glands to muscarine and choline derivatives despite their anatomically sympathetic innervation. The evanescent action of acetylcholine suited its potential role as a chemical transmitter along with adrenaline. The sweat glands were the exception in sympathetic innervation which fact finally led to the introduction by Dale (1933) of the terms adrenergic and cholinergic.

The ultimate step in demonstration of chemical transmission by intervention of acetylcholine was taken by Loewi (1921). With specific reference to sweat glands this step was taken by Dale and Feldberg (1934) who demonstrated the release of acetylcholine from the cat foot pad on stimulation of its sudomotor supply. Cholinergy in transmission from sympathetic nerve to sweat glands was firmly established for the cat and by implication for man who's sweat glands similarly are "sensitive to pilocarpine and atropine and indifferent to adrenaline".

II. Evidences of adrenergic control of sweat glands

Until fairly recently the sweat glands of cat and man generally were held to be indifferent to adrenaline or noradrenaline. In the meantime much work culminating in that of LOVATT EVANS and SMITH (1956) had shown the glands of the horse (and of cattle and sheep) to be under humoral control by adrenaline. The present account is not concerned with these latter species.

First mention of adrenaline in relation to sweat secretion in the cat foot pad is that of LANGLEY (1901) who injected extracts of supra-renal bodies into the pads and in one cat observed a trace of sweating. ELLIOTT (1905) found no effect on sweat glands. DIEDEN (1916) likewise found no effect in the normal pad, but after denervation there was profuse secretion. This was presumably an example of the law of denervation (CANNON 1939) in action, exemplification of which in relation to lachrymal glands and the dilator pupillae was already at hand in the work of LANGLEY (1901) and ELLIOTT (1905). LANGLEY and UYENO (1922) obtained sweating by injection into the pad of adrenaline, but warned against the method and dismissed the result, for injected saline was at least as effective if not more so. An action of adrenaline introduced by other routes was not seen until FROELICH and ZAK (1932) by subcutaneous injection produced a secretion that was not blocked by atropine. The response was strong after denervation. In the same year, HARA (1932) recorded what are now known as secretory potentials from the cat foot pad in response to intravenous injection of pilocarpine and adrenaline. Atropine blocked the action of pilocarpine, but not that of adrenaline. HARA raised the question of double innervation.

PATTON (1949) blocked transmission to the foot pad sweat glands completely by means of atropine, but failed to achieve any block by means of the adrenergic blocker, dibenamine. This result is difficult to credit for the congeneric phenoxybenzamine blocks completely the neural and humoral, both by acetylcholine and noradrenaline, control of the cat sweat glands (LLOYD 1963). Also HAIMOVICI (1948) and SONNENSCHEIN (1949) had shown dibenamine block of neosynephrine and adrenaline induced sweating in man.

In the years about 1950 there was a flurry of observation on sweating produced in man by adrenaline and similar substances (HAIMOVICI 1948, 1950; SONNENSCHEIN 1949; KISIN 1949; WADA 1950; SONNENSCHEIN et al. 1951; BARNETT 1951; CHALMERS and KEELE 1951, 1952; RANDALL and KIMURA 1955; COLLINS et al. 1959). These followed the demonstrations by ICHIHASHI and KASHIWABARA (cited by KUNO 1956). In general it was found that adrenaline, noradrenaline and related compounds produced secretion; that atropine did not block the action; that dibenamine did.

To return to the object proper of this account (sweating in the cat foot pad) Nakamura and Hatanaka (1958) caused sweating by systemic injection of adrenaline (the sciatic nerve being cut to obviate central effects) which was sensitive to dihydroergotamine, but not to atropine. They paid particular attention to exaggeration of the sweating response to adrenaline (and nicotine and mecholyl) following denervation, this being another demonstration of denervation supersensitivity (Cannon 1939). Finally Lloyd (1959, 1963), observing impedance changes (cf. 1959) and potential changes (cf. 1961), demonstrated the secretory action of nerve impulses, of intravenous acetylcholine, adrenaline, noradrenaline. Further it was shown that atropine blocked the action of nerve impulses and of acetylcholine, but not that of adrenaline and noradrenaline, and that the adrenergic blocker phenoxybenzamine prevented response to any of these sources of excitation.

Histochemistry has made its contribution to the study of sweat gland innervation. Thus Hurley et al. (1953) demonstrated the presence of cholinesterase in nerve fibers surrounding the secretory tubules of eccrine sweat glands in man. There was no cholinesterase about the apocrine glands which are generally conceded to be adrenergically innervated. In the horse, who's sweat glands are apocrine and activated by circulating adrenaline, amine oxidase is present and cholinesterase absent (Hellmann 1955). More germane to the present concern are observations on the sweat glands of the cat. The nerve endings surrounding these give a strong positive reaction for cholinesterase (Hellmann 1952, 1955). But more important in the present framework of discussion is the fact that the sweat glands of the foot pad showed definite if slight amine oxidase activity (Hellmann 1955). The result implies that adrenergy as well as cholinergy may be involved in transmission to sweat glands in the cat.

Ideas concerning the physiological role of adrenaline or noradrenaline and of adrenergy in the sympathetic control of sweat gland activity are varied. One view, held prominently by those who have studied sweat secretion in man, is that these substances play no physiological role in transmission. Rothman (1954) proposed an auxiliary role assuming that adrenergic agents contract the myoepithelial elements to express preformed droplets and thus cause "sweating" without actually acting on the secretory cells. That this is not so is shown by the fact that adrenaline and noradrenaline will cause electrical signs of sweat formation and duct filling after a prolonged rest period at which time the ducts are empty (Lloyd 1959, 1962). Also the action is prolonged in eccrine glands (Lloyd 1959, 1963; Weiner and Hellmann 1960) whereas expulsion of sweat by myoepithelial contraction would take place in one or more short bursts (Hurley and Shelley 1954). Finally, there is the possibility that noradrenaline plays a role in transmission (Lloyd 1963, 1964).

III. Concerning duality of innervation

The possibility of adrenergic innervation inevitably is raised by the fact, now well-established, that noradrenaline activates sweat glands of the cat foot pad. It is raised necessarily despite the well-established and generally accepted fact of cholinergy. Hence in considering the possibility of adrenergy one must speak of duality of innervation. Duality may take one of two forms: cholinergic and adrenergic elements may lie in parallel, or cholinergic and adrenergic steps may lie in series. Certain it is that the sweat glands are not divided into two groups, however intermingled, the one group cholinergically, the other adrenergically innervated, for atropine block is complete. Furthermore, MELLINKOFF and SONNENSCHEIN (1954) identify the sweat glands responding to acetylcholine and adrenaline. The possibility that each gland cell, or a majority of gland cells, receives both cholinergic and adrenergic endings, which would constitute a parallel arrangement, is not so easily dismissed, but the evidence is more in favor of a serial arrangement.

Recently the hypothesis of serial relationship in which the action of liberated acetylcholine is to liberate noradrenaline in sympathetic postganglionic fibers has come to the forefront of attention in another connection (BURN and RAND 1959). The fibers concerned, unlike those to sweat glands, are ordinarily considered adrenergic rather than cholinergic. KOELLE (1962) proposed a variant of this hypothesis, adapted to purely cholinergic functions, in which the action of liberated acetylcholine is firstly to act on the presynaptic terminals themselves to liberate further quanta of acetylcholine.

One step in establishing a serial relationship between the actions of acetylcholine and noradrenaline is to show that acetylcholine has an action on prejunctional structures. Evidence for such an action in ganglia and neuromuscular junctions is to be found in a number of studies (MASLAND and WIGTON 1940; DEMPSHER and RIKER 1957; RIKER et al. 1957). There is to be considered also the widespread axon-reflex sweating induced by acetylcholine and other drugs with nicotine-like action (COON and ROTHMAN 1939, 1941; ROTHMAN and COON 1940), and the pilomotor action of acetylcholine (COON and ROTHMAN 1940) and nicotine (ROTHMAN and COON 1939). It is important to note that these actions on presynaptic structures and the six examples of acetylcholine-like action in sympathetic systems assembled by BURN (1961) are nicotinic in nature whereas the direct action of acetylcholine and nerve impulses on sweat glands in atropine-sensitive and muscarinic in nature.

Perhaps more convincing evidence of presynaptic action of acetylcholine with specific reference to the sweat glands of the cat is to be found in the fact that cholinergic agents no longer act following degeneration of the nerve supply (LUCHSINGER 1877; NAWROCKI 1878; VULPIAN 1878; OTT 1879; LANGLEY 1922). Loss of response to acetylcholine and pilocarpine is seen also in man after degeneration of the postganglionic nerves (Kahn and Rothman 1942; Janowitz and Grossman 1950, 1951; Chalmers and Keele 1952). Admittedly there is some ambiguity about the reaction of the cat sweat glands following nerve degeneration (Burn 1922, 1925; Simeone et al. 1951; Nakamura and Hatanaka 1958), nevertheless Simeone et al. working in Cannon's laboratory, clearly state that the cat sweat glands, acted on by pilocarpine, provide the exception to Cannon's law of denervation.

The converse of failure to act following degeneration of nerve supply has been the uniform finding with respect to adrenaline action on sweat glands (Dieden 1916; Froelich and Zak 1932; Hara 1932; Nakamura and Hatanaka 1958). In each instance a supersensitivity to adrenaline developed, which is in accord with the law of denervation (Cannon 1939), and suggests a postjunctional action.

Thus a reasonable case can be advanced, on the basis of response to denervation, for the hypothesis of duality of sweat gland innervation and for a serial rather than a parallel relationship between acetylcholine and noradenaline, with acetylcholine acting prejunctionally, noradrenaline postjunctionally. At least this hypothesis has the merit of reconciling most of the discrepancies concerning the law of denervation with respect to sweat glands. A definitive restudy of sweat gland denervation undoubtedly is in order.

From the electron microscope comes evidence concerning cholinergic and adrenergic endings that may in the future have germanity to the problem of sweat gland innervation. Richardson (1962) has drawn a distinction between granular and agranular vesicles in nerve endings in the vas deferens. Agranular vesicles are seen in the endings of neuromuscular junctions (Birks et al. 1960) which act by release of acetylcholine in discrete multi-molecular packets. Both types are found in some, but not in other endings in the vas deferens (Richardson 1962). In the iris, however, where cholinergic and adrenergic endings can be distinguished, the cholinergic endings are found to contain agranular vesicles, whereas adrenergic endings contain both types (Richardson 1964). Moreover, catechol amines are contained in the granular fraction (Blashko et al. 1955). These observations have interest in relation to the hypothesis of dual transmission and, by implication, to the specific question of the innervation of sweat glands. Obviously the innervation of these glands is a challenging problem for electron microscopy.

IV. Action of adrenergic blocking agents on neural and humoral control of sweat glands

In an effort further to assess the possible role of adrenergy in the process of transmission to sweat glands one turns to the study of adrenergic blocking agents. To date five of these have been investigated: dibenamine (Patton 1949), dihydroergotamine (NAKAMURA and HATANAKA 1958), phenoxybenzamine (LLOYD 1963), bretylium and guanethidine (LLOYD 1964).

Interpretation of experiments involving adrenergic blockers may be equivocal. For instance, phenoxybenzamine was found by BOYD et al. (1963) completely to block several pelvic nerve-effector (cholinergic) systems. On the other hand UVNÄS (1948) found dibenamine to block sympathetic salivary secretion, but not chorda secretion. Bretylium and guanethidine exert a curare-like (antinicotinic) action (BOURA and GREEN 1959; DIXIT et al. 1961) but, with one reported exception (the guinea-pig pelvic nerve-rectum preparation [BOYD et al. 1963] which was partially blocked) these agents are without effect upon mammalian muscarinic cholinergic junctions. In examplification, BOYD et al. (1962) found a clean-cut block by bretylium and guanethidine of the sympathetic inhibition of the rabbit colon, the parasympathetic action being unaltered, and also that the sympathetic inhibitory response of the ileum was replaced by a motor response, all of which would not have happened if these substances had blocked the muscarinic action of released acetylcholine. In large doses, BW 467C60 briefly blocks cholinergic mechanisms (BOURA and GREEN 1963). BURN (1961) interprets the effect of bretylium and guanethidine on adrenergic transmission as the block of a nicotinic acetylcholine action of releasing noradrenaline (cf. also BURN and FROEDE 1963).

In commenting on their study of the cholinergic blocking action of adrenergic agents BOYD et al. (1963) conclude that "The use of autonomic drugs as analytical tools is not reliable unless combined with both histochemical examination of the preparation and assay of the substances released upon stimulation of the nerves". As already noted there is some histochemical evidence favoring a role for adrenergy in sweat gland innervation (HELLMANN 1955). An attempt to show release by the sudomotor fibers of noradrenaline, in the manner of the DALE-FELDBERG (1934) demonstration of the release of acetylcholine, would be futile for sources other than the sweat glands exist. Stimulation of the presynaptic fibers liberates in the superior cervical ganglion acetylcholine (FELDBERG and GADDUM 1934) and an adrenaline-like substance (BÜLBRING 1944).

Dihydroergotamine was shown by NAKAMURA and HATANAKA (1958) to block the sudorific action of injected adrenaline which action was, however, unaffected by atropine.

Dibenamine and phenoxybenzamine are perhaps the most controversial agents. Nevertheless, their action, or lack of action, is worth recording. PATTON (1949) found no effect of dibenamine on transmission to sweat glands. Phenoxybenzamine (LLOYD 1963), on the other hand, completely blocks neural transmission, the action of intravenously injected acetylcholine and the action of intravenously injected noradrenaline.

The actions of bretylium and guanethidine are similar, one to the other (LLOYD 1964). As indicated by change in the action potential response of sweat glands to motor nerve stimulation block begins to develop in about one min. Response is halved in about 10 min. Rate of decline decreases approximately exponentially and responses reach a final low amplitude in 50 to 90 min. At such times block is up to 98 percent complete. Occasionally a maximum is reached in fifteen minutes followed by partial remission of block to an intermediate value. Discontinuity, of greater or lesser degree, in the early development of block clearly indicates a duality of blocking action. One action would be of short latency, brief and possibly anticholinergic, the other of longer latency, enduring and possibly antiadrenergic.

The actual situation with respect to adrenergic blocking agents is perhaps not so equivocal as would seem from the conclusion of Boyd et al. (1963). High requisite dosage and relative brevity of action appear likely to distinguish between cholinergic and adrenergic blockade (Boura and Green 1963). And then there is evident duality in the blocking process at the junctions to sweat glands that could accommodate action at two stages of transmission, one cholinergic, the other adrenergic.

Considering all the evidence presented one concludes that a dual hypothesis, however unpalatable, encompassing serial action of acetylcholine and noradrenaline in that order, at the present time best accommodates the facts concerning transmission from postganglionic sympathetic nerve fibers to sweat glands in the cat foot pad.

Acknowledgement

This investigation was supported by a P.H.S. research grant, NB-02816, from the National Institute of Neurological Diseases and Blindness.

References

Barnet, A. J. (1951). *Nature* (Lond.) **167**, 482.
Birks, R., Huxley, H. E. & Katz, B. (1960). *J. Physiol.* (Lond.) **150**, 134.
Blashko, H., Hagen, P. & Welch, A. D. (1955). *J. Physiol.* (Lond.) **129**, 27.
Boura, A. L. A. & Green, A. F. (1959). *Brit. J. Pharmacol.* **14**, 536.
Boura, A. L. A. & Green, A. F. (1963). *Brit. J. Pharmacol.* **20**, 36.
Boyd, G., Gillespie, J. S. & Mackenna, B. R. (1962). *Brit. J. Pharmacol.* **19**, 258.
Boyd, H., Burnstock, G., Campbell, G., Jowett, A., O'Shea, J. & Wood, M. (1963). *Brit. J. Pharmacol.* **20**, 418.
Bülbring, E. (1944). *J. Physiol.* (Lond.) **103**, 55.
Burn, J. H. (1922). *J. Physiol.* (Lond.) **56**, 232.
Burn, J. H. (1925). *J. Physiol.* (Lond.) **60**, 365.
Burn, J. H. (1961). *Brit. Med. J.* **1**, 1623.
Burn, J. H. & Froede, H. (1963). *Brit. J. Pharmacol.* **20**, 378.
Burn, J. H. & Rand, M. J. (1959). *Nature* (Lond.) **184**, 163.
Cannon, W. B. (1939). *Am. J. Med. Sci.* **198**, 737.
Chalmers, T. M. & Keele, C. A. (1951). *J. Physiol.* (Lond.) **114**, 510.
Chalmers, T. M. & Keele, C. A. (1952). *Brit. J. Dermatol.* **64**, 45.
Collins, K. J., Sargent, F. & Weiner, J. S. (1959). *J. Physiol.* (Lond.) **148**, 592.

COON, J. M. & ROTHMAN, S. (1939). *Proc. Soc. Exp. Biol. Med.* **42**, 231.
COON, J. M. & ROTHMAN, S. (1940). *J. Pharmacol. exp. Therap.* **68**, 301.
COON, J. M. & ROTHMAN, S. (1941). *J. Pharmacol. exp. Therap.* **73**, 1.
COYNE, P. (1878). *C. R. Acad. Sci. Paris* **86**, 1276.
CUTINHO (1874), quoted by HENDERSON, V. E. & ROEPKE, M. H. (1937). *Physiol. Rev.* **17**, 373.
DALE, H. H. (1933). *J. Physiol.* (Lond.) **80**, 10 P.
DALE, H. H. & FELDBERG, W. (1934). *J. Physiol.* (Lond.) **82**, 121.
DEMPSHER, J. & RIKER, W. F. (1957). *J. Physiol.* (Lond.) **139**, 145.
DIEDEN, H. (1916). *Z. Biol.* **66**, 387.
DIXIT, B. N., GULATI, O. D. & GOKHALE, S. D. (1961). *Brit. J. Pharmacol.* **17**, 372.
DIXON, W. E. (1906). *Brit. Med. J.* **2**, 1807.
ELLIOTT, T. R. (1905). *J. Physiol.* (Lond.) **32**, 401.
EVANS, C. L. & SMITH, D. F. G. (1956). *Proc. roy. Soc.* B **145**, 61.
FELDBERG, W. & GADDUM, J. H. (1934). *J. Physiol.* (Lond.) **81**, 305.
FROELICH, A. & ZAK, E. (1932). *Arch. Exp. Path. Pharmak.* **168**, 620.
GOLTZ, F. (1875). *Arch. ges. Physiol.* **11**, 52.
HAIMOVICI, H. (1948). *Proc. Soc. Exp. Biol. Med.* **68**, 40.
HAIMOVICI, H. (1950). *J. Appl. Physiol.* **2**, 512.
HARA, Y. (1932). *Jap. J. Med. Sci. III Biophysics* **2**, 215.
HELLMANN, K. (1952). *Nature* (Lond.) **169**, 113.
HELLMANN, K. (1955). *J. Physiol.* (Lond.) **129**, 454.
HERMANN, L. & LUCHSINGER, B. (1878). *Arch. ges. Physiol.* **17**, 310.
HUNT, R. & TAVEAU, R. DE M. (1906). *Brit. Med. J.* **2**, 1788.
HURLEY, H. J. & SHELLEY, W. B. (1954). *J. Invest. Dermatol.* **22**, 143.
HURLEY, H. J., SHELLEY, W. B. & KOELLE, G. B. (1953). *J. Invest. Dermatol.* **21**, 139.
JANOWITZ, H. D. & GROSSMAN, M. I. (1950). *J. Invest. Dermatol.* **14**, 453.
JANOWITZ, H. D. & GROSSMAN, M. I. (1951). *Experientia* (Basel) **7**, 275.
KAHN, H. & ROTHMAN, S. (1942). *J. Invest. Dermatol.* **5**, 431.
KISIN, E. E. (1948). *Vestnik. Venerol. Dermatol.* **5**, 27.
KOELLE, G. B. (1962). *J. Pharm. Pharmacol.* **14**, 65.
KUNO, Y. (1956). *Human Perspiration.* Springfield: C. C. Thomas.
LANGLEY, J. N. (1901). *J. Physiol.* (Lond.) **27**, 237.
LANGLEY, J. N. (1922). *J. Physiol.* (Lond.) **56**, 110.
LANGLEY, J. N. & UYENO, K. (1922). *J. Physiol.* (Lond.) **56**, 206.
LLOYD, D. P. C. (1959.) *Nature* (Lond.) **184**, 277.
LLOYD, D. P. C. (1960). *J. Gen. Physiol.* **43**, 713.
LLOYD, D. P. C. (1961). *Proc. Nat. Acad. Sci.* **47**, 351.
LLOYD, D. P. C. (1962). In *Eccrine sweat glands and eccrine sweating* (ed. by MONTAGNA, W., ELLIS, R. A. & SILVER, A. F.), p. 127. Oxford: Pergamon Press.
LLOYD, D. P. C. (1963). *J. Physiol.* (Lond.) **169**, 116 P.
LLOYD, D. P. C. (1964). *J. Physiol.* (Lond.) **175**, 74 P.
LOEWI, O. (1921). *Arch. ges. Physiol.* **189**, 239.
LUCHSINGER, B. (1877). *Arch. ges. Physiol.* **15**, 482.
MASLAND, R. L. & WIGTON, R. S. (1940). *J. Neurophysiol.* **3**, 269.
MELLINKOFF, S. M. & SONNENSCHEIN, R. R. (1954). *Science.* **120**, 997.
NAKAMURA, Y. & HAPANAKA, K. (1958). *Tohoku. J. Exp. Med.* **68**, 225.
NAWROCKI, F. (1878). *Centralbl. Med. Wiss.* **16**, 97.
NOTHNAGEL, G. (1894). *Arch. Pharmacie.* **232**, 261.
OTT, I. (1879). *J. Physiol.* (Lond.) **2**, 42.
OTT, I. & WOOD FIELD, G. B. (1878). *J. Physiol.* (Lond.) **1**, 193.
PATTON, H. D. (1948). *J. Neurophysiol.* **11**, 217.
PATTON, H. D. (1949). *Proc. Soc. Exp. Biol. Med.* **70**, 412.
RANDALL, W. C. & KIMURA, K. K. (1955). *Pharmacol. Rev.* **7**, 365.
RANVIER, L. (1879). *C. R. Acad. Sci.* (Paris) **89**, 1120.

Richardson, K. C. (1962). *J. Anat.* **96**, 427.
Richardson, K. C. (1964). *Amer. J. Anat.* **114**, 173.
Richter, C. P. & Whelan, F. (1943). *J. Neurophysiol.* **6**, 191.
Riker, W. F., Roberts, J., Standaert, F. G. & Fugimori, H. (1957). *J. Pharmacol. exp. Therap.* **121**, 286.
Rothman, S. (1954). *Physiology and Biochemistry of skin.* Chicago: U. Chicago Press.
Rothman, S. & Coon, J. M. (1939). *Arch. Dermatol. Syph.* **40**, 999.
Rothman, S. & Coon, J. M. (1940). *J. Invest. Dermatol.* **3**, 79.
Simeone, F. A., Mentha, C. & Rodriques, H. A. (1951). *Am. J. Physiol.* **165**, 356.
Sonnenschein, R. R. (1949). *Proc. Soc. Exp. Biol. Med.* **71**, 654.
Sonnenschein, R. R., Kobrin, H., Janowitz, H. D. & Grossman, M. I. (1951). *J. Appl. Physiol.* **3**, 573.
Uvnäs, B. (1948). *Acta. physiol. scand.* **15**, 362.
Vulpian, A. (1878). *C. R. Acad. Sci.* (Paris) **87**, 311.
Wada, M. (1950). *Science* **111**, 376.
Waller, A. D. (1901). *Proc. roy. Soc.* **69**, 171.
Waller, A. D. (1904). *Proc. roy. Soc.* **73**, 92.
Wang, G. H. & Lu, T. W. (1930). *Chinese J. Physiol.* **4**, 393.
Weiner, J. S. & Hellmann, K. (1960). *Biol. Rev.* **35**, 141.

EFFECTS OF BARBITURATES AND LACK OF OXYGEN ON THE MONOSYNAPTIC REFLEX PATHWAY OF THE CAT SPINAL CORD

by Yngve Løyning

General anaesthetics or lack of oxygen block the monosynaptic reflex in the spinal cord if the level of anaesthesia or oxygen deprivation are great enough. Lack of oxygen, whether induced by breathing gas of low oxygen content (hypoxia, anoxia) or by stopping ventilation or circulation (asphyxia), gives an initial period of enhanced reflex activity before block occurs (see Kirstein 1951). Many investigators have expressed different opinions about the differential sensitivities of the various parts of the spinal reflex pathway to both anaesthetics and lack of oxygen, about the site of blocking, and the mechanism of their action. Some of the differences can presumably be explained by differences in experimental techniques. The present investigation was designed to establish the principal site of action of barbiturates and hypoxia and their effects before block occurs on the various components of the monosynaptic reflex pathway within the cat spinal cord.

A short-acting barbiturate, thiamylal sodium, was slowly injected intravenously in doses of 10—20 mg/kg and the effects were studied on monosynaptic reflexes in the lower lumbar spinal segments. The cats were already lightly anaesthetized with pentobarbital sodium, immobilized with gallamine triethiodide (Flaxedil), and artificially ventilated. (For further description of methods used, see Løyning, Oshima and Yokota 1964).

This dose of thiamylal gave a marked reduction in the monosynaptic reflex without producing the fall in the systemic blood pressure and oxygen tension in the spinal cord that occurred with greater doses or more rapid injections. Fig. 1 shows the typical changes caused by 10 mg/kg thiamylal in potentials recorded intracellularly from a motoneurone. Four to five minutes after the injection the monosynaptic excitatory potential (EPSP), evoked by stimulation of the afferent nerve, was decreased to 60—89% of the control amplitude (14 observations: mean = 79%, S.E. = 7%). Complete recovery usually was observed within 30 minutes (Fig. 1). No changes were observed in the time course of the EPSP and in the potentiation seen after tetanic stimulation. Furthermore, the typical bimodal reduction in EPSP size normally seen for stimulation frequencies of 0.3—50 per second (CURTIS and ECCLES 1960; LØYNING, OSHIMA and YOKOTA 1964) was the same during thiamylal anaesthesia, apart from a slightly greater depression at the higher frequencies of 20—50 impulses per second. There were no changes in such characteristics of the motoneurone membrane as the firing

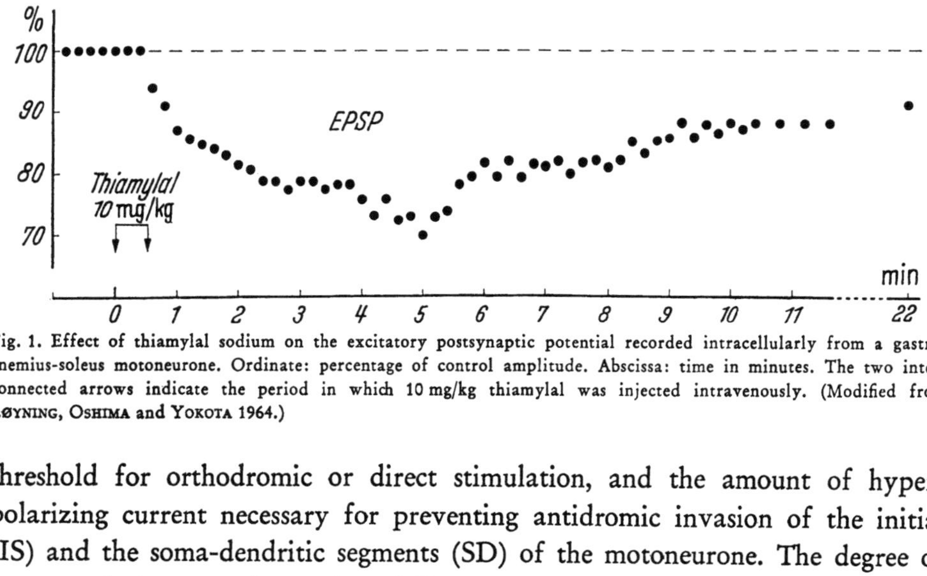

Fig. 1. Effect of thiamylal sodium on the excitatory postsynaptic potential recorded intracellularly from a gastrocnemius-soleus motoneurone. Ordinate: percentage of control amplitude. Abscissa: time in minutes. The two interconnected arrows indicate the period in which 10 mg/kg thiamylal was injected intravenously. (Modified from LØYNING, OSHIMA and YOKOTA 1964.)

threshold for orthodromic or direct stimulation, and the amount of hyperpolarizing current necessary for preventing antidromic invasion of the initial (IS) and the soma-dendritic segments (SD) of the motoneurone. The degree of accommodation was also unaltered (LØYNING, OSHIMA and YOKOTA 1964).

Two methods were employed to test the effects of thiamylal on the afferent nerve terminals. The excitability of the terminals was tested directly by stimulating them with a brief current pulse of constant submaximal strength through a microelectrode situated in a motor nucleus and measuring the size of the evoked compound action potential (ANP) recorded peripherally in the corresponding nerve (WALL 1958). By this method, motoneurones were also stimulated directly, and the size of the evoked potential recorded from the ventral root (VRP) was used as an indication of the excitability of these cells. The other method employed was to alternate this stimulation through the microelectrode with the recording within the motor nucleus of extracellular focal potentials evoked by orthodromic stimulation of the corresponding afferent nerve. These potentials consist of an initial triphasic action potential (IP), generally considered to represent the activity in the afferent nerve terminals, followed by a negative synaptic potential (FSP) (Brooks and Eccles 1947). A depolarization of spinal presynaptic terminals has been shown to produce an increase in the peripherally recorded action potential, thus indicating an increase of the excitability of the afferent fibres and terminals (Eccles, Magni and Willis 1962). However, both components of the focal potential are reduced in size, which indicates that the increase in excitability has not brought about a detectable increase in the invasion of terminals for orthodromic activation (Løyning, Oshima and Yokota 1964). Hyperpolarization of the terminals has the opposite effect (Eccles and Rall 1951).

Fig. 2 shows the effect of 10 mg/kg thiamylal on the excitabilities of the

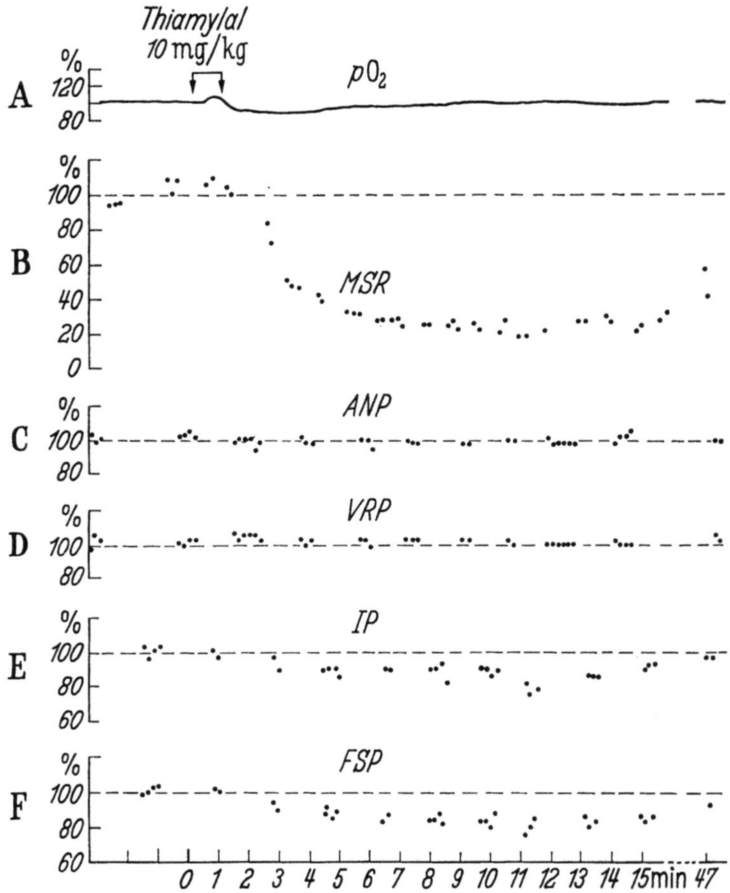

Fig. 2. Effects of thiamylal on the monosynaptic reflex (MSR), on the excitabilities of the afferent nerve terminals and the motoneurone somata, and on the orthodromic focal potential in the motor nucleus involved in the reflex. A NaCl-filled micropipette was placed extracellularly in the posterior biceps-semitendinosus motor nucleus and used for both recording and stimulating purposes (see text). A. Oxygen tension recorded in spinal cord, as a percentage of control value. B. Monosynaptic reflex (MSR) recorded from the seventh lumbar ventral root in response to stimulation of the posterior biceps-semitendinosus nerve; as a percentage of the control value. C, D. Antidromic "nerve potential" (ANP) recorded in posterior biceps-semitendinosus nerve and potential recorded from ventral root (VRP) in response to stimulating through the microelectrode (0.1 msec, 65 V); both as percentage of control responses before the thiamylal injection. E, F. Presynaptic action potential (IP) and negative synaptic potential (FSP) recorded in the motonucleus in response to the same stimulus as in B. Abscissae — time in minutes. The two interconnected arrows indicate the period in which 10 mg/kg thiamylal was injected. (Modified from Løyning, Oshima and Yokota 1964.)

afferent nerve terminals and the motoneurone somas as judged by the amplitudes of the "nerve potential" (ANP) and the ventral root potential (VRP) respectively, and on the focal potential consisting of the initial presynaptic action potential (IP) and the following synaptic potential (FSP). Also illustrated are the effects on the ventral root monosynaptic reflex (MSR), and on the oxygen tension (pO_2) recorded polarographically with open-ended platinum electrodes in the fourth lumbar segment of the spinal cord. This dose of thiamylal resulted in a decrease of the monosynaptic reflex (MSR) to 20% of its control size, and a concomitant decrease in both components of the focal potential. The excitability of the nerve terminals and the motoneurone somas were not changed, indicating that the action potentials in the terminals were depressed for reasons other than depolarization. These results were confirmed in 16 similar trials. Injection of pentobarbital sodium (10 mg/kg) produced similar results. The reduction in amplitude of the presynaptic action potential (IP), whether fewer terminals were invaded or the individual spike was smaller, may be brought about by a restriction of ion movements across the membrane, a mechanism postulated for the action of local anaesthetics (see also Løyning, Oshima and Yokota 1964) and also suggested as a possible mechanism for the action of general anaesthetics (see Frank and Sanders 1963). Curtis and Phillis (1960) have shown that procaine administered electrophoretically near cat spinal motoneurones increased the firing threshold without altering the resting and postsynaptic potentials.

From these results it is clear that barbiturates in these doses produce a reduction in the monosynaptic reflex that is more likely due to an action on the presynaptic terminals than to a postsynaptic effect on the motoneurone membrane. The postsynaptic effects reported by other investigators, such as an increase in the firing threshold (Brooks and Eccles 1947; Somjen and Gill 1963), and in the degree of accommodative increase in firing threshold to a linearly rising current (Sasaki and Otani 1962), were obtained with greater doses of barbiturates (25—120 mg/kg) than were used in the present experiments. Barbiturates injected rapidly, and in doses greater than 20 mg/kg, significantly reduce the systemic blood pressure and the oxygen tension in the spinal cord, and may thus also produce an hypoxia effect.

On the basis of these results it is postulated that the principal site of action of barbiturates in doses which reduce the monosynaptic reflex by up to 80%, is at the afferent terminals. It is suggested that the mode of action is similar to that of local anaesthetics: a reduction in the spike size and in the output of transmitter, and consequently a diminution in the synaptic potential.

The same experimental technique was used to study the effects of oxygen lack on the monosynaptic reflex pathway (Eccles, Løyning and Oshima, unpublished results). Oxygen lack was induced by ventilating the cat either with 5% oxygen in nitrogen, or with pure nitrogen. The oxygen tension in the spinal cord decreased rapidly after the onset of hypoxia, usually stabilized within one minute, and was then unaffected by minor blood pressure changes. At the return of air ventilation, the oxygen tension rose sharply, often with an overshoot before it returned to the initial level.

Fig. 3 illustrates the effects of two minutes ventilation with pure nitrogen on

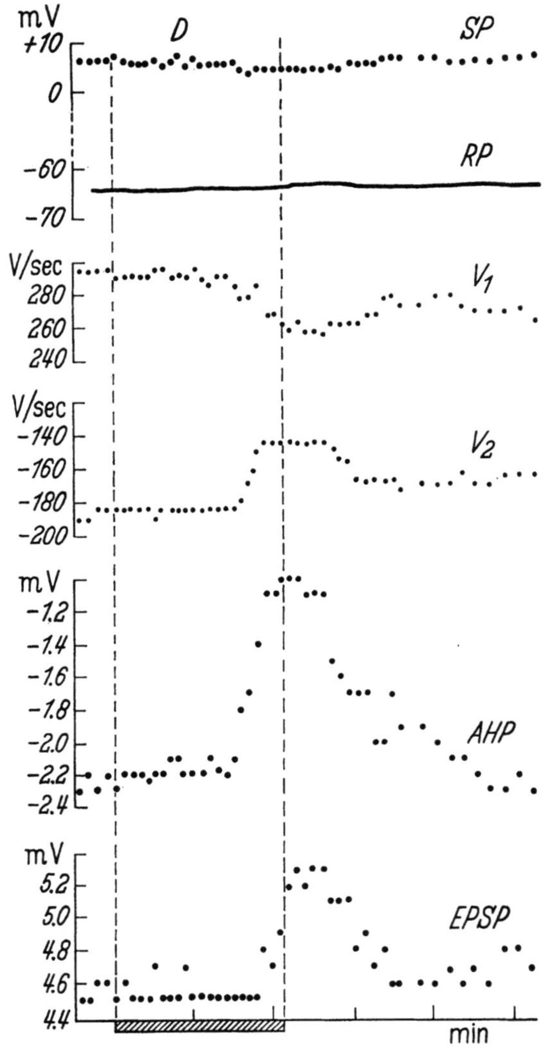

Fig. 3. Effects of ventilation with nitrogen for two minutes and five seconds on the following intracellular records from a gastrocnemius motoneurone: the resting potential (RP); the amplitude of the EPSP and of the after-hyperpolarization (AHP) of the antidromic spike potential; the amplitude of the spike measured as overshoot (SP), and its rate of rise (V_1) and fall (V_2) measured from differentiated records of the spike potential. The vertical broken lines connected with the bar at the bottom of the graph indicate the period of nitrogen ventilation. (Note that an evoked EPSP preceded the antidromic spike potential to aid firing the cell.)

potentials recorded intracellularly from a gastrocnemius motoneurone. The graph shows the size of the resting potential and the amplitude of the excitatory post-synaptic potential (EPSP), and the following characteristics of the antidromic spike potential: the overshoot (SP), rate of rise (V_1) and fall (V_2), and the amplitude of the after-hyperpolarization (AHP). This short exposure to nitrogen produced no significant change in the resting potential, a slight reduction in the overshoot of the spike, a moderate reduction in the rate of spike rise and a somewhat greater reduction in its rate of fall, a marked reduction in the after-hyperpolarization of the spike, and a clear increase in the EPSP. Ventilation with 5% oxygen in nitrogen for up to 17 minutes produced similar, but somewhat less pronounced effects. In addition, a degree of depolarization was observed which depended upon the resting potential level.

Under normal circumstances intracellular records can be obtained from motoneurones for several hours. (The records shown in Fig. 3 are from a cell observed for nearly three hours.) During exposures to hypoxia however, the cells were easily lost in a few minutes before any changes could be observed. Sudden reductions in resting potential occurred frequently and these were most probably due to movements of the tissue in relation to the microelectrode caused by swelling of the cells which has been shown to occur during hypoxia (van Harreveld and Biersteker 1964). A stepwise and considerable depolarization (up to 50% reduction in resting potential) was reported by Kolmodin and Skoglund (1959) to occur in cat motoneurones within one minute of asphyxia. They discussed and dismissed the possibility that this depolarization could be partly movement artefacts. On the other hand, Nelson and Frank (1963) who also observed sudden large and smaller changes in the cat motoneurone resting potential during anoxia, regarded them as movement artefacts. They claimed that when they had eliminated them, the resting potential was remarkable insensitive to lack of oxygen, showing at the most a depolarization of 2—5 mV after 4—5 minutes anoxia. Washizu (1960) observed in toad motoneurones a gradual depolarization of about 10 mV after twenty to thirty minutes of anoxia.

Analysis of 15 selected hypoxia trials which were free from major movement artifacts, and in which the intracellular recordings were from cells with a resting potential between 60 to 80 mV, has shown that the degree of depolarization produced by ventilation with 5% oxygen for 6—15 minutes depended upon the resting potential level. Those cells with a high initial resting potential showed little if any depolarization, and those with a lower resting potential showed more depolarization during hypoxia, the maximum rate observed being 1.8 mV per minute. It is possible that the cells with the lower resting potentials which showed less resistance to depolarization by hypoxia, were damaged by the impalement. These resting potentials could however, recover completely after hypoxia and be maintained at the same level for more than an hour. In addition, the spike potential of these cells were normal with an overshoot of 10—20 mV. Ito and Oshima (1964) have shown that the sodium pump operates effectively in cells with lower resting potential levels, as judged by the ability to extrude electrophoretically injected sodium ions. They also found that there was less depolarization following injection of sodium ions in cells with higher resting potentials, and sometimes a slight hyperpolarization in cells with the highest resting potentials. It appears that a reduction in intracellular potassium concentration, whether it be caused by hypoxic reduction in the active potassium influx or by injection of sodium ions, produces a depolarization only in cells already somewhat depolarized. A similar relationship between intracellular potassium concentration and its effect on the resting potential has been demonstrated for the squid giant axons (Baker, Hodgkin and Shaw 1962; Baker, Hodgkin and Meves 1964). Ito and Oshima (1964) have estimated that the intracellular sodium concentration would increase at a rate of 10 mM per minute if the sodium pump stopped immediately. If potassium is lost from the cell at the same rate, the cell would be depolarized at a minimum rate of 1.8 mV per minute. This was the maximum rate of depolarization observed during exposure to 5% oxygen in nitrogen.

According to the ionic hypothesis of excitation (Hodgkin 1951, 1958; Hodgkin and Huxley 1952 a, b) the rate of rise and overshoot of the spike potential are determined chiefly by the rapid increase in sodium conductance and the equilibrium potential for sodium, and the rate of fall and the degree of after-hyperpolarization are chiefly determined by the increased potassium conductance and equilibrium potential for potassium, with some overlapping action of these factors at the crest of the spike potential. The changes observed in the components of the spike potential during hypoxia may thus be due either to changes in the ion conductances during activation or to reductions in the equilibrium potentials for sodium and potassium, e. g. by a slowing of the metabolically dependant "sodium-potassium pump", or to both. The methods employed in these studies do not however, allow a discrimination between these two possible mechanisms.

The increase in amplitude of the excitatory postsynaptic potential was not accompanied by changes in the half decay time of the potential. This effect was most likely due to a presynaptic action of hypoxia in increasing the amount or efficiency of the transmitter which was released. The effects of hypoxia on the excitability of the afferent nerve terminals, and on the motor nucleus focal potential comprising the presynaptic action potential and the synaptic potential, were studied with the same technique as used in the thiamylal studies. The excitability showed a slight initial increase followed by a decrease almost to inexcitability, at which time the monosynaptic reflex was also blocked, indicating that the terminals were gradually depolarized until the spike generating mechanism was inactivated. Before block occurred, the synaptic potential showed an increase of about 70% at a time when both the excitability of the terminals and the presynaptic action potential were only slightly increased. Therefore, it appears that the amount or efficiency of the transmitter is increased during hypoxia for other reasons than a greater invasion of the afferent terminals or a hyperpolarization of the terminals with greater individual spike sizes which has been shown to be correlated with a greater transmitter output (Hagiwara and TASAKI 1958; TAKEUCHI and TAKEUCHI 1962). Studies of the effect of hypoxia on neuromuscular transmission in the isolated rat phrenic nerve-diaphragm preparation have shown that before block of conduction occurs in the nerve terminals, there is also here an increase in the excitatory postsynaptic potential. This is due to both a presynaptic action, as revealed by an increase in the quantal content of the endplate potential, and to a postsynaptic action producing an increase in the amplitude of the spontaneous miniature endplate potentials (HUBBARD and LØYNING, unpublished data).

LLOYD and MCINTYRE (1949) have suggested that the ability of the somas of the motoneurone to generate impulses in response to presynaptic volleys fails first during asphyxia, and that the subsequent failure takes place at the branching point of the parent afferent fibres, rather than in the branches or terminals. The present experiments have indicated that the most probable site of block of the monosynaptic reflex during hypoxia is in the presynaptic terminals, or in the actual process of synaptic transmission, since the presynaptic action potential was not diminished at the time when the synaptic potential disappeared. Furthermore, the motoneurone soma was still excitable to direct stimulation when the reflex was blocked, as was also shown by BROOKS and ECCLES (1947). Upon restoration of the oxygen supply after returning to air ventilation, the observed changes during hypoxia could recover completely. A new block or reduced excitability due to a hyperpolarization in the recovery period from asphyxia or anoxia, as reported by LLOYD (1953) and WASHIZU (1960), was not observed.

The observed initial increase in the EPSP and the reduction in the after-hyperpolarization of the spike potential, which would shorten the refractory period and increase the ability to repetitive firing of motoneurones, may explain the enhanced reflex activity commonly observed initially during hypoxia (see KIRSTEIN 1951).

References

BAKER, P. F., HODGKIN, A. L. & MEVES, H. (1964). *J. Physiol.* (Lond.) **170**, 541.
BAKER, P. F., HODGKIN, A. L. & SHAW, T. I. (1962). *J. Physiol.* (Lond.) **164**, 355.
BROOKS, C. MCC. & ECCLES, J. C. (1947). *J. Neurophysiol.* **10**, 349.
CURTIS, D. R. & ECCLES, J. C. (1960). *J. Physiol.* (Lond.) **150**, 374.
CURTIS, D. R. & PHILLIS, J. W. (1960). *J. Physiol.* (Lond.) **153**, 17.
ECCLES, J. C., MAGNI, F. & WILLIS, W. D. (1962). *J. Physiol.* (Lond.) **160**, 62.
ECCLES, J. C. & RALL, W. (1951). *J. Neurophysiol.* **14**, 353.
FRANK, J. B. & SANDERS, H. D. (1963). *Brit. J. Pharmacol.* **21**, 1.
HAGIWARA, S. & TASAKI, I. (1958). *J. Physiol.* (Lond.) **143**, 114.
HODGKIN, A. L. (1951). *Biol. Rev.* **26**, 339.
HODGKIN, A. L. (1958). *Proc. Roy. Soc.* B. **148**, 1.
HODGKIN, A. L. & HUXLEY, A. F. (1952 a). *J. Physiol.* (Lond.) **116**, 449.
HODGKIN, A. L. & HUXLEY, A. F. (1952 b). *J. Physiol.* (Lond.) **117**, 500.
ITO, M. & OSHIMA, T. (1964). *Proc. Roy. Soc.* B. **161**, 132.

KIRSTEIN, L. (1951). *Acta physiol. scand.* 23, suppl. 80, 1.
KOLMODIN, G. M. & SKOGLUND, C. F. (1959). *Acta physiol. scand.* 45, 1.
LLOYD, D. P. C. (1953). *J. gen. Physiol.* 36, 673.
LLOYD, D. P. C. & MCINTYRE, A. K. (1949). *J. gen. Physiol.* 32, 409.
LØYNING, Y., OSHIMA, T. & YOKOTA, T. (1964). *J. Neurophysiol.* 27, 408.
NELSON, P. G. & FRANK, K. (1963). *J. Physiol.* (Lond.) 205, 208.
SASAKI, I. & OTANI, T. (1962). *Jap. J. Physiol.* 12, 383.
SOMJEN, G. G. & GILL, M. (1963). *J. Pharmacol.* 140, 19.
TAKEUCHI, A. & TAKEUCHI, N. (1962). *J. gen. Physiol.* 45, 1181.
VAN HARREVELD, A. & BIERSTEKER, P. A. (1964). *Amer. J. Physiol.* 206, 8.
WALL, P. D. (1958). *J. Physiol.* (Lond.) 142, 1.
WASHIZU, Y. (1960). *J. Physiol. Soc. Jap.* 22, 23.

MONOAMINES AND SPINAL REFLEXES

by ANDERS LUNDBERG

There have been many reports that monoamines influence transmission in the spinal cord. Systemic administration of adrenaline (A) and noradrenaline (NA) may give both facilitatory and inhibitory effects on spinal reflexes (for references see MCLENNAN 1963, p. 84) but since these catecholamines in all likelihood do not penetrate the blood brain barrier it is possible that the effects are secondary to circulatory changes. The same indirect action probably also applies to experiments with systemic injection of 5-hydroxytryptamine (5-HT). Some effects have been found with topical application of monoamines on the spinal cord (cf. MCLENNAN 1963), but CURTIS, PHILLIS and WATKINS (1961) found no effect of electrophoretically administered NA and 5-HT on single cells in this tissue.

A strong impetus for a renewed interest in the effect of monoamines on the spinal cord has been given by investigations with the histochemical fluorescence technique of FALCK (1962) and FALCK, HILLARP, THIEME and TORP (1962). CARLSSON, FALCK, FUXE and HILLARP (1964) have shown that there are synaptic terminals of descending pathways which contain NA or 5-HT. In part this termination is within autonomic nuclei, but there is also a diffuse termination in the gray matter (not on blood vessels). Another important line of investigation has been the histochemical analysis of monoamines in the spinal cord and their disappearance after transection of the cord (MAGNUSSON and ROSENGREN 1963; CARLSSON, MAGNUSSON and ROSENGREN 1963). It has also been shown that in the frog and mouse electrical stimulation of the spinal cord may cause liberation of 5-HT (ANDÉN, CARLSSON, HILLARP and MAGNUSSON 1964).

DOPA (L-3, 4 dihydroxyphenylalanine) and 5-HTP (5-hydroxytryptophan) are intermediate metabolites in the synthesis of the catecholamines (dopamine

[DA], NA and A) and of 5-HT, respectively. Hence DOPA and 5-HTP can act as precursors of these monoamines, and after entry in monoaminergic terminals induce increased synthesis and overflow of transmitters. CARLSSON et al. (1963) found that DOPA and 5-HTP could change spinal reflexes. Results from an electrophysiological analysis of their effect will now be discussed.

The effect of DOPA

In the acute spinal cat DOPA depresses transmission from the FRA (flexor reflex afferents) to primary afferents (Fig. 1), motoneurones (Fig. 2) and to ascending pathways (ANDÉN, LUNDBERG, ROSENGREN and VYKLICKÝ 1963). This action is selective in that neither dorsal root potentials (DRP) evoked from group I afferents nor the Ia excitatory inhibitory pathways to motoneurones are changed. Excitatory and inhibitory pathways from the FRA are depressed to the same degree. The effect is fully reversible after 1—2 hours (Fig. 1), and

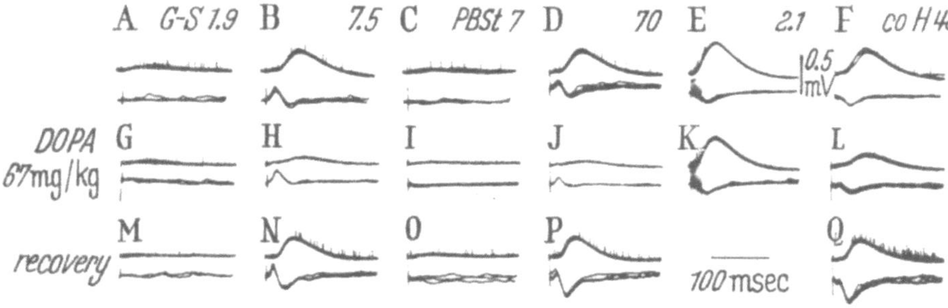

Fig. 1. The effect of DOPA on dorsal root potentials (DRP). The DRPs (upper traces) were recorded in the most caudal dorsal rootlet in L 6. The rootlet was cut 15 mm from the entry into the cord and placed on two electrodes, one close to the entry zone and the other at the cut end. The lower traces in all records are from the cord dorsum in L 7. Records A—F were taken before, and the corresponding records G—L 10 minutes after, intravenous injection of DOPA (67 mg/kg). The lower records M—G were taken 90 minutes after the injection of DOPA and show the recovery. The nerves from gastrocnemius-soleus (G—S), posterior biceps-semitendinosus (PBSt) and contralateral hamstring (coH) were stimulated as indicated above each column of records. Note that DOPA depresses the DRPs from ipsi- and contralateral high threshold muscle afferents, but not the DRP evoked from group I muscle afferents (E and K). (From an experiment by ANDÉN, LUNDBERG and VYKLICKÝ.)

can at any step be relieved by the α receptor antagonist phenoxybenzamine (Fig. 2) and chlorpromazine, but not by the β antagonist nethalide. The selective action, and the quick reversibility, after administration of an α receptor antagonist exclude the possibility that the depression is secondary to circulatory changes.

Although many amino acids act on nerve cells (CURTIS and WATKINS 1960), the experimental findings suggest that DOPA does not act directly. Catecholamines are formed from DOPA by decarboxylation. Inhibition of the decarboxylase (by meta-hydroxybenzylhydrazine) prevents the action of DOPA. The catecholamines are at least in part inactivated by monoamine oxidase. Inhibition of this enzyme (by nialamide) gives a tenfold potentiation of the effect of a small dose of DOPA (ANDÉN, JUKES and LUNDBERG 1964). It is therefore suggested that DOPA acts through formation of catecholamines. The available evidence suggests that the catecholamine is NA; A is not formed from DOPA in the spinal

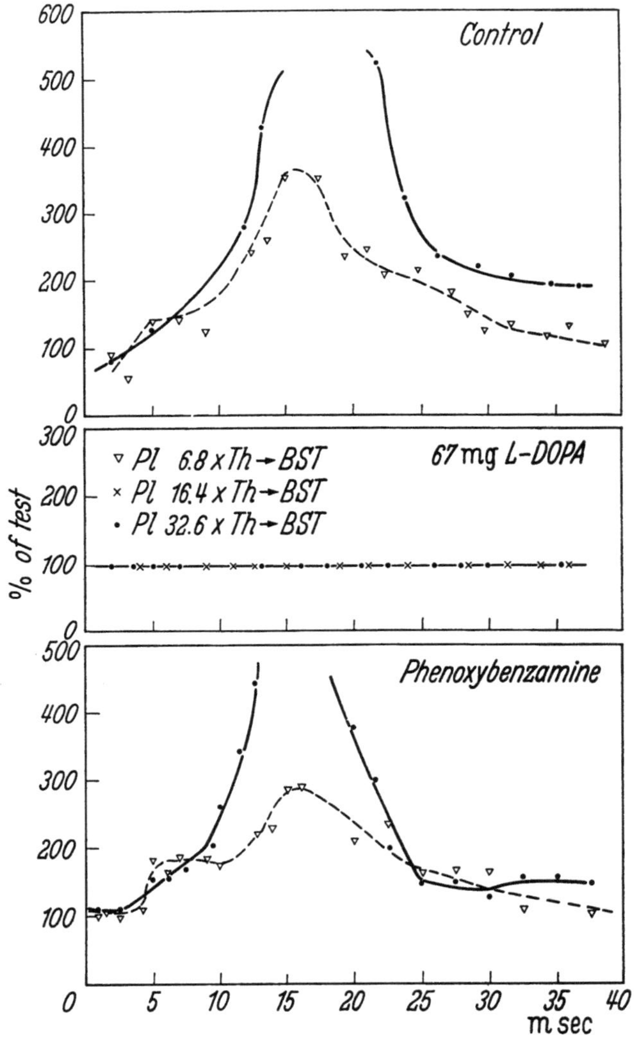

Fig. 2. The effect of DOPA on the synaptic action from high threshold muscle afferents to motoneurones. Monosynaptic test reflexes from the nerve to posterior biceps-semitendinosus (BST) were conditioned by single volleys in the nerve to plantaris (Pl); the conditioning stimulus strengths are given in multiples of threshold strength. 100% on the ordinate represents the unconditioned test reflex, the conditioned response is expressed in % of the unconditioned with time interval on the abscissa. The upper curves were obtained before, and the curve in the middle graph 10 minutes after, the injection of DOPA, 67 mg/kg. The curves in the lower graph were obtained 15 minutes later and 5 minutes after the injection of the α receptor antagonist phenoxybenzamine (20 mg/kg). (From an experiment by ANDÉN, LUNDBERG and VYKLICKÝ.)

cord and it cannot be DA since DOPA is ineffective after reserpine, under which conditions there is normal formation of DA but no formation of NA.

Our hypothesis is that DOPA acts by liberation of NA from the synaptic terminals of a noradrenergic descending pathway with an inhibitory action on transmission from the FRA. Since some segmental actions from the FRA are not depressed (cf. ANDÉN, JUKES, LUNDBERG and VYKLICKÝ 1964), it is likely that this inhibition is exerted at an interneuronal level and not through primary afferent depolarization. Our hypothesis is summarized in the tentative scheme of Fig. 3. If this hypothesis is correct it would be expected that DOPA would exert no action in a chronic spinal cat. However, four weeks after spinal transection there is still some effect, but even at this time some DOPA decarboxylase activity remains in the spinal cord. (ANDÉN, MAGNUSSON and ROSENGREN 1964.)

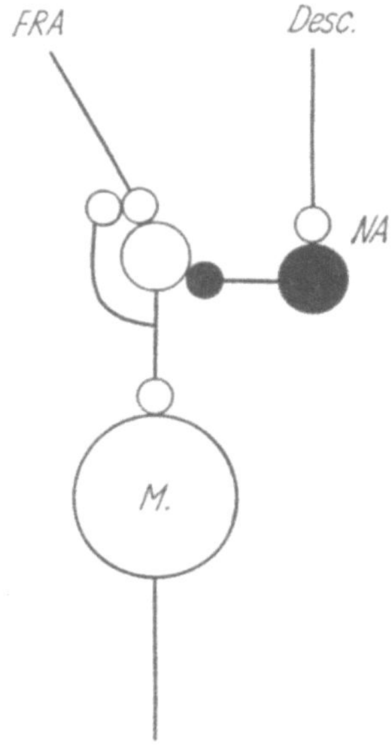

Fig. 3. Tentative diagram of connections from the noradrenergic (NA) descending pathway inhibiting transmission from the FRA. A single interneurone in the diagram may represent a chain. Excitatory interneurones and those evoking primary afferent depolarization are drawn with open circles, inhibitory neurones are drawn with filled circles

The effect of 5-HTP

The intravenous injection of 5-HTP (25—100 mg/kg) also depresses transmission from the FRA to primary afferents, motoneurones and ascending pathways (ANDÉN, JUKES and LUNDBERG 1964). There is full recovery in 3—5 hours. In addition, 5-HTP gives an increased excitability in flexor and extensor motoneurones, there is regularly an increase of the monosynaptic reflexes and often a discharge in the ventral root. The effect of 5-HTP is not prevented by the α receptor blocker, phenoxybenzamine, which reverses the action of DOPA; hence these substances act on different receptors. The 5-HT antagonist BOL (2-bromo *d*-lysergic acid diethylamide) partially reverses the effect of 5-HTP.

After administration of a decarboxylase inhibitor there is no effect by large doses of 5-HTP (200 mg/kg). This indicates that 5-HTP acts entirely through liberation of 5-HT.

The effect of reserpine

Reserpine is well known to release monoamines from the stores in the presynaptic terminals. In itself reserpine has no effect on transmission from the FRA, but when given after administration of the monoamine oxidase inhibitor, nialamide, it evokes a profound inhibition of transmission from the FRA (ANDÉN, JUKES and LUNDBERG 1964). The effect of reserpine may be due both to the release of NA and 5-HT.

Comments

In conjunction with the histochemical experiments of Carlsson et al. (1964), the present findings indicate that there is a noradrenergic, and possibly also a 5-hydroxytryptaminergic, descending pathway inhibiting transmission from the FRA. In addition, there may be a 5-hydroxytryptaminergic pathway giving excitation to α motoneurones. These monoaminergic pathways probably take origin in the lower brain stem (Dahlström and Fuxe 1964) and experiments are planned to find out if any of the known descending systems controlling transmission from the FRA (Eccles and Lundberg 1959; Holmqvist and Lundberg 1959; Lundberg and Vyklický 1963) are monoaminergic. Of particular interest in this connection is the finding that cells in the raphe nuclei contain 5-HT (Dahlström and Fuxe 1964); the centres that maintain the tonic decerebrate inhibition of transmission from the FRA are located in the medial part of the medulla and lower pons (Holmqvist and Lundberg 1959; Engberg and Lundberg, unpublished results). With respect to the evidence for chemical transmission discussed above, there is the curious situation that the cells to which the postulated transmission occurs have not been investigated directly. Their location in the spinal cord is not even known. A more direct approach has been customary in this field (Eccles 1964), but, although experiments with micro-application are highly desirable, there is the difficulty that the cells may be hard to find.

The depression by DOPA and 5-HTP of transmission in the short latency paths from the FRA has revealed new reflex connections from these afferents. (Andén, Jukes, Lundberg and Vyklický 1964; Jankowska, Jukes, Lund and Lundberg 1964.)

References

Andén, N.-E., Carlsson, A., Hillarp, N.-Å. & Magnusson, T. (1964). *Life Sciences* **3**, 473.
Andén, N.-E., Jukes, M. G. M. & Lundberg, A. (1964). *Nature* (Lond.) **202**, 1222.
Andén, N.-E., Jukes, M. G. M., Lundberg, A. & Vyklický, L. (1964). *Nature* (Lond.) **202**, 1344.
Andén, N.-E., Lundberg, A., Rosengren, E. & Vyklický, L. (1963). *Experientia* (Basel) **19**, 654.
Andén, N.-E., Magnusson, T. & Rosengren, E. (1964). *Experientia* (Basel) **20**, 328.
Carlsson, A., Falck, B., Fuxe, K. & Hillarp, N.-Å. (1964). *Acta physiol. scand.* **60**, 112.
Carlsson, A., Magnusson, T. & Rosengren, E. (1963). *Experientia* (Basel) **19**, 359.
Curtis, D. R., Phillis, J. W. & Watkins, J. C. (1961). *J. Physiol.* (Lond.) **158**, 296.
Curtis, D. R. & Watkins, J. C. (1960). *J. Neurochem.* **6**, 117.
Dahlström, A. & Fuxe, K. (1964). *Acta physiol. scand.* **62**, Suppl. 232, 5.
Eccles, J. C. (1964). *The Physiology of Synapses.* Berlin: Springer-Verlag.
Eccles, R. M. & Lundberg, A. (1959). *J. Physiol.* (Lond.) **147**, 565.
Falck, B. (1962). *Acta physiol. scand.* **56**, Suppl. 197, 1.
Falck, B., Hillarp, N.-Å., Thieme, G. & Torp, A. (1962). *J. Histochem. Cytochem.* **10**, 348.
Holmqvist, B. & Lundberg, A. (1959). *Arch. ital. Biol.* **97**, 340.
Jankowska, E., Jukes, M. G. M., Lund, S. & Lundberg, A. *Nature* (Lond.) (in press)
Lundberg, A. & Vyklický, L. (1963). *Acta physiol. scand.* **59**, Suppl. 213, 91.
Magnusson, T. & Rosengren, E. (1963). *Experientia* (Basel) **19**, 229.
McLennan, H. (1963). *Synaptic Transmission.* Philadelphia: W. B. Saunders Co.

WATER METABOLISM OF DESERT RUMINANTS

by W. V. MACFARLANE

There is a logarithmically linear relationship between water turnover and body weight amongst mammals (ADOLPH 1949; RICHMOND, LANGHAM and TRUJILLO 1962). This generalisation requires some modification since there are exceptions to the body weight rule. The kangaroo rat for instance turns over less water than rats of the same size (RICHMOND, TRUJILLO and MARTIN 1960) while it is clear that cattle *(B. indicus* or *B. taurus)* turn over two or three times more water than other ruminants of comparable weight. These appear to be inherent physiological differences between major groups of mammals.

Another variable which enters into patterns of water usage is the ambient temperature or any other form of heat load. Some of the input of heat is dissipated by evaporation of the secretion of the nasal or sweat glands so that water turnover becomes an index of evaporative cooling. Life in dry hot regions appears to increase the total body water content and the volume of extracellular fluids, as well as the rate of water turnover in ruminants. This aspect of water metabolism in relation to desert adaptation has been explored. Water distribution in the bodies of ruminants has been measured in central Australia and in Kenya (MACFARLANE 1964) and the water turnover was estimated from the dilution of tritium by free water ingested, food water and metabolic water.

Measurements were made under field conditions, with minimal disturbances of feeding habits or of the amount of exposure to heated air and radiation.

Methods

The animals — comprising camels *(C. dromedarius)*, cattle *(B. taurus, B. indicus)*, sheep *(Ovis aries* — Karakul, blackhead Persian [Somali], Dorper, Merino), goats (*Capra hircus*) and kangaroos (*Macropus rufus*) — were numbered and weighed. A polythene tube with obturator was tied into the jugular vein of all animals to allow accurate introduction of markers and withdrawal of blood at timed intervals.

Thiocyanate spaces were determined in the morning, and tritium equilibration with body water began at that time. A dose of 15 ml/kg sodium thiocyanate and 0.4 mg/kg T 1824 was injected through the jugular cannula. This was washed in with 2 mC of tritium oxide (TOH) for cattle, 3 mC for camels and 400 μC for sheep and goats.

The T 1824 space was determined from a sample taken 5 min after the injection of the marker solution, and the thiocyanate space 20 min. These points were shown in sheep (MACFARLANE, MORRIS, HOWARD and BUDTZ-OLSEN 1959) to give estimates of extracellular distribution as satisfactory as extrapolation to zero time. The 20 min sample appears to give a comparable estimate of the thiocyanate distribution in cattle and camels.

Tritium distribution was estimated at 6 hr from the time of injection in the case of sheep, and the relatively small Boran cattle and goats. By that time equilibration was complete. In camels, equilibration was slower so no water was given the animals overnight and further blood was taken after 24 hr. The tritium concentration reached equilibrium between 8 and 16 hr after injection. Amongst *B. taurus* cattle there was equilibration at about 8 hr. Water distribution in the camel appears to be slower than in cattle. The delay in equilibration is greater in dehydrated animals, and probably derives from the rate of secretion of salivary and pancreatic fluids, as well as from the rate of turnover between alimentary tract and blood stream.

Tritium concentration was determined by scintillometry. Samples of blood or urine were sublimed to give water free of contaminants and were taken to dryness to eliminate the possibilitiy of separation of TOH and HOH (VAUGHAN and BOLING 1961). Two ml of sublimed water were added to 14 ml of scintillation fluid, made up of naphthalene 100 g, 1-4-bis-2-(4-methyl-5-phenyl-oxazolyl) benzene 250 mg, 2-5-diphenyloxazole 10 g, and dioxane to 1 l. Samples were counted in a Packard Tri-Carb spectrometer.

The thiocyanate was determined by the method of BOWLER (1944).

In Australia, Merino sheep of the Peppin strain were studied during winter on latitude 35 °S at Canberra (a cool temperate climate) as well as in the tropics. At Cunnamulla (28 °S) the dry subtropical conditions were interrupted in late summer by heavy rains so that the sparse *Astrebla* pastures were replaced for a short time by succulent herbage. Measurements were made there in both dry and wet summer periods. Within the tropics (21 °S) at Julia Creek in N. W. Queensland there is no winter rain, the summer is long and hot and there is a short monsoonal period yielding an average annual rainfall of 16 inches in 2 or 3 months. Peppin Merinos adapted to the dry tropics were investigated at Julia Creek before the rains. Maximum summer temperatures ranged from 30 ° to 45 °C.

Cattle, camels and kangaroos were maintained on dry country south of Alice Springs (24 °S) in the Northern Territory. Although there was little grazing, some scrub and trees *(Acacia, Casuarina,* and *Hakea)* were used as food by the cattle and camels. Kangaroos were maintained in a partly shaded enclosure and were fed lucerne hay. All animals had access to water although the cattle and camels had to walk more than a mile to reach it. Measurements were made during late summer when maximum air temperatures ranged from 38 to 45 °C dry bulb and the humidities were between 20 and 30 per cent during the hot part of the day. There was virtually no cloud protection from the sun but there were periodic dust storms.

The water turnovers of three breeds of sheep were compared at Katumani in Kenya, on the equator at 4000 feet elevation. The sheep grazed green *Cynodon* pasture. Air temperatures were low (up to 28 °C) but the radiation loads were high. At a higher altitude identical twin cattle were compared, as they grazed temperate pastures, at 7500 feet, on the equator.

In northern Kenya near Wajir (2 °N., elevation 600 feet) livestock of Somali nomads were measured for water metabolism. In the *Acacia* scrub of the semi-desert herdsmen tended camels, Boran cattle, fat-tail Somali (so called Persian) sheep and goats, which cropped *Acacia* and grazed the few legumes growing in the red sand. Maximum air temperatures ranged from 33 to 38 °C but humidity during the daylight hours did not rise above 30 per cent and there was little cloud. The observations were made at the beginning of the dry season in the sun and there was no shortage of food since the previous year had had an unusually high rainfall of 28 inches which had produced unprecedented amounts of vegetation. The usual range of precipitation is 2 to 8 inches annually.

The animals were watered by the herdsmen who hauled water from wells 30 feet deep. Water was offered every two or three days. At night the animals were enclosed in a thornbush boma to protect them from lion or human predators.

Results

Water distribution

The *total water content* of the same Peppin strain of Merino sheep was greater in winter than in summer in any one environment (MACFARLANE et al. 1959) and it was greater in the tropics than in temperate regions (Fig. 1). The mean

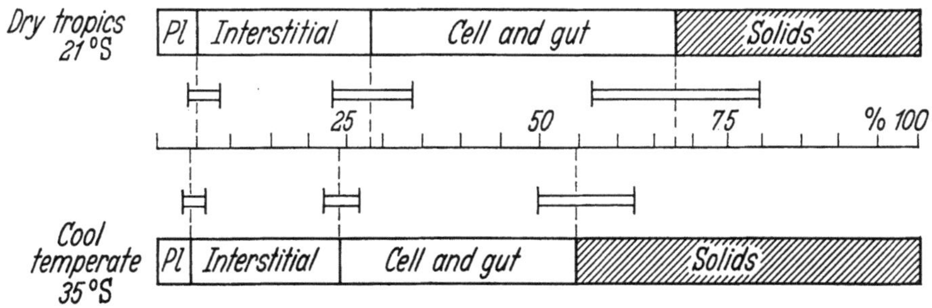

Fig. 1. Body fluid distribution in adult Peppin Merino sheep, measured during summer in a cool temperate climate and in the dry tropics. The range and mean of plasma, extracellular and total body water volumes are shown. Each space was larger in the tropical than in the temperate zone sheep

body water of summer Merinos in the temperate regions was 55 per cent, compared with 68 per cent of body weight for tropical sheep of comparable strain and age. A period of feeding on freshly grown and succulent herbage following rain in summer, increased the content of water in all body spaces of Merino sheep, as well as the turnover of water (Table 1).

Camels and cattle in hot arid regions also had high water contents ranging from 66 to 85 per cent of body weight. *B. indicus* cattle in the cool highlands of Kenya with body water of 68.0 per cent of body weight carried less water than similar animals on the hot plains of the Northern Frontier District (NFD) of Kenya, which averaged 76.9 per cent water.

Table 1

	Latitude	Mean max T°C	Thiocyanate space per cent body weight	Body Water per cent body weight	Water turnover	
					ml/kg/24hr	ml/kg$^{0.82}$/24hr
Merinos, Australia						
Canberra	35°	9	24.2	56.2	63	127
Cunnamulla (dry)	28°	40	23.0	56.8	93	186
(wet)	28°	34	26.7	62.8	130	255
Julia Creek	21°	42	28.9	77.1	135	255
Kenya sheep						
Wajir (dry)	3°	36				
Blackhead Somali			29.2	67.5	107	197
Katumani (warm)	1°	27				
Dorper			22.8	67.9	90	177
Merino			23.8	66.5	95	182
Karakul			24.8	78.6	114	211

The average values for groups of 6 to 12 adult sheep were obtained in cool, hot and dry environments. Extracellular and total body water increased in parallel with the water turnover.

Plasma volume and thiocyanate spaces were greater in the tropics and desert than in cool regions. Fig. 1 illustrates this for sheep during one summer, but in sheep at Julia Creek (21 °C) during five other seasons, a wider spectrum of extracellular spaces was found, reaching 52 per cent of body weight without oedema. In the NFD the thiocyanate space for sheep averaged 29.2 per cent, for Boran cattle 28.8 per cent and for goats 25.5 per cent of body weight. These spaces are one third higher than those for comparable animals in cool areas. This fluid space in Borans of the Kenya Highlands averaged 14.2 per cent of body weight, for instance, and in Dorpers 22.8 per cent.

The extracellular and total body water of identical twin cattle grazing together agreed within 4 to 7 per cent and the turnover (ml/kg$^{0.82}$/24 hr) to within 1.4 per cent.

When there was a high water content in the food at Cunnamulla with consequent high turnover of water (and maximum air temperatures below 35 °C), all fluid spaces increased, relative to those of the same sheep in a dry hot season (Table 1). A doubling in water turnover brought about by shearing in summer also increased the extracellular fluid volume by 13 per cent.

Water turnover

In Table 2 the water turnover of several genera of ruminants living in hot arid country are compared. The daily turnover per unit weight was least in camels and greatest in *B. taurus* cattle. When converted to water turnover per kg $^{0.82}$, non-lactating camels, goats and sheep behaved similarly. The values ranged from 174 to 197 ml/kg$^{0.82}$/24 hr and the regression of water turnover on body weight was linear (Fig. 2).

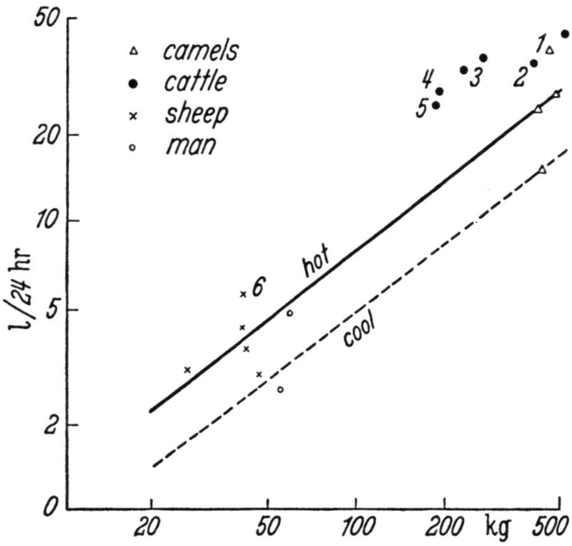

Fig. 2. The daily water turnover of most ruminants is related in a linear logarithmic fashion to the body weight. In a cool environment (when evaporative fluid loss is minimal) the regression line lies parallel with but below that for groups of sheep and camels in hot dry regions (maximum daily temperature above 40° C). — There are exceptions to this linear relationship, however, due to both innate and adaptive functions. The points above the line for hot climates are (1) lactating camels (2) *B. indicus* (Boran) cattle in the cool temperate climate of the Kenya highlands (3) *B. taurus*, also in Kenya highlands (4) *B. taurus* (Shorthorn) in semi-desert (5) *B. indicus* (Boran) in semi-desert (6) Merinos, in the sub-tropics, feeding on succulent new-grown pasture

There was a major deviation, however, by both species of cattle from the exponential weight index of turnover. The bovidae turned over nearly twice as much water as camels or sheep living in the same environment (Macfarlane, Morris and Howard 1963). *B. taurus* identical twins in the Highlands of Kenya turned over 364 and 359 ml/kg$^{0.82}$/24 hr, Boran twins 261 and 261 ml/kg$^{0.82}$ per 24 hr. In the NFD desert, however, Borans subjected to the hot environment showed a turnover of 336 to 360 ml/kg$^{0.82}$/24 hr while Shorthorn *B. taurus* in Australian hot arid conditions ranged from 316 to 416 ml/kg$^{0.82}$/24 hr, compared with 184 ml/kg$^{0.82}$/24 hr for camels.

In the Kenya highlands at Katumani, it was possible to compare three breeds of sheep living on warm well-watered country. The mean water turnover of Dorper sheep (Table 1) was less than that of Merinos and 27 per cent less than that of Karakul. The skin and body temperatures of these black sheep were raised in the sun, and respiratory rates were faster than those of Merinos or Dorpers.

Merino sheep in Australia increased water turnover proportionally to heat load. The dry tropical flock used more than the subtropical, and that again, than the temperate climate groups of sheep. At Cunnamulla a selected group producing 12 per cent more wool than controls, turned over 13 per cent more water during the dry season.

On the desert, camels, goats and sheep were exposed to the sun. Kangaroos, however, kept to the shade, as is their custom, and they fed on lucerne chaff

Table 2

	Water content percent body weight	Water turnover	
		ml/kg/24hr	ml/kg$^{0.82}$/24hr
Camels			
Alice Springs (2 lactating)	72.5	92	278
Alice Springs (2 dry)	67.3	61	184
Wajir (4 dry)	—	61	188
Cattle			
Alice Springs (4 Shorthorn steers)	75.6	148	382
Wajir (4 Boran bulls)	76.9	135	346
Sheep			
Cunnamulla (9 Merinos, control)	57.4	88	174
Cunnamulla (10 Merinos, high yield)	56.3	98	197
Wajir (12 Blackhead Persian)	67.5	107	197
Goats			
Wajir (4 Somali)	69.0	95	185

Water content and turnover of ruminants living in hot, dry areas. The Australian observations were made during summer, and the Kenya observations, after rains, in the equatorial semi-desert of the Northern Frontier District.

rather than on native grasses. The mean turnover of 88 ml/kg/24 hr for red kangaroos in the heat of summer indicates that they fall into the same water metabolism category as sheep and are probably not exceptional.

Discussion

The water content of ruminants has one component additional to those found in monogastric animals. The rumen contents are 85 per cent water, or more, and up to 18 per cent of the body weight may comprise water in the alimentary tract (HUNGATE, PHILLIPS, MCGREGOR, HUNGATE and BEUCHNER 1959). This allows a greater total water content of the body to develop than is found in healthy monogastric animals, though this makes the *in vivo* determination of the proportion of body fat difficult.

Camels studied have shown a total body water range from 67 to 72 per cent of body weight, but Shorthorn cattle in the desert reached 85 per cent. Desert Borans averaged 77 per cent (MACFARLANE 1964).

Merino sheep in good condition reached 78 per cent water but in the same flock there was a range of 20 per cent in the water component of body weight.

Assuming that water makes up a constant 67 per cent of cells, the high values of total body water must represent increases of gut fluids and of extracellular fluid. When the latter was measured it was found to account for 5 to 8 per cent of body weight gain, less than half of the greater water content of tropical ruminants, relative to temperate area types. An increment in gut water by more than 5 per cent of body weight is therefore likely in these animals. The activity of salivary glands may be important in adjusting rumen volume, and maintaining higher levels of alimentary water.

The finding of a high body water in desert animals automatically reduces the possible fat content. Well nourished desert Boran, Somali sheep and camels carry specialised fat depots, but they are poorly provided with fat elsewhere. This would reduce insulation, while allowing space for extracellular fluid.

Increases of the extracellular fluid volume in sheep can follow exposure to heat, cold and starvation (MORRIS, HOWARD and MACFARLANE 1962). Apparently healthy animals in the tropics or desert during summer show an expansion of the extracellular spaces. It seems likely that a high rate of water turnover during evaporative cooling leads to a greater renal retention of water and electrolytes than is found in the steady state of unheated animals. Although starvation increases the extracellular fluid this was not evident in the ruminant groups studied. The mechanism by which the regulation of extracellular volume is reset to a higher value in the desert is not clear. The ruminant kidney has anomalous responses to vasopressin and aldosterone (KINNE, MACFARLANE and BUDTZ-OLSEN 1961) and during exposure to heat, sodium is often excreted rather than — as in man — retained. This would make expansion of the extracellular space difficult.

Water turnover is high in several circumstances. Lactation, sweating and respiratory cooling all increase the demand for water (Table 2). Pasture with more than 80 per cent of water in the plants, leads the animals to consume large amounts of water in order to obtain a satisfying supply of energy from carbon and nitrogen compounds. In the desert the increased turnover of water followed very much the same exponential relation to weight as that found in cool climates. ADOLPH (1949) suggested that weight $^{0.88}$ was the likely function for free water intake, while RICHMOND et al. (1962) found the exponent to be 0.80. Our observations led to the 0.82 power of the body weight as the relationship of water turnover to body weight for mammals in hot areas.

There is a clear exception to this amongst the bovidae. Cattle appear, on both arid and well-watered pasture, to pass more water in urine and faeces than sheep or camels. This is so even when cattle are deprived of water in summer (unpublished observation).

Within species and genera there are functional groups. *B. indicus* types of cattle can reduce water loss from the intestine more than *B. taurus* (QUARTERMAN, PHILLIPS and LAMPKIN 1957). Zebu cattle correspondingly drink less than European breeds in the same environment (HORROCKS and PHILLIPS 1961;

HUTCHINSON 1961). Boran cattle were found to turn over less water on cool pasture than European types, but Borans in the desert used 52 per cent more than Borans in the temperate highlands. The physiological demand for water by the Boran is high though less than that of *B. taurus* breeds.

Amongst sheep a similar difference of water metabolism was found between the Dorper (a cross between Dorset Horn and Blackhead Somali, which is well adapted to a hot arid land) and the Karakul. The black wool of Karakuls absorbs more radiant energy than Dorpers, but this seems insufficient to account for the 27 per cent difference between them in water turnover.

The inherent water metabolism in some genera of mammals differs from other genera (e. g. *Rattus* and *Dipodomys, Bos* and *Ovis).* There are also genetically determined differences between breeds within a species, and selected groups within breeds (e. g. selected and unselected Merinos, where the high yielding groups used water more than controls).

The genetic or inherent control of water turnover is illustrated by the close similarity of turnover by identical twins. This function was more closely regulated in the twin pairs than body weight.

Conclusion

There is evidence from these observations that the rate of water turnover has intrinsic controls, which differ amongst genera. At the individual level, identical twins have closely similar water distribution and turnover patterns, so that there is a genetic background to these functions. Within a species, physiological strains use water at different rates. Different species of cattle and breeds of sheep living in the same environment have consistent differences in water use patterns.

Besides the intrinsic regulation of water, however, there is extrinsic modification of water metabolism. The nature of the foodstuff taken and its water content affects water turnover. An increase of heat load increases water spaces and water consumption in all ruminant types.

The close correspondence of plasma, interstitial and total body water volume with water turnover suggests that volume regulation is in part determined by the rate of passage of water through the animals.

Acknowledgements

Colleagues in many places have helped with field measurements, and their organisation or physical efforts were essential components of the investigations. —
Dr. G. MOULE and Mr. A. BELL at Julia Creek; Mr. C. H. S. DOLLING at Cunnamulla; Dr. B. HOWARD and Mr. R. J. H. MORRIS in Canberra; Mr. J. WHITTEN and Mr. B. D. SIEBERT at Alice Springs; Dr. E. RUSSELL, Dr. W. J. A. PAYNE,

Mr. J. MacDonald, Mr. J. Gold, Nelson Mwangi, Mohamed Ogle and their assistants in Kenya.

The support of the Wool Trust Fund and the Rural Credits Fund of the Commonwealth Bank, Australia, is gratefully acknowledged.

References

Adolph, E. F. (1949). *Science* **109**, 579.
Bowler, R. B. (1944). *J. Biochem.* **38**, 385.
Horrocks, D. & Phillips, G. D. (1961). *J. Agric. Sci.* **56**, 379.
Hungate, R. E., Phillips, G. D., McGregor, A., Hungate, D. P. & Beuchner, H. K. (1959). *Science* **130**, 1192.
Hutchinson, H. J. (1961). Republic of Tanganyika, Ministry of Agriculture Annual Report, p. 51.
Kinne, R., Macfarlane, W. V. & Budtz-Olsen, O. E. (1961). *Nature* (Lond.) **192**, 1084.
Macfarlane, W. V. (1964). In *Handbook of Physiology*, vol. **4**, *Adaptation to Environment*, p. 509. Washington: Amer. Physiol. Soc.
Macfarlane, W. V., Morris, R. J. H. & Howard, B. (1963). *Nature* (Lond.) **199**, 270.
Macfarlane, W. V., Morris, R. J. H., Howard, B. & Budtz-Olsen, O. E. (1959). *Aust. J. Agric. Res.* **10**, 269.
Morris, R. J. H., Howard, B. & Macfarlane, W. V. (1962). *Aust. J. Agric. Res.* **13**, 320.
Quarterman, J., Phillips, G. D. & Lampkin, G. H. (1957). *Nature* (Lond.) **180**, 552.
Richmond, C. R., Langham, W. H. & Trujillo. T. T. (1962). *J. Cell. Comp. Physiol.* **59**, 45.
Richmond, C. R., Trujillo, T. T. & Martin, D. W. (1960). *Proc. Soc. exp. Biol. Med.* **104**, 9.
Vaughan, B. E. & Boling, E. A. (1961). *J. Clin. Lab. Med.* **57**, 159.

SOME APPLICATIONS OF INPUT-OUTPUT TECHNIQUE

by A. K. McIntyre

Although in the broadest sense almost any neurophysiological investigation could be regarded as an exercise in input-output relations, studies undertaken with the specific object of measuring the extent and nature of postsynaptic action at various levels of input have yielded results of particular value. Especially noteworthy amongst such investigations have been those upon the stretch-reflex arc; input-output studies of other synaptic systems have also provided interesting information. It is proposed to survey briefly the contributions which some of these studies have made to our understanding of synaptic and reflex mechanisms.

The stretch-reflex arc and its afferent limb

The delicate grading of the extent of myotatic reflex response to differing degrees of muscle-stretch was first observed by Liddell and Sherrington (1924); but detailed input-output analysis was not carried out until progress in electrical recording methods made it possible to recognize the discharge of stretch-reflex impulses in the axons of motoneurones. The first quantitative input-output study of this system was carried out by Lloyd (1943 a) in his examination of the growth of ventral root reflex discharge in response to graded and monitored afferent volleys in a corresponding dorsal root. The early and apparently monosynaptic segmental ventral root reflex discharge, first recorded by Eccles and Pritchard (1937) and confirmed as monosynaptic by Renshaw (1940), was later identified by Lloyd (1943 b, 1943 c) as a manifestation of activity in the stretch-reflex arc. Lloyd's input-output curves for segmental monosynaptic reflex discharge, and similar ones for monosynaptic facilitatory action made it clear, amongst other things, that the afferent limb for this reflex consists of large fibres of low threshold (designated Group I) which include the very largest in the dorsal roots, and that impulses in fibres of higher threshold (Groups II and III) exert no action in monosynaptic tempo upon motoneurones. Because of the key position which the stretch-reflex arc has continued to hold in the analysis of synaptic and reflex function, these and other related pioneer studies involving it have been of particular value in building a firm foundation upon which our present and greatly expanded knowledge rests. Developments such as the use of intracellular microelectrodes with which such prodigious advances have been made, notably by Sir John Eccles himself, owe a great deal to these careful and revealing earlier investigations.

The plateau in Lloyd's input-output curves for the segmental monosynaptic reflex response, though demonstrating monosynaptic ineffectiveness of impulses in dorsal root fibres of higher threshold than those of the Group I band, did not exclude the presence within the latter of fibres concerned with actions other than the stretch-reflex. Such duality of Group I function was in fact demonstrated later by Laporte and Lloyd (1952), who disclosed the presence in muscle nerves of Group I fibres lacking monosynaptic action upon motoneurones, but responsible for other reflex effects (notably the lengthening reaction) through pathways including interneurones. Correspondingly, the Group I band was shown by Hunt and Kuffler (1951) to contain fibres from muscle-spindle (IA) and tendon-organ (IB) receptors. Little difference in threshold to electrical stimulation of the two types of Group I fibre would be expected from the subsequent careful analysis of single muscle-afferent units by Hunt (1954); Laporte and Lloyd (1952) had in fact noted that thresholds of Group I fibres yielding the two different reflex patterns were closely similar. In consonance with this, Hunt showed that excitatory action in the monosynaptic reflex arc of *triceps surae* continued to grow until the recorded Group I input volley reached its maximum (Hunt 1955). Thus the Group IA fibres, generally conceded to be those executant for the stretch reflex (Hunt and Perl 1960), appear from these results to be distributed throughout the Group I range; an essentially similar distribution would be expected for IB tendon-organ fibres of autogenetic inhibitory function.

However, apparently conflicting results have been obtained in other investigations. Input-output curves published by RALL (1955) for *triceps surae* motoneurones suggest that a considerable fraction of the higher threshold Group I fibres (presumably of tendon-organ origin) are without monosynaptic excitatory action; and BRADLEY and ECCLES (1953), and ECCLES, ECCLES and LUNDBERG (1957) have described a splitting of Group I volleys into fast and slow components, with corresponding threshold differences. They concluded from the patterns of response impressed upon motoneurones by graded Group I input volleys that the fast component consists largely of impulses in fibres serving muscle-spindles, and the slow component chiefly impulses in tendon-organ (IB) fibres. Group I volleys of double contour are indeed commonly observed upon stimulating nerves supplying muscles of the thigh; and in this situation LAPORTE and BESSOU (1957), by analysis of receptor units identified as to type, provided partial confirmation of BRADLEY and ECCLES' interpretation. They found in animals showing such splitting of the Group I volley that the fast component was largely composed of spindle impulses with a few from tendon-organs; while the slower deflexion was a mixture with impulses from tendon-organs predominating, but including many from muscle-spindles as well. ECCLES et al. (1957) also described splitting in some preparations of Group I volleys from muscle-nerves of the leg such as *triceps surae*, and their input-output measurements upon single impaled motoneurones indicated, in agreement with RALL (1955), that monosynaptic excitatory action reaches a ceiling with afferent volleys considerably less than maximal for Group I fibres.

The lack of agreement between these results and those of HUNT (1955) may be more apparent than real, since microelectrode technique may not provide representative sampling of a motoneurone pool, and variations in the spread of I A afferent input must be expected for individual neurones of the pool. Furthermore, the use of motoneurone response-pattern as the criterion of engagement in the input volley of tendon-organ afferent fibres might be misleading because of the possible need for summation in the di- or trisynaptic pathway through which such fibres exert their actions. Be that as it may, however, the crucial importance to a number of spinal reflex investigations of knowing how the two components of the Group I afferent fibre population are recruited with graded increases in strength of stimulation suggested the need for further study.

SUMNER (1961) carried out an analysis of Group I fibres of *triceps surae* in much the same manner as HUNT (1954), but measuring threshold within the muscle nerve of each afferent fibre isolated, in addition to determining its conduction velocity and defining its receptor of origin. By relating the threshold of each unit to the threshold and Group I maximal strength of stimulation for the whole nerve in each preparation, the pooled results could be plotted. In Fig. 1, the size of Group I afferent volley with increasing strength of stimulation is represented on the abscissa as percentage of maximum; the ordinate shows amount of Group IA input (open circles) and IB (filled circles) calculated from the unit thresholds and also expressed as percentage of maximum. If IA and IB thresholds were equally distributed, obviously all the points would lie along the broken line. However, it is apparent that there is in fact a considerable difference in threshold, though conduction velocity measurements, which closely matched those of Hunt (1954), did not suggest the likelihood of such a distinct

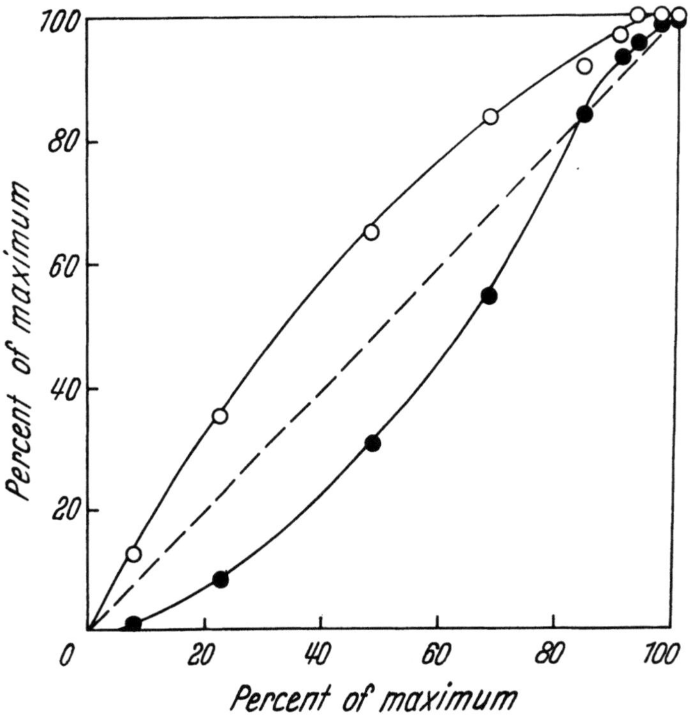

Fig. 1. Curves showing proportion of IA spindle and IB tendon organ fibres in Group I volleys of graded amplitude, constructed from threshold measurements of 128 IA and 88 IB units in nerve of cat *gastrocnemius medialis*. *Abscissa:* Group I volley-size in percentage of maximum. *Ordinate:* number of IA (open circles) and IB (filled circles) recruited, also in percentage of maximum. [From Sumner, A. J. (1961), Proc. Univ. Otago Med. Sch. 39. 3.]

threshold differentiation. IA fibres in this sample have thresholds distributed almost throughout the Group I range; volleys greater than about 8% of Group I maximum would engage IB as well as IA fibres, though the latter would predominate. Obviously it would be very difficult to obtain by electrical stimulation any clear segregation of the two fibre types, except with stimuli only slightly above threshold, as found for reflex differentiation by Laporte and Lloyd (1952). It is likely, of course, that sharper threshold differentiation can be achieved in certain individual preparations; and conversely, in some of Sumner's animals there was considerably less threshold difference than that shown in the pooled results of Fig. 1.

Some other applications of input-output technique

Many other examples may be quoted of the use of this technique as a means of identifying those afferent fibres in a mixed input channel which are responsible for a particular action. For instance, by this means it was shown that Group I fibres are not concerned in bringing about forelimb-hindlimb interactions in the long spinal reflex system, and that different afferent fibre-populations are responsible for facilitatory and inhibitory actions within this system (LLOYD and MCINTYRE 1948). Similarly, the late and apparently polysynaptic discharges of chromatolytic motoneurones in response to muscle afferent volleys were shown to be the result of Group I impulses operating through disordered monosynaptic links (DOWNMAN, ECCLES and MCINTYRE 1953; ECCLES, LIBET and YOUNG 1958; MCINTYRE, BRADLEY and BROCK 1959). The still unexplained participation in the flexor reflex of impulses in myelinated afferent fibres distributed almost throughout the range of those present in cutaneous nerves was also revealed by input-output experiments (BERNHARD and WIDÉN 1953; LLOYD 1957; MARK and STEINER 1958).

In addition to helping in the identification of afferent fibres concerned in a given action, input-output studies can also provide information of relevance to the mechanisms operating at synaptic junctions interposed between input and output neurones. Comparison of input-output curves of monosynaptic facilitation and discharge of motoneurones shows very clearly the need for summation in the stretch-reflex arc (LLOYD 1943 a). Input-output curves for synapses in afferent pathways stand in contrast: the smallest measurable input usually secures some postsynaptic discharge, as in the linkage between Group I afferent fibres and the cells of CLARKE's column (LLOYD and MCINTYRE 1950). A somewhat similar relationship has been found for the synapses of the lateral geniculate body (BISHOP and MCLEOD 1954). The experiments of MARK and STEINER (1958) on the pathways for cutaneous afferent fibres to cerebral cortex provide a good example of input-output relations for an afferent system. Their experiments demonstrated that impulses in the very largest fibres of lowest threshold in skin nerves have access to cortical somatosensory receiving areas I and II of the cerebral cortex, both by the dorsal column-lemniscal system and by the pathway described by MORIN (1955), and that the synaptic linkages on these pathways must be extremely powerful. Fig. 2 gives an example of an input-output curve for this afferent projection system. The most striking features of such curves are their steep rise and tendency to form a plateau at input values exceeding 20 to 50 percent of maximum. Commonly, the curves rise even more steeply than the one illustrated in Fig. 2, though in the latter, cortical response attains nearly 90 percent of its maximum with input but 50 percent of maximum. The form of these curves reveals the existence of amplification systems in the synaptic machinery of these pathways, an arrangement which appears to favour the transfer of minimal signal-input from receptors of the skin to the cortex. This contrasts with the strong inhibitory actions of massive input volleys in the somatosensory pathways to cerebral cortex (Marshall 1941; Eccles 1964). Mark and Steiner (1958) demonstrated that the plateau usually seen in their curves is not because the fibres of higher threshold fail to make connexions with second-order neurones of the pathway as in the case of the dorsal-to-ventral root monosynaptic reflex, but is a consequence of occlusive and inhibitory pro-

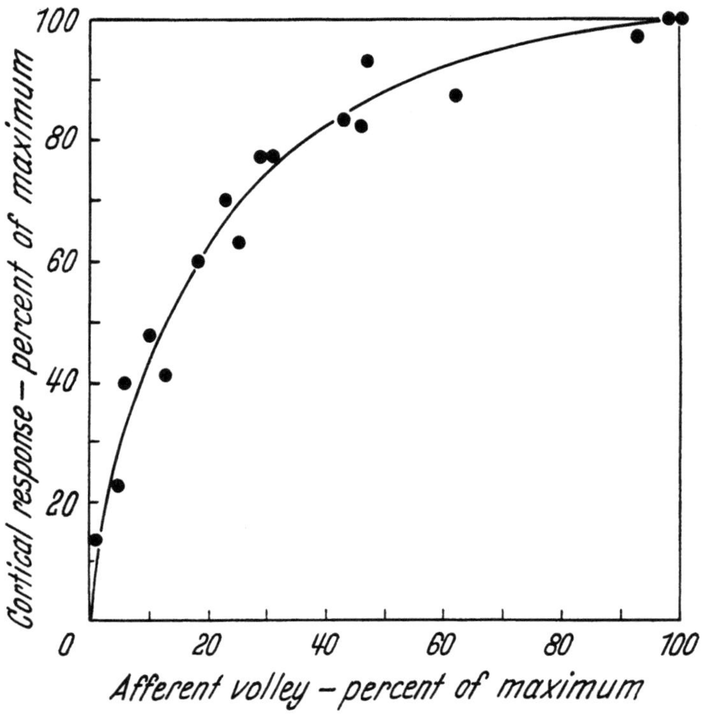

Fig. 2. Input-output relation for volleys in superficial radial nerve and evoked response in contralateral somatic area II of cerebral cortex (cat under pentobarbitone anaesthesia). *Abscissa:* size of monitored superficial radial nerve volley in percentage of maximum (A $\alpha\,\beta$ fibres). *Ordinate:* amplitude of initial positive wave of evoked cortical response, also in percentage of maximum

cesses brought about by prior action of impulses in faster fibres of lower threshold. It looks as if these great input channels to the cerebral cortex are equipped with an automatic volume control mechanism.

Central actions of minimal input volleys

The apparent existence of an amplifying mechanism for small input volleys in afferent synaptic systems poses the question: what is the minimum input necessary to secure a detectable output response? Because of the very extensive overlap of receptor terminals in the skin, even the smallest natural stimulation will set up impulses in many afferent fibres (HUNT and McINTYRE 1960); and electrical stimulation of a nerve trunk even at threshold must usually excite a number of fibres. However, the large size, discreteness and extreme mechanical sensitivity of Pacinian corpuscles together make it possible to initiate afferent volleys restricted to one or a few impulses. A situation convenient for the exposure and stimulation of Pacinian corpuscles is in the region of the crural interosseous membrane where many such corpuscles are supplied by the interosseous nerve (HUNT 1961); their impulses have ready access to the cerebral cortex by the dorsal column-lemniscal system (McINTYRE 1962). Individual impulses can be set up at will by applying very weak mechanical pulses to an exposed corpuscle and monitoring them in the nearby nerve. In this way, it has recently been found that the minimal input — a single impulse — from this type of receptor is capable of eliciting a detectable response in the cerebral cortex (HOLMAN, McINTYRE and VEALE, unpublished).

In the case of the stretch-reflex arc, with which this survey began, it is exciting to note that at long last the refinement of determining the minimal input-to-output relationship has been achieved. KUNO (1964) has recently succeeded in recording from single impaled motoneurones the monosynaptic excitatory action of single impulses in appropriately selected I A afferent fibres. He has not only demonstrated quantal transmitter-release at mammalian monosynaptic junctions, but has been able to estimate the value of "m" — the mean number of quantal units released upon a motoneurone by a single impulse in one stretch-reflex afferent fibre, a value which turns out to be one. A fitting climax to the investigations of monosynaptic reflex action heralded by ECCLES and PRITCHARD's observation of 27 years ago!

References

BERNHARD, C. G. & WIDÉN, L. (1953). *Acta physiol. scand.* **29**, suppl. 106, 42.
BISHOP, P. O. & McLEOD, J. G. (1954). *J. Neurophysiol.* **17**, 387.
BRADLEY, K. & ECCLES, J. C. (1953). *J. Physiol.* (Lond.) **122**, 462.
DOWNMAN, C. B. B., ECCLES, J. C. & McINTYRE, A. K. (1953). *J. comp. Neurol.* **98**, 9.
ECCLES, J. C. (1964). *The physiology of synapses.* 1st ed. p. 233. Berlin: Springer-Verlag.
ECCLES, J. C. & PRITCHARD, J. J. (1937). *J. Physiol.* (Lond.) **89**, 43 P.
ECCLES, J. C., ECCLES, R. M. & LUNDBERG, A. (1957). *J. Physiol.* (Lond.) **136**, 527.
ECCLES, J. C., LIBET, B. & YOUNG, R. R. (1958). *J. Physiol.* (Lond.) **143**, 11.
HUNT, C. C. (1954). *J. gen. Physiol.* **38**, 117
HUNT, C. C. (1955). *J. gen. Physiol.* **38**, 813.
HUNT, C. C. (1961). *J. Physiol.* (Lond.) **155**, 175.
HUNT, C. C. & KUFFLER, S. W. (1951). *J. Physiol.* (Lond.) **113**, 298.
HUNT, C. C. & McINTYRE, A. K. (1960). *J. Physiol.* (Lond.) **153**, 99.
HUNT, C. C. & PERL, E. R. (1960). *Physiol. Rev.* **40**, 538.
KUNO, M. (1964). *J. Physiol.* (Lond.) **175**, 81.
LAPORTE, Y. & BESSOU, P. (1957). *J. Physiol.* (Paris) **49**, 1025.
LAPORTE, Y. & LLOYD, D. P. C. (1952). *Amer. J. Physiol.* **169**, 609.

Liddell, E. G. T. & Sherrington, C. S. (1924). *Proc. roy. Soc.* B. **96**, 212.
Lloyd, D. P. C. (1943 a). *J. Neurophysiol.* **6**, 111.
Lloyd, D. P. C. (1943 b). *J. Neurophysiol.* **6**, 293.
Lloyd, D. P. C. (1943 c). *J. Neurophysiol.* **6**, 317.
Lloyd, D. P. C. (1957). *J. gen. Physiol.* **41**, 297.
Lloyd, D. P. C. & McIntyre, A. K. (1948). *J. Neurophysiol.* **11**, 455.
Lloyd, D. P. C. & McIntyre, A. K. (1950). *J. Neurophysiol.* **13**, 39.
McIntyre, A. K. (1962). *J. Physiol.* (Lond.) **163**, 46.
McIntyre, A. K., Bradley, K. & Brock, L. G. (1959). *J. gen. Physiol.* **42**, 931.
Mark, R. F. & Steiner, J. (1958). *J. Physiol.* (Lond.) **142**, 544.
Marshall, W. H. (1941). *J. Neurophysiol.* **4**, 25.
Morin, F. (1955). *Amer. J. Physiol.* **183**, 245.
Rall, W. (1955). *J. cell. comp. Physiol.* **46**, 413.
Renshaw, B. (1940). *J. Neurophysiol.* **3**, 373.
Sumner, A. (1961). *Proc. Univ. Otago Med. Sch.* **39**, 3.

INTRACELLULAR RECORDING FROM NEURONES OF THE RETICULAR FORMATION

by F. Magni and W. D. Willis

The reticular formation of the brain stem has received a great deal of attention from anatomists and from physiologists in recent years (see reviews by Brodal 1957, and by Rossi and Zanchetti 1957; see also the symposium edited by Jasper et al. 1958). It has become increasingly clear that the reticular formation is not a diffusely organized system, but rather that it includes a number of discrete nuclei which subserve quite diverse functions. Unfortunately, many of the physiologic studies of the reticular formation have neglected this fact, at least to the extent that there was no attempt to associate electrical activity in the reticular formation with particular types of neurones.

In the experiments to be described, reticular neurones were classified according to the direction of projection of their axons. The preparations were cerebellectomized and were either anaesthetized with pentobarbital sodium or decerebrated. Single reticular neurones were impaled with glass microelectrodes, using standard techniques (Brock, Coombs and Eccles 1952). Their axons were stimulated either in the spinal cord or in the mesencephalon, and the antidromically conducted action potential as recorded from the soma-dendritic region (Brock, Coombs and Eccles 1953) served to identify the type of neurone under investigation. The effect of stimulation of various pathways afferent to the reticular formation could then be determined for each type of neurone. The full papers describing the techniques and the findings in more detail may be consulted elsewhere (MAGNI and WILLIS 1963, 1964 a, b).

I. Results

1. *Types of reticular neurones.* A total of 520 neurones of the lower pontine and rostral medullary reticular formation were identified by antidromic activation. Many of these neurones displayed spike potentials having a simple rising phase (eg., Figs. 1 A, 2 H, 4 A), whereas others had spike potentials that showed an inflection on the rising phase comparable to the initial segment-soma dendritic delay of spinal cord motoneurones (Fig. 3 A; cf. BROCK, COOMBS and ECCLES 1953). However, if the microelectrode was not actually within the soma-dendritic region in those neurones having simple spike potentials, it must have been within a portion of the axon which was very close to the cell body in order to record postsynaptic potentials. Further, in many of these neurones the postsynaptic potentials had very brief times to peak, suggesting little or no electronic decrement (eg., in Fig. 4 H, the time to peak of the excitatory postsynaptic potential was about 1.2 msec).

Three types of reticular neurones were investigated: those with axons descending into the spinal cord; those with axons ascending into the mesencephalon; and those with axons projecting in both directions. Indirect evidence for the presence of the last mentioned type of reticular neurone was found by BRODAL's group (BRODAL 1957), and such neurones have been described histologically by SCHEIBEL and SCHEIBEL (1958). An example of electrical activity recorded from this type of neurone is given in Fig. 1.

The antidromically conducted action potentials in Fig. 1 A—C were recorded following stimulation of the reticulospinal tract at the level of the first lumbar segment of the spinal cord. A small excitatory postsynaptic potential (EPSP) was also evoked by spinal cord stimulation; this is best seen in the high gain record taken at threshold for the axon (Fig. 1 C). Antidromic action potentials were also produced by stimuli applied by a concentric bipolar electrode whose tip was within the mesencephalon (Fig. 1 D and E). The tip position was verified histologically by placing an electrolytic lesion through the electrode; it was in the region of the central tegmental tract (Fig. 1 G). Mesencephalic stimulation below threshold for the ascending axon produced a small EPSP which often triggered action potentials (Fig. 1 E and F). The approximate position of the neurone is indicated in Fig. 1 H.

2. *Effects of cortico-reticular volleys.* The action of various types of volleys afferent to the reticular formation was studied once a reticular neurone was identified. Of particular interest was the effect of volleys evoked by stimulation

of the cerebral cortex. In many experiments, the cortical areas marked by the rectangles in Fig. 2 I were stimulated in succession during the course of impalement of a reticular neurone; in addition, the contralateral pre- and postcruciate gyri were also excited. The records in Figs. 2 and 3 were taken in the same preparation from reticulospinal neurons located within a few millimetres of each other.

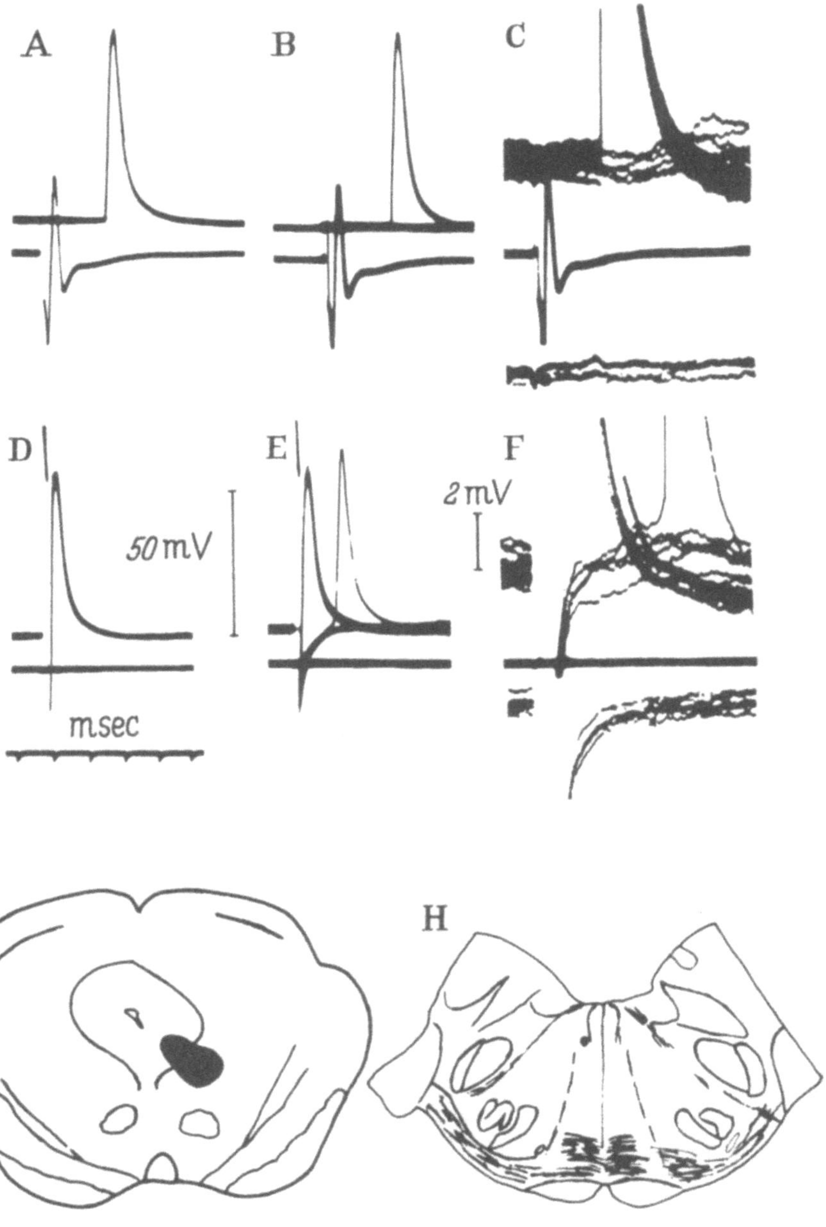

Fig. 1. Reticular neurone with axons projecting both rostrally and caudally. See text for description of records. The time scale applies to A—F. The 50 mV potential scale is for A, B, D and E, while the 2 mV scale is for C and F. (From Magni and Willis 1963.)

All of the cerebral cortical areas stimulated produced an EPSP in the neurone shown in Fig. 2. The largest effect was produced by excitation of the temporal cortex (Fig. 2 F). An action was also noted following stimulation of the occipital cortex (Fig. 2 G).

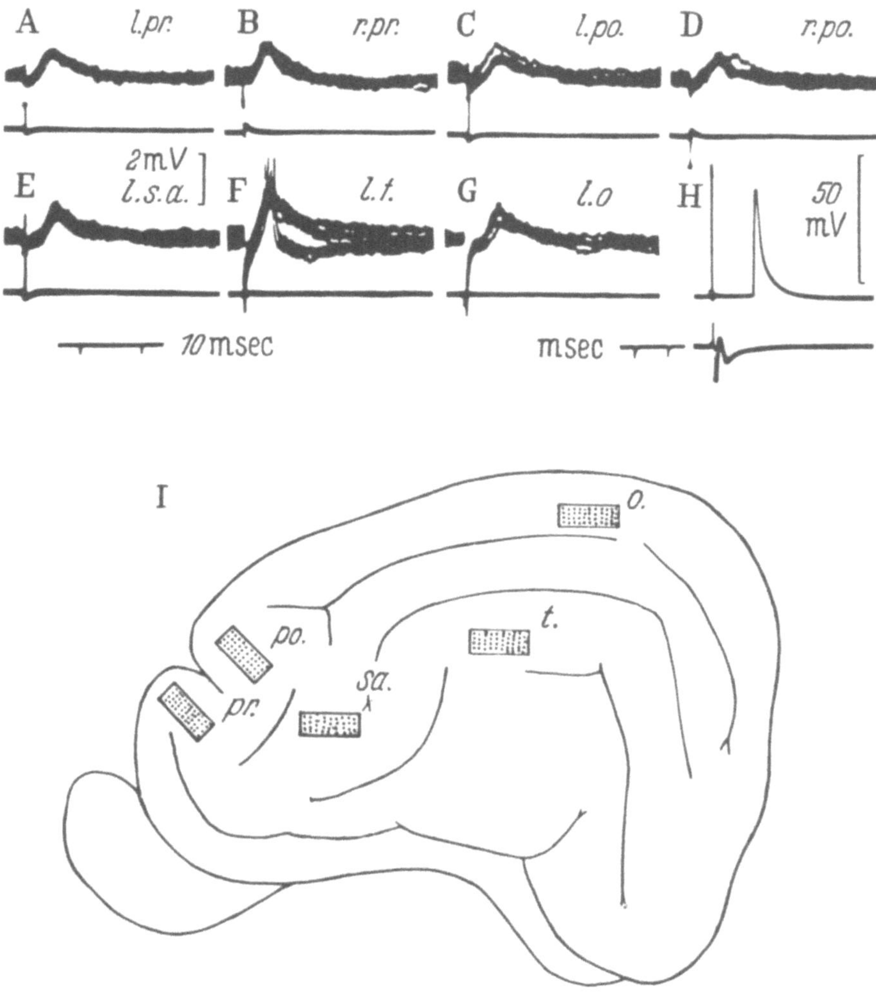

Fig. 2. EPSPs from various cerebral cortical regions as recorded in a reticulospinal neurone. The EPSPs are seen in A—G, and the antidromic action potential following spinal cord stimulation is in H. The 10 msec timer and the 2 mV potential scale apply to A—G, while the msec and 50 mV scales are for H. The diagram in I shows the cortical areas stimulated. Abbreviations: pr., precruciate gyrus; po., postcruciate gyrus; s. a., anterior suprasylvian gyrus; t., temporal lobe; o., occipital lobe. The letters l. and r. refer to the left and right sides of the body. (From Magni and Willis 1964 a.)

The temporal and occipital areas apparently lacked excitatory connexions to a neighboring neurone (Fig. 3 G and H). Stimulation of other parts of the cerebral cortex, however, produced large EPSPs, which often caused a repetitive discharge of the neurone (Fig. 3 C—F). Stimuli applied within the region of the central tegmental tract of the mesencephalon also had a strong excitatory action (Fig. 3 B). It should be noted that the EPSPs in Fig. 3 B—F had latencies of less than 1 msec, indicating monosynaptic connexions both from the cerebral cortex and from a pathway traversing the mesencephalic tegmentum. The EPSPs in Fig. 2 had latencies of several msec; these may or may not have been monosynaptic.

Almost all reticulospinal neurones received EPSPs from 1 or more areas of the cerebral cortex (87% of 359 neurones), and so did most neurones having

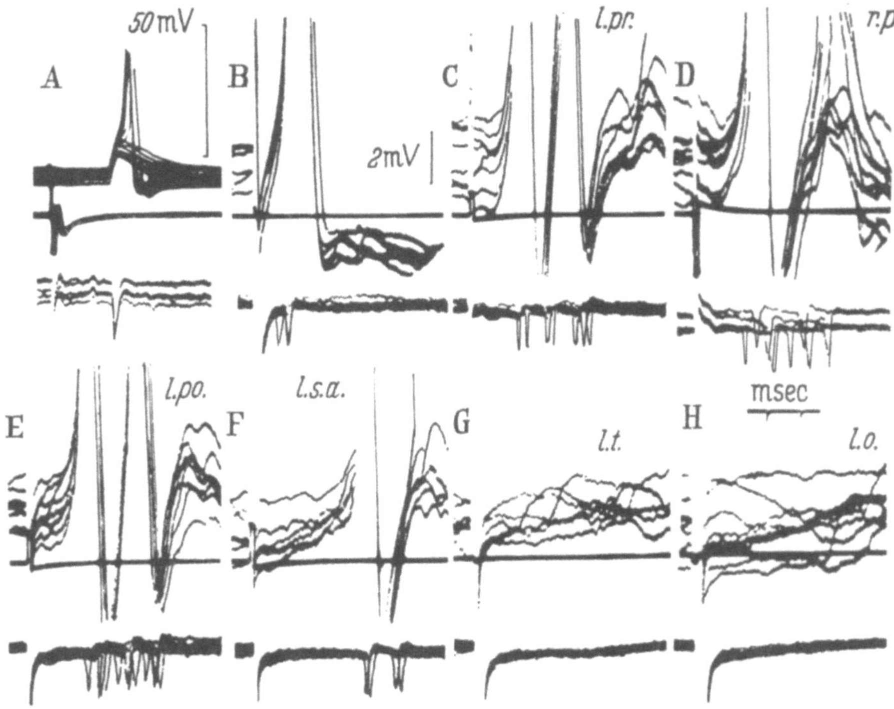

Fig. 3. Monosynaptic EPSPs in a reticulospinal neurone from a mesencephalic tegmental path and from the cerebral cortex. The antidromic action potential from spinal cord stimulation is seen in A. The EPSP from mesencephalic stimulation is in B, while the effects of activation of various regions of the cerebral cortex are seen in C—H. The msec time scale applies to all records. The 50 mV potential scale is for A, and the 2 mV scale for the other records. The extracellular field potential in the third trace of A was taken at the higher gain, as were the other field potentials in B—H. (From Magni and Willis 1964 a.)

axons projecting both to the spinal cord and to the mesencephalon (67% of 15 neurones). However, only about one fifth of reticular neurones with just ascending axons received EPSPs from cortical stimulation (19% of 48 neurones). This finding is consistent with the histologic evidence of Rossi and Brodal (1956), but there may have been a danger of sampling error, since the number of ascending reticular neurones impaled was relatively small. There was no obvious difference in the strength of the excitatory connexions of the sensorimotor areas of the two sides of the cerebral cortex onto individual reticular neurones.

It ist interesting that very few inhibitory postsynaptic potentials were produced in reticular neurones by corticofugal volleys. All of the ones observed followed EPSPs (eg., Fig. 5 D).

3. *Effects of a pathway in the mesencephalic tegmentum.* As already mentioned, stimulation within the region of the central tegmental tract of the mesencephalon often results in monosynaptic excitation of reticular neurones (Figs. 1 E and F, 3 B). Another example of such an action is given in Fig. 4. The EPSPs evoked by stimuli of several grades of strength are illustrated in Fig. 4 E—H. The tip of the stimulating electrode was shown to be at the point marked in Fig. 4 I.

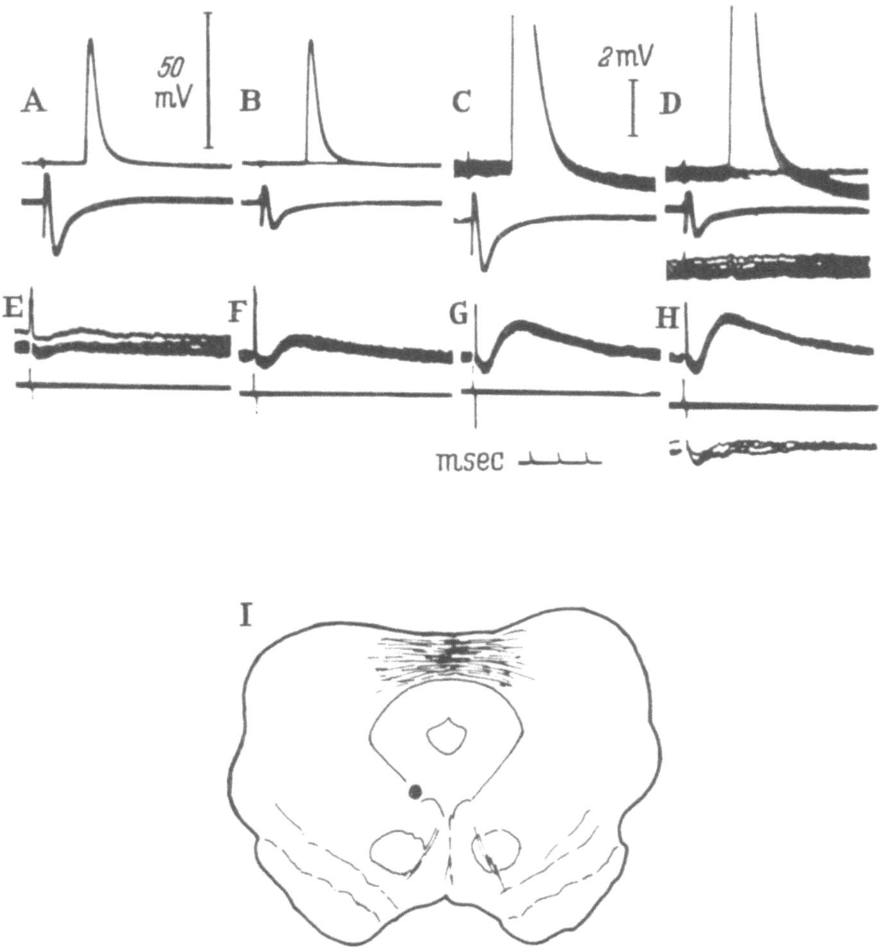

Fig. 4. EPSPs in a reticulospinal neurone following stimulation of the region of the central tegmental tract. A—D show the antidromic action potential resulting from cord stimulation. The other records and the diagram are described in the text. The msec time scale is for A—H. The 50 mV potential scale is for A and B, while the 2 mV scale applies to C—H. (From MAGNI and WILLIS 1964b.)

In a few reticular neurones, inhibitory postsynaptic potentials (IPSPs) were produced by stimulation in approximately the same part of the mesencephalic tegmentum. These IPSPs either followed EPSPs or occured in isolation. Their latency indicated a disynaptic pathway.

The frequency of occurrence of synaptic action from this mesencephalic pathway upon reticular neurones seemed to be greater in the case of reticulospinal neurones (26% of 376 neurones) than of ascending reticular neurones (3% of 86 neurones). However, it was more difficult to study such an effect in the latter neurones, since the same stimulus would also produce antidromic invasion.

4. *Effects of peripheral nerve stimulation.* The action of volleys set up in peripheral nerves of the forelimb was investigated, but only reticulospinal neurones were found in sufficient numbers for the results to be reported. The nerves

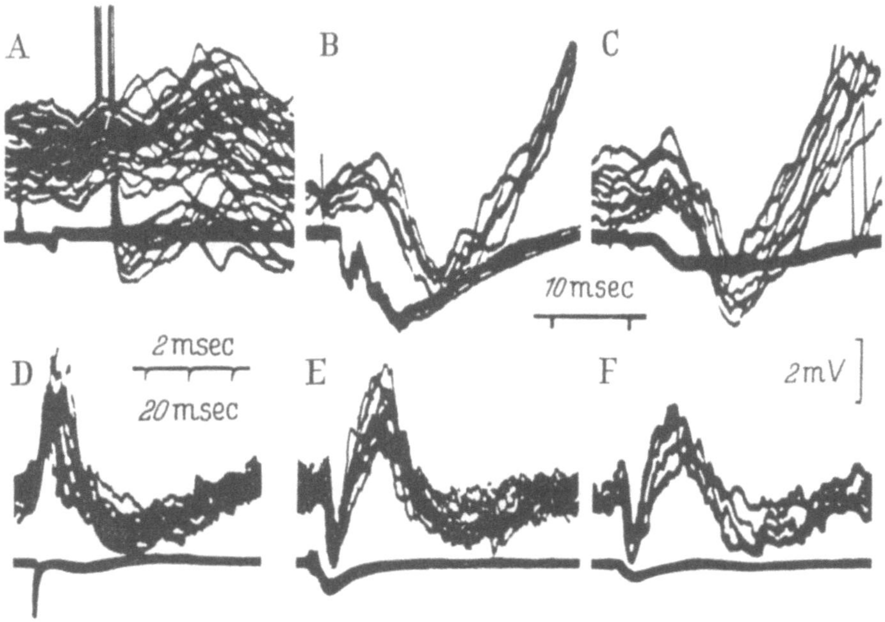

Fig. 5. Complex response in a reticulospinal neurone evoked by peripheral nerve volleys. A shows the antidromic action potential from spinal cord stimulation. B, C, E and F are described in the text. D is the EPSP-IPSP sequence produced by stimulation of the left sensorimotor cortex. The 2 mV potential scale is for all the records. The timer beneath record A is 2 msec when applied to A and 20 msec when applied to D—F. The 10 msec timer is for B and C. (From MAGNI and WILLIS 1964b.)

employed were the deep radial (largely a muscular nerve) and the superficial radial (a cutaneous nerve). All of the reticulospinal neurones had axons descending below the first lumbar segment.

Postsynaptic potentials were recorded in response to stimulation of forelimb nerves in 48% of 122 reticulospinal neurones. In most instances, the postsynaptic potentials had a simple waveform, being eithers EPSPs or IPSPs. However, in many neurones there was a diphasic sequence of postsynaptic potentials, and in one neurone a triphasic sequence.

Fig. 5 illustrates the IPSP-EPSP-IPSP sequence recorded from a reticulospinal neurone in response to cutaneous and to muscular nerve stimulation. The action of the superficial radial nerve volley is seen in full in Fig. 5 E; the initial part of the potential shifts is better seen in the record of Fig. 5 B, taken on a faster sweep. Similarly, the potentials evoked by stimulation of the deep radial nerve are illustrated in Fig. 5 F and C.

II. Discussion

Obviously, this study represents just a small part of the major effort required to relate electrical activity within the reticular formation to the behaviour of particular kinds of reticular neurones. No attempt was made to study reticular neurones which project onto the cerebellum. Nor could groups of reticulospinal or ascending reticular neurones be further subdivided according to the function of their terminals within the spinal cord or the brain. A further difficulty emerged from the confirmation of the histologic evidence (Brodal 1957; Scheibel and Scheibel 1958) for the presence of reticular neurones with axons projecting both caudally and rostrally (Fig. 1). It was impossible to exclude the presence of axons passing in the opposite direction from neurones proved to have axons projecting either rostrally or caudally. Thus, the classification of reticular neurones followed in this study becomes somewhat arbitrary. Nevertheless, by adhering primarily to the positive findings, one can draw some useful conclusions from the evidence so far obtained.

Reticulospinal neurones almost invariably receive excitatory connexions from the cerebral cortex. The cortical areas involved include the sensorimotor regions, but some cortico-reticular fibres may also arise in the temporal and in the occipital cortex (Fig. 2). The latencies of the EPSPs in some reticulospinal neurones (eg., Fig. 3) establish that at least part of the cortico-reticular pathway from the sensorimotor regions is monosynaptic. The sensorimotor projections to the reticular formation from the two sides of the brain appear to be equally effective. All of these conclusions are in agreement with and extend the observations of Rossi and Brodal (1956). Corticofugal connectivity to ascending reticular neurones was less impressive, but it was present.

It has now been established that there is a pathway traversing the region of the central tegmental tract of the mesencephalon which contains fibres that monosynaptically excite many reticular neurones (Fig. 4). The origin of this pathway is at present obscure, but it, together with the cortico-reticulospinal path, presumably forms an important part of the extrapyramidal system.

The finding that nearly half of reticulospinal neurones having axons projecting to the lumbosacral spinal cord received afferent connexions from peripheral nerves of the forelimb indicates that a spino-reticulospinal loop is involved in the transmission of information between the limbs. The pathway must involve a fairly complicated synaptic organization to explain the di- and triphasic sequences of postsynaptic potentials which it may evoke in reticular neurones (Fig. 5) without the assistance of cerebellar circuits.

It is hoped that further studies along the lines begun in the present experiments will help unravel some, at least, of the patterns of connectivity of the reticular formation.

Acknowledgements

The research was sponsored jointly by the Office of Scientific Research, OAR, through the European Office, Aerospace Research, United States Air Force, under Grant EOAR 62-9, and by the Rockefeller Foundation. During the investigation Dr. Willis was a Postdoctoral Research Fellow of the National Institute of Neurological Diseases and Blindness, United States Public Health Service.

References

Brock, L. G., Coombs, J. S. & Eccles, J. C. (1952). *J. Physiol.* (Lond.) **117**, 431.
Brock, L. G., Coombs, J. S. & Eccles, J. C. (1953). *J. Physiol.* (Lond.) **122**, 429.
Brodal, A. (1957). *The Reticular Formation of the Brain Stem.* Edinburgh: Oliver and Boyd.
Jasper, H. H. et. al. (eds.) (1958). *Reticular Formation of the Brain.* Boston and Toronto: Little, Brown & Co.
Magni, F. & Willis, W. D. (1963). *Arch. ital. Biol.* **101**, 681.
Magni, F. & Willis, W. D. (1964 a). *Arch. ital. Biol.* **102**, 418.
Magni, F. & Willis, W. D. (1964 b). *Arch. ital. Biol.* **102**, 434.
Rossi, G. F. & Brodal, A. (1956). *J. Anat.* **90**, 42.
Rossi, G. F. & Zanchetti, A. (1957). *Arch. ital. Biol.* **95**, 199.
Scheibel, M. E. & Scheibel, A. B. (1958). In *Reticular Formation of the Brain* (ed. by Jasper, H. H. et. al.) p. 31. Boston and Toronto: Little, Brown & Co.

PHARMACOLOGICAL STIMULATION OF THE DEEP REGIONS OF THE HUMAN BRAIN BY INTRAVENTRICULAR INJECTIONS

by S. Obrador

The practice of Neurosurgery provides occasional opportunities for recording some neurophysiological observations. In fact, since the beginning of this speciality, many neurosurgeons have dedicated a large part of their efforts to the development of experimental and clinical neurophysiology. It was my privilege to spend some time in the Laboratory of Physiology of Sir Charles Sherrington in Oxford, and to work under the guidance of Sir JOHN ECCLES at the beginning of my training.

Most of the neurophysiological studies on the human brain have been carried out by electrical stimulation of exposed nervous tissue, such as the cerebral cortex, or by implantation of electrodes within different subcortical levels. Another way to study the deep regions of the human brain is the direct injection of drugs into the ventricular system. This field was also opened by the great neurosurgeon HARVEY CUSHING (1932), who observed the autonomic effects which were induced by intraventricular injections of pituitrin and pilocarpine. During the last years FELDBERG (1963) and his co-workers have analyzed the actions of many drugs applied directly to the nervous tissue of the inner surface of the brain of cats by intraventricular injection.

We have developed a method that allows percutaneous and frequent injections of drugs into the ventricular system of man. A special metal chamber with a silicone rubber diaphragm is screwed in the skull and connected to a catheter placed in the frontal horn of the lateral ventricle (OBRADOR 1964). This arrangement is well tolerated by the patient, and the apparatus can be kept in place many weeks or longer if necessary. The drugs injected into the cerebral fluid that fills the metal chamber pass through the foramen of MONRO into the third ventricle and normally continue through the aqueduct to reach the fourth ventricle.

We have studied so far the effects induced by many stimulant and depressant drugs in different cerebral conditions in which it seemed suitable to perform such tests. I shall only refer here very briefly to the main effects obtained by stimulant drugs in comatose patients, and in a few parkinsonian and other patients with central pain, before stereotaxic or psychosurgical procedures were carried out.

A great variety of autonomic responses were obtained with some drugs, e. g. nausea, vomiting, sweating, vasodilatation or vasoconstriction, tachycardia, piloerection, changes of blood pressure and respiration rate, hyperthermia, hiccup, lacrimation, pupillary changes, micturition and increase of bronchial secretion. These autonomic effects appeared from several minutes to one hour after the intraventricular injection and lasted several hours in a few cases. Adrenaline (0.25 mgm), acetylcholine (0.5—1 mgm), prostigmine (0.25 mgm), pituitrin (5 U), cocaine (1 mgm), morphine (2 mgm), tubocurarine (6—11 mgm), chlorpromazine (2—5 mgm) and promethazine (2—5 mgm) were the drugs which showed a more marked central autonomic action.

Some of these drugs were very dangerous by intraventricular application in the above-mentioned doses because of severe depressant action on respiration (cocaine, morphine, tubocurarine) or blood pressure (chlorpromazine, promethazine). Such action was usually associated with a generalized and marked diminution of central reactivity, that lasted in some cases for several hours and even required the use of different therapeutic measures (respirator, stimulant drugs, etc.).

In patients with coma secondary to severe head injuries, operated brain tumors or other cerebral conditions we tried intraventricular injections of various stimulant drugs as a therapeutic measure. Different degrees of excitation with arousal and an increase of reactivity and alertness was often observed in these unfortunate individuals having profound signs of decortication. Several minutes to one or two hours after the intraventricular injection of some drugs there appeared more alert expression of the face with gestures, sighs, yawning, increase of the ocular and head movements, mastication and swallowing, tendency to vocalization, increase of emotional response with smiles, crying and in some cases even the performance of responses to simple commands together with a more lively spontaneous activity. This increase of reactivity usually lasted several hours.

The more effective stimulant drugs in our experience so far are tubocurarine (6—11 mgm), caffeine and theophylline (25—75 mgm), methylglutarimide (20 to 50 mgm) and benzedrine (5 mgm). Stimulant effects appear sometimes with other drugs, such as lobeline (2.5 mgm), methylphenidate (5—10 mgm), and nikethamide (25 mgm). Most of these stimulant drugs produce also moderate autonomic responses and a tendency towards some degree of desynchronisation in the EEG recording, but often these changes were not very marked.

Intraventricular tubocurarine is a strong central stimulant but it also has a marked convulsive action that has been demonstrated in animal studies (Feldberg 1963), and confirmed by clinical observation and the EEG in human cases. For this reason this drug seems unsuitable for therapeutic purposes in man.

The actions of these stimulant drugs applied centrally to the inner surface of the human brain are uncertain, however, and not very clear in some cases, perhaps because there are lesions of the central structures necessary to produce the arousal responses. We have some evidence that in tumours of the third ventricle the intraventricular injection of stimulant drugs is ineffective.

Finally, my intention to record very briefly these pharmacological observations is to emphasize once more that the neurosurgeon has the opportunity of making some studies which perhaps may be useful, not only for the future development of therapeutic procedures, but also for the knowledge of how the human brain responds to a direct and deep stimulation.

References

Cushing, H. (1932). *Papers relating to the pituitary body, hypothalamus and parasympathetic nervous system.* Springfield, Ill.: Charles C. Thomas.

Feldberg, W. (1963). *A pharmacological approach to the brain from its inner and outer surface.* London: Edward Arnold.

Obrador, S. (1964). *J. Neurosurg.* In the Press.

THE NEUROSECRETORY CELLS OF THE MAMMALIAN NEUROHYPOPHYSIS

by W. J. O'CONNOR

In many species, certain cells are believed to be *neurosecretory;* their embryological origin is that of nervous tissue, they have morphological characteristics of nerve cells but they produce hormones which are released into the blood stream (ORTMANN 1960). An example in mammals is the supraoptic nucleus, whose cells, situated in the hypothalamus and with characteristics of nerve cells, are believed to liberate the hormones of the posterior lobe of the pituitary, e. g. the antidiuretic hormone.

The supraoptic nuclei contain large cells from which non-myelinated axons extend across the floor of the third ventricle through the pituitary stalk into the neural lobe of the pituitary where the axons form a network of fibres. Cells of similar appearance in the paraventricular nuclei also contribute axons to the network in the pars nervosa. This account was first presented completely by FISHER, INGRAM and RANSON (1938) in their study of the experimental production of diabetes insipidus and is based on the classical methods of neurohistology — staining with silver, degeneration of axons following lesions in the pituitary stalk and also retrograde degeneration in the cells of the nuclei (see also O'CONNOR 1947).

Since about 1945 further studies of the morphology of these cells have considered largely of the description, from light or electron microscopic examination, of inclusions within the cells. These inclusions are regarded as secretory in nature (see SLOPER 1958; ORTMANN 1960) and are distributed throughout the cells; while they are found both in the cell bodies and in the axons, the greatest amounts are found in the endings of the axons in the pars nervosa. This corresponds with the fact that the pharmacological activities (antidiuretic, oxytocic etc.) characteristic of extracts of the posterior lobe are also present, but in lesser degree, in extracts of the region of the hypothalamus containing the supraoptic nuclei; thus LEDERIS (1962) found that the posterior lobe in dogs, contains about 8,000 milli-units of antidiuretic activity and the hypothalamus only about 200 m-u. At present physiologists interested in neurosecretion are debating whether the hormones are produced in the cell bodies and transported along the axons into the pars nervosa or whether they are formed in all parts of the cells including the axons (see HELLER and CLARK 1962).

Electrophysiological properties of the cells of the supraoptic nucleus

The most complete electrophysiological investigation of neurosecretory cells is by KANDEL (1964) of the cells of the preoptic nucleus of the goldfish. This nucleus is analogous to the supraoptic nucleus of mammals, although its hormonal polypeptides and their probable function are different. By intracellular recording the cells had a membrane potential of 50 mV, action potentials similar to those in nerve cells occurred spontaneously or could be set up antidromically by stimulation of the pituitary gland; these neurosecretory cells thus had the electro-physiological properties of ordinary nerve cells in the brain of the goldfish.

In mammals, intracellular records have not been reported. In anaesthetised rabbits and cats CROSS and GREEN (1959) and BROOKS, USHIYAMA and LANGE (1962) using extracellular microelectrodes in the region of the supraoptic nuclei, have recorded action potentials which were similar to those from units in other parts of the hypothalamus where there are no neurosecretory cells. An obvious hope would be to identify neurosecretory cells by antidromic impulses set up by stimulation of the pituitary stalk or neural lobe but neither group appear to have achieved this.

The neurosecretory cells can apparently be excited by electrical stimulation; HARRIS (1947) showed that in conscious rabbits stimulation by means of an electrode implanted in the region of the pituitary stalk caused antidiuresis and contraction of the uterus. However, since current spread obviously occurred in these experiments, the point of stimulation could have been ordinary nerve structures of the hypothalamus rather than the neurosecretory cells themselves.

Both CROSS and GREEN (1959) and BROOKS et al. (1962) found that the activity of units in the region of the supraoptic nuclei was modified by the intracarotid injection of hypertonic solutions of sodium chloride, the discharge being usually increased but sometimes decreased. The reason for testing the effect of hypertonicity was the work of VERNEY (1947) which is discussed below. However VERNEY's work was on conscious dogs whereas the unit recordings were in anaesthetized rabbits or cats; the release of antidiuretic activity is abnormal in anaesthetized animals (for discussion see LEDSOME, LINDEN and O'CONNOR 1962) and there is no satisfactory evidence that it is altered by changes in the osmotic pressure of the blood.

Natural stimulation of the neurosecretory cells

Stimulation by hypertonic blood

Release of antidiuretic hormone determined by the osmotic pressure of the blood has been studied in detail by VERNEY (1947). He injected hypertonic solutions into the carotid artery of conscious dogs during water diuresis; the result was inhibition of urine flow which was clearly shown to be due to the release into the blood of antidiuretic hormone. The significance of this finding in relation to the control of water balance has been considered by VERNEY (1947) and many other authors including O'CONNOR (1962) but here it is discussed with particular reference to the function of the neurosecretory cells.

By making detailed studies of the arterial systems in the brains of their dogs, JEWELL and VERNEY (1957) were able to establish that for release of antidiuretic hormone to occur, blood made hypertonic by an intracarotid injection must reach the anterior part of the hypothalamus, the effective area including the supraoptic and paraventricular nuclei: on the other hand if the injected material reached the pars nervosa of the pituitary but not the hypothalamus there was no release of antidiuretic hormone. It was not possible to decide whether the cells of the supraoptic nuclei were themselves stimulated by the increase in osmotic pressure or whether the osmoreceptors were elsewhere in the hypothalamus, the neurosecretory cells then being activated by impulses in axons making synaptic connections with them.

In some of the experiments of VERNEY (1947), sodium chloride was infused into one carotid artery in amounts which would produce an increase of only 1% in the osmotic pressure of the blood and this produced release of antidiuretic hormone at 0.06 milli-unit per min, which was just enough to inhibit water diuresis. This slow rate of release of hormone is believed by VERNEY (1947) to be all that is involved in the normal control of urine volume. However in dogs large amounts of sodium chloride given by stomach tube can cause a maintained release of antidiuretic hormone of 1.4 milli-units per min (O'CONNOR 1950) and during brief periods of emotional stress as much as 20 milli-units is released in a minute (O'CONNOR and VERNEY 1942). The normal control of the volume of the urine is thus by release of hormone at a rate only about 1% of that which has been observed under some experimental circumstances.

Suckling

The mechanisms involved in the release of milk during suckling have been discussed by LINZELL (1959). Intravenous injection of extracts of the posterior lobe in lactating animals produces discharges of milk due to contraction of myoepithelial cells in the mammary glands: both vasopressin and oxytocin have this action but oxytocin is the more effective. During suckling release of hormones occurs from the neurohypophysis partly due to stimulation by the young of afferent nerves from the nipples and partly due to conditioned responses of the mother to the process of suckling or the circumstances in which she is milked. In either case it must be supposed that the immediate cause of the release of hormones is the arrival of impulses at synaptic endings on the cells of the supraoptic or paraventricular nuclei. Milk ejection in suckling is mainly due to the release of oxytocic activity, but suckling also produces inhibition of water diuresis so that some vasopressin-like activity is also released.

Emotion

Emotional stress, copulation, stimulation of the vagina have all been shown to cause release of antidiuretic activity. In the bitch, release of 1—20 milli-units of antidiuretic activity by emotional stress (O'CONNOR and VERNEY 1942) is accompanied by about 20 times as much oxytocic activity as indicated by uterine contraction (ABRAHAMS and PICKFORD 1954). O'CONNOR and VERNEY (1945) found that the presence of adrenaline in the circulation prevented the release of antidiuretic hormone by emotion. Adrenaline injected intravenously or adrenaline released from the animal's own suprarenal glands were equally effective and in consequence inhibition of water diuresis by emotion was obtained more consistently in dogs after denervation of the suprarenals. It is not known how adrenaline acts to prevent the release of the antidiuretic hormone.

The site of release of the hormones

In all of the above instances stimulation of the neurosecretory cells occurs in the hypothalamus at the cell bodies, and it is usually assumed that the release of hormones occurs from the endings in the neural lobe of the pituitary in consequence of the passage of impulses along the axons. There is, however, no evidence which compels this interpretation. It is true that after removal of the posterior lobe or section of the pituitary stalk the amounts of antidiuretic activity released by any stimulus is greatly reduced, but there is also retrograde degeneration of the cells of the supraoptic nuclei; loss of function is parallel to the loss of cells (O'CONNOR 1947) and so does not prove that release of hormone occurs from the posterior lobe. Indeed the fact that there is a small release of hormones after removal of the posterior lobe might suggest that release can occur from surviving cell bodies deprived of their axons.

Prolonged deprivation of water greatly decreases the vasopressin content of the posterior lobe in rats, but only if the rats are kept without water for about a week, and then they stop eating. Loss of the polypeptides from the posterior lobe is just as likely to be a non-specific effect of the disturbed protein metabolism of starvation as a specific effect of dehydration. After only 2—3 days without water, when the animals are still eating, there is no reduction in the antidiuretic activity in the posterior lobe (DICKER and NUNN 1957).

The main reason for the accepted belief that release occurs from the posterior lobe is the large amounts of oxytocin and vasopressin found in extracts of the posterior lobe. However, the smaller amounts found in the hypothalamic portions of the neurosecretory cells are ample to provide for ordinary releases. Thus the hypothalamus of the dog contains about 200 milli-units of antidiuretic activity (LEDERIS 1962) which although much less than the 8,000 milli-units in the posterior lobe, seems ample to act as the store the release of 0.1 milli-unit per min involved in the normal control of the urine volume. As far as existing evidence goes it is possible to suggest that the axons of the neurosecretory cells in the mammal rarely carry impulses; ordinarily when the cell body is stimulated release of hormones could occur from precursor material in the cell bodies with the axon playing no part, in which case the cell bodies would be acting like any secreting cell. Perhaps it is only under experimental circumstances when a fast rate of release is attained that impulses are propagated along the axons and invoke release from the large stores of hormone in the posterior lobe.

References

ABRAHAMS, V. C. & PICKFORD, M. (1954). *J. Physiol.* (Lond.) **126**, 329.
BROOKS, C. McC., USHIYAMA, J. & LANGE, G. (1962). *Amer. J. Physiol.* **202**, 487.
CROSS, B. A. & GREEN, J. D. (1959). *J. Physiol.* (Lond.) **148**, 554.
DICKER, S. E. & NUNN, J. (1957). *J. Physiol.* (Lond.) **136**, 235.
FISHER, C., INGRAM, W. R. & RANSON, S. W. (1938). *Diabetes insipidus and the neurohormonal control of water balance: a contribution to the structure and function of the hypothalamico-hypophyseal system.* Ann Arbor: Edwards Bros.
HARRIS, G. W. (1947). *Phil. Trans.* B. **232**, 385.
JEWELL, P. A. & VERNEY, E. B. (1957). *Phil. Trans.* B. **240**, 197.
HELLER, H. E. & CLARK, R. B. (Ed.) (1962). *Third International Congress on Neurosecretion.* Memoirs of the Society for Endocrinology, **12**. London: Academic Press.
KANDEL, E. R. (1964). *J. gen. Physiol.* **47**, 691.
LEDERIS, K. (1962). In *Third International Congress on Neurosecretion.* Memoirs of the Soc. for Endocrinology, 12 (ed. by HELLER, H. E. & CLARK, R. B.) p. 227. London: Academic Press.
LEDSOME, J. R., LINDEN, R. J. & O'CONNOR, W. J. (1962). *J. Physiol.* (Lond.) **159**, 87.
LINZELL, J. L. (1959). *Physiol. Rev.* **39**, 534.
O'CONNOR, W. J. (1947). *Quart. J. exp. Physiol.* **34**, 29.
O'CONNOR, W. J. (1950). *Quart. J. exp. Physiol.* **36**, 21.
O'CONNOR, W. J. (1962). *Renal Function.* London: Edward Arnold.
O'CONNOR, W. J. & VERNEY. E. B. (1942). *Quart. J. exp. Physiol.* **31**, 393.
O'CONNOR, W. J. & VERNEY, E. B. (1945). *Quart. J. exp. Physiol.* **33**, 77.
ORTMANN, R. (1960). In *Handbook of Physiology,* Section 1. Neurophysiology. Vol. 2, (ed. by FIELD, J.) p. 1039. Washington: American Physiological Society.
SLOPER, J. C. (1958). *Int. Rev. Cytol.* **7**, 337.
VERNEY, E. B. (1947). *Proc. Roy. Soc.* B **135**, 25.

PROPRIOCEPTIVE AND EXTEROCEPTIVE PROJECTIONS TO THE PERICRUCIATE CORTEX OF THE CAT

by OLOV OSCARSSON

Despite many investigations the exact extent of the primary motor and sensory areas of the cat cerebral cortex has remained a matter of dispute, and so have the projections paths to these areas and their modality components. The recent findings that large muscle spindle afferents project to the cerebral cortex (Oscarsson and Rosén 1963) and that cutaneous afferents project to both motor and sensory cortices through short-latency paths (Oscarsson and Rosén 1964) help to clarify some of the problems.

Projection paths

There are two short-latency paths to the somatic areas of the cat cerebral cortex. They are both components of the medial lemniscus-thalamocortical system which relays in the ventrobasal complex of the thalamus (Morin 1955; Mark and Steiner 1958; Busch 1961; Andersson 1962; Norsell and Voorhoeve 1962). One component of the medial lemniscus originates from the cuneate and gracile nuclei and is activated from fibres ascending in the dorsal funiculus. The other component consists of axons from the lateral cervical nucleus and is activated by impulses in the spinocervical tract which is located in the dorsomedial part of the lateral funiculus (Lundberg and Oscarsson 1961; Norsell and Voorhoeve 1962). Tactile information is carried by both lemniscal components but that activated from the dorsal funiculus is the only path available for some sensory modalities. Information from Group I muscle afferents reaches the cerebral cortex from the forelimbs but not the hindlimbs and is mediated exclusively by fibres ascending in the dorsal funiculus (Oscarsson and Rosén 1963). Information from specific joint receptors and from Pacinian corpuscles presumably also reaches the cerebral cortex exclusively through the latter pathway (Oscarsson 1958; Oswaldo-Cruz and Kidd 1964; McIntyre 1962).

The degree of segregation of the two lemniscal components in the ventrobasal complex has been recently investigated (Landgren, Nordwall and Wengström 1965) and the relay of information from large muscle spindle afferents demonstrated in a region dorsomediorostral to the cutaneous relay (Mallart 1964). There is a conspicuous modality segregation in the further projection of the lemniscal pathways to the cerebral cortex. Information from primary endings of muscle spindles and from specific joint receptors reaches the first but not the second somatic area (Andersson 1962; Oscarsson and Rosén 1963), whereas information from Pacinian corpuscles seems to reach the second but not the first somatic area McIntyre 1962). Furthermore, as will be described below, different modalities have separate projection areas in the first somatic area.

Projection areas

Fig. 1 shows the projection areas related to the contralateral forelimb of the cat and investigated with the evoked-potential technique in preparations with either the dorsal funiculus or the spinocervical tract interrupted (Oscarsson and Rosén 1963, 1964, and unpublished). The hatched areas refer to regions in which large surface-positive potentials with short latency were evoked by stimulation of skin and muscle nerves. No potentials or very small ones were recorded after interruption of both pathways.

Cutaneous afferents evoke potentials in two distinct areas (Fig. 1, A) with only slightly different extent in the two types of preparation. In experiments with intact spinocervical tract these areas were more circumscribed and the potentials often larger in the rostral region. In experiments with intact dorsal funiculus the areas were sometimes confluent in their lateral part and the potentials usually most prominent in the caudal area. The latency was approximately the same in both areas: on stimulation of the superficial radial nerve about 4.5 msec in preparations with intact dorsal funiculus and about 5.5 msec in preparations with intact spinocervical tract. The longer latency is due to the additional synaptic relay in the latter pathway.

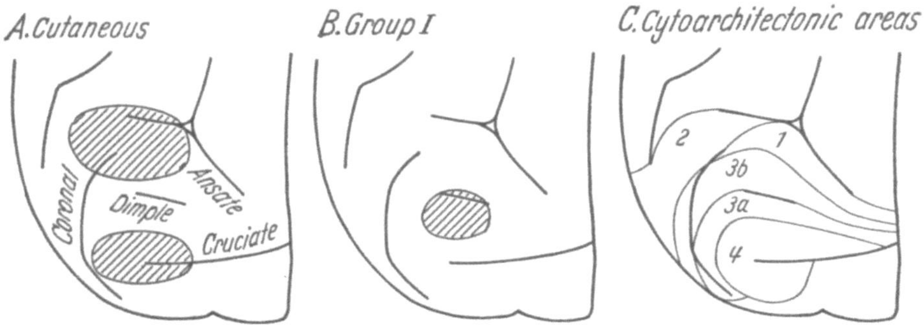

Fig. 1. Dorsal view of rostral pole of right hemisphere. The ansate, coronal, and cruciate sulci and the postcruciate dimple are indicated in A. Short-latency surface-positive potentials are evoked in the hatched areas by volleys in cutaneous afferents (A) and by volleys in Group I muscle afferents (B) of forelimb nerves. C, approximate extent of the cytoarchitectonic areas 1, 2, 3a, 3b, and 4 according to HASSLER and MUHS-CLEMENT (1964)

Information from Group I muscle afferents reaches the cerebral cortex exclusively through the dorsal funiculus. The projection area is located in the rostral part of the postcruciate gyrus, mainly rostral to the postcruciate dimple as shown in Fig. 1, B. The potential maximum has an eccentric position in this area, within, or just in front of the dimple. This is also the site where cortical cells activated by Group I volleys are found most readily (OSCARSSON and ROSÉN, unpublished). Typical findings on intracellular recording from such a cell are shown in Fig. 2. The EPSPs are large and of long duration, and have a short latency (G) indicating monosynaptic excitation from corticothalamic fibres. The potentials appear with a small ingoing volley (I—L) demonstrating that little spatial summation is needed in the projection path. There is often convergence of excitation from a wide variety of muscle nerves (A—E) and a cutaneous volley is sometimes effective but then after a longer latency (F, H).

The projection of information from joint receptors to the area caudal to the postcruciate dimple has been described by MOUNTCASTLE (1957). Cells activated from these receptors are intermingled with cells activated from tactile receptors, but have a tendency to occur more often in the rostral part of the area. This is the part bordering the Group I projection. It is remarkable that Mountcastle and Powell (1959) found a reverse localization in the monkey where cells activated from joint receptors are especially common in the caudal part of the postcentral gyrus.

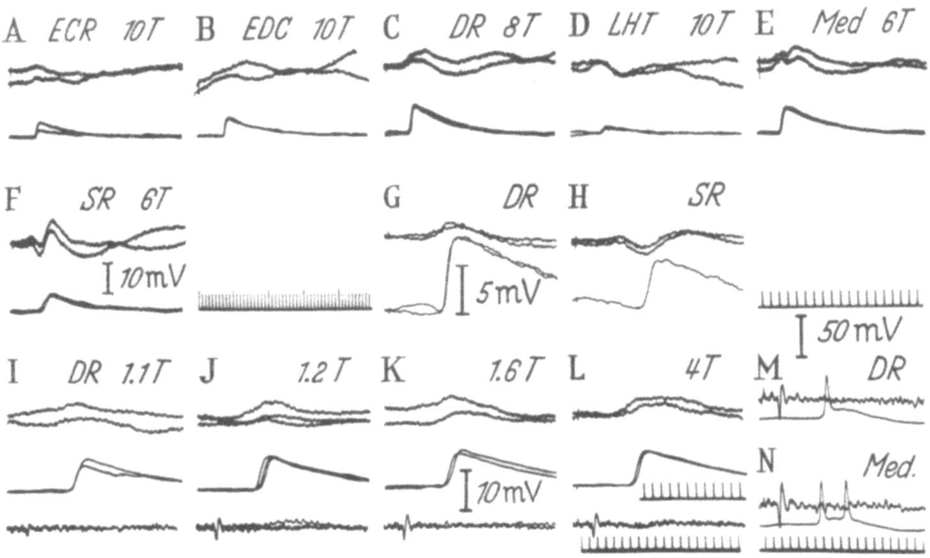

Fig. 2. Intracellular recording from cell located close to the postcruciate dimple and activated from Group I afferents. Pentobarbitone anaesthesia. Upper traces in A—L record surface potentials (positivity upwards) at the microelectrode site. Lower traces in I—L and upper traces in M—N record ingoing volleys from dorsal funiculus at C 3. Stimulated nerves and stimulation strengths in multiples of nerve threshold are indicated. ECR, nerve to extensor carpi radialis; EDC, nerve to extensor digitorum communis; DR, remaining portion of deep radial nerve; LHT, nerve to long head of triceps; Med, median nerve; SR, superficial radial nerve. A—L obtained at a low membrane potential (about 30 mV) and M—N immediately after the penetration at a membrane potential of 60 mV. Time marker, msec. Note different sweep speeds in A—F, G—H, I—L, and M—N. Two upper traces in I—L at slightly faster speed than the lower trace. Voltage scales to refer to microelectrode recording

Attempts to correlate physiological observations with the anatomy of the cat cerebral cortex have been hampered by the lack of sufficiently detailed anatomical data. The difficulties have been mitigated by the recent histological studies of Hassler and Muhs-Clement (1964). The approximate extent of the cytoarchitectonic areas around the cruciate sulcus, as given by them, is indicated in Fig. 1, C. It appears that the cutaneous projections shown in Fig. 1, A, correspond reasonably well to the forelimb regions of the classical motor (area 4) and sensory (areas 1, 2 and 3 b) areas. The Group I projection has its maximum in area 3 a but extends into the motor cortex.

Functional significance

It can now be regarded as established that cutaneous afferents project through short-latency paths to discrete regions which represent the primary motor and sensory areas. The projection to the former area does not necessarily indicate a sensory function of the motor cortex. The information from cutaneous afferents might be used in the integration of cortical motor activity. In this connection it is of interest that the cutaneous projection areas also receive information from high threshold muscle afferents (OSCARSSON and ROSÉN, unpublished). Several ascending pathways are activated from the combination of cutaneous afferents, Group II and III muscle afferents, and high threshold joint afferents. These afferents evoke the flexion reflex at the segmental level and the tracts they activate have been suggested to inform about flexion reflex patterns (LUNDBERG 1959; HOLMQVIST, LUNDBERG and OSCARSSON 1960; LUNDBERG, NORSELL and VOORHOEVE 1963). The possibility that projections of cutaneous and high threshold muscle afferents to the cerebral cortex also signal flexion reflex patterns should be considered. Such information would obviously be of value in the co-ordination of motor functions.

Large muscle spindle afferents project to a region which is separated from the two cutaneous projection areas and located partly caudal to the classical motor area. It has been suggested that the Group I projection serves as a feed-back channel for the regulation of movements elicited from the cerebral cortex (OSCARSSON and ROSÉN 1963). Kinesthesis and position sense might be primarily related to the projection from joint receptors as suggested by MOUNTCASTLE and POWELL (1959).

References

ANDERSSON, S. A. (1962). *Acta physiol. scand.* **56**, Suppl. 194.

BUSCH, H. F. M. (1961). *An anatomical analysis of the white matter in the brain stem of the cat.* (Thesis.) Van Gorcum & Comp. N. V. Assen.

HASSLER, R. & MUHS-CLEMENT, K. (1964). *J. f. Hirnforschung* **6**, 377.

HOLMQVIST, B., LUNDBERG, A. & OSCARSSON, O. (1960). *Arch. ital. Biol.* **98**, 60.

LANDGREN, S., NORDWALL, A. & WENGSTRÖM, C. (1965). *J. Physiol.* (Lond.) **176**, 12 P.

LUNDBERG, A. (1959). Symposia and Special Lectures, XXI Int. Physiol. Congr. Buenos Aires. p. 100.

LUNDBERG, A., NORSELL, U. & VOORHOEVE, P. (1963). *Acta physiol. scand.* **59**, 462.

LUNDBERG, A. & OSCARSSON, O. (1961). *Acta physiol. scand.* **51**, 1.

MALLART, A. (1964). *C. R. Acad. Sc. Paris* **259**, 1215.

MARK, R. F. & STEINER, J. (1958). *J. Physiol.* (Lond.) **142**, 544.

MCINTYRE, A. K. (1962). *J. Physiol.* (Lond.) **163**, 46.

MORIN, F. (1955). *Amer. J. Physiol.* **183**, 245.

MOUNTCASTLE, V. B. (1957). *J. Neurophysiol.* **20**, 408.

MOUNTCASTLE, V. B. & POWELL, T. P. S. (1959). *Johns Hopk. Hosp. Bull.* **105**, 173.

NORSELL, U. & VOORHOEVE, P. (1962). *Acta physiol. scand.* **54**, 9.

OSCARSSON, O. (1958). *Arch. ital. Biol.* **96**, 199.

OSCARSSON, O. & ROSÉN, I. (1963). *J. Physiol.* (Lond.) **169**, 924.

OSCARSSON, O. & ROSÉN, I. (1964). *J. Physiol.* (Lond.) **172**, 28 P.

OSWALDO-CRUZ, E. & KIDD, C. (1964). *J. Neurophysiol.* **27**, 1.

THE SODIUM PUMP OF MAMMALIAN NERVE CELLS

by T. OSHIMA

Interest in active transport processes of mammalian nerve cells stems from their high rate of oxygen consumption (HOLMES 1930; HELLER and ELLIOT 1955; KOREY and ORCHEN 1959), the high vulnerability to oxygen lack (VAN HARREVELD 1946; BROOKS and ECCLES 1947; LLOYD 1953; GELFAN and TARLOV 1955; KOLMODIN and SKOGLUND 1959) and high activity of sodium- and potassium-stimulated adenosine-triphosphatase (BONTING, SIMON and HAWKINS 1961; BONTING, CARAVAGGIO and HAWKINS 1962). These data suggest a very rapid active transfer of ions across the nerve cell membrane which maintains the intracellular ionic concentrations.

The first demonstration of active cation transport in motoneurones was made by COOMBS, ECCLES and FATT (1955 a). They injected sodium, tetramethylammonium or choline ions into cat motoneurones by passing depolarizing currents through intracellular microelectrodes which were filled with solutions of salts of the appropriate cation species. Changes thereby induced in motoneuronal potentials indicated that in the intracellular fluid there was a transient increase of the selected cations which accompanied a decrease of potassium and an increase of chloride. There would also be movement of water, balancing the osmolarity between the inside and the outside of the cell. After a sodium injection these changes were gradually restored to the initial levels within ten minutes, while injections of tetramethylammonium or choline ions were followed by only a slow and partial recovery. The fast rate of recovery from the sodium injection was ascribed to the action of the sodium pump, because the recovery was so specific to the sodium injection, and because the presumed sodium efflux after the injection was in opposition to the electrochemical gradient for this ion species.

Recently sodium injections have been repeated, and the results compared with those of other alkaline cations (ARAKI, ITO, KOSTYUK, OSCARSSON and OSHIMA 1962, 1965; ECCLES, ECCLES and ITO 1964 a, b; ITO and OSHIMA 1964 a, b, c). The ionic events occurring after the sodium injection were satisfactorily analysed particularly in those motoneurones which maintained a high resting potential around —80 mV. The resting potential at such a high level was much less sensitive to the intracellular ionic changes produced by the sodium injection than at a relatively low level, and therefore the complication due to an altered resting potential could be eliminated from the measurements of motoneuronal potentials during the recovery phase of sodium injection (ITO and OSHIMA 1964 a, b). It was then possible to calculate the rate constant of the recovery process which characterized active sodium extrusion from motoneurones.

Fig. 1 illustrates the changes produced by a sodium injection in three types of motoneuronal potentials: spike, after-potential and inhibitory postsynaptic potential (IPSP). Immediately after the injection (the second row of Fig. 1)

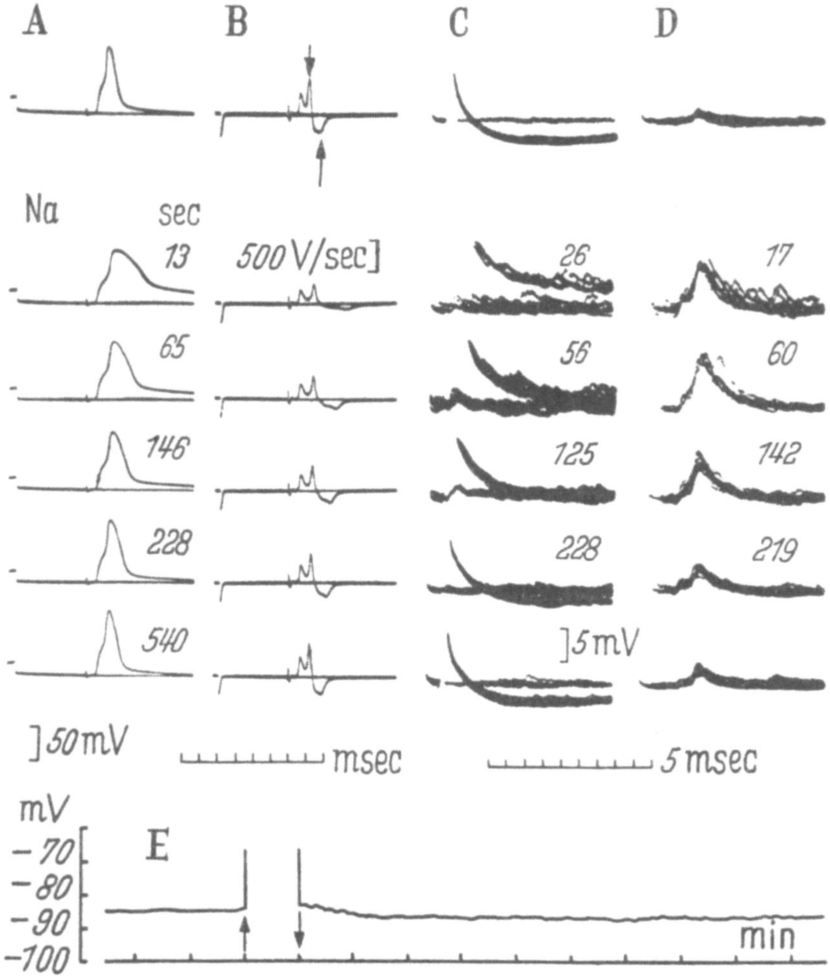

Fig. 1. Effects of a sodium injection into a gastrocnemius-soleus motoneurone. Column A, antidromic spike; B, electrically-differentiated records of the spike; C, after-potential of the spike, and D, the polysynaptic IPSP induced by a volley from skin afferent fibres in peroneal nerve. Note that the IPSP was a depolarizing potential before the injection, presumably due to chloride ions which diffused from the microelectrode (2 M NaCl). The antidromic stimulus was set just at threshold for the L7 ventral root, so that about half the sweeps in A—C were without antidromic invasion. Immediately after the records of the top row were taken, sodium ions were injected by a current of 5×10^{-8} A for 60 sec. The subsequent changes in each potential are illustrated from the second to the sixth rows. The approximate time after the injection is indicated in sec on the right of each record; otherwise it is given by the entry in the records to the left. msec scale applies to A and B, and 5 msec to C and D. E gives time course of the resting potential, upward and downward arrows indicating the period of sodium injection

there was a reduction of the spike amplitude (A) with concomitant decreases of the rates of rise and fall of the spike (B), conversion into depolarization of the after-hyperpolarization that followed the spike (C), and an increase of the IPSP in the depolarizing direction (D). The resting potential was depolarized a little and recovered rapidly within one or two minutes after the injection (E). Recovery processes in the other potentials were much slower (the second to the sixth rows), and completed within 6.4 ± 1.2 min (mean and standard deviation estimated from thirty-seven injections by passage of currents of 5×10^{-8} A for 60 sec).

The ionic hypothesis developed earlier for nerve and muscle fibres (cf. HODGKIN and KATZ 1949; HODGKIN 1951, 1958) and also for motoneurones (COOMBS et al. 1955 a, b) provided satisfactory explanations for the above-described effects of the sodium injection. The rising slope of the spike is generated by the inward sodium current that flows across the membrane during the phase of high sodium permeability. When the intracellular sodium concentration (Na_i) is increased after the sodium injection, the net inward sodium current at the spike initiation decreases, and so would result in a slower rising slope of the spike. The maximum rate of rise of the spike (V_2) was measured in the differentiated record (indicated by a downward arrow in Fig. 1 B), and was taken to be linearly related to Na_i. The maximum rate of fall of the spike (V_4, upward arrow in Fig. 1 B), and the amplitude of the after-hyperpolarization were assumed to be measures of the intracellular potassium concentration (K_i), because they are both generated by enhanced potassium permeabilities and so would be controlled by changes in K_i. The effects on the IPSP have been attributed to an intracellular increase of chloride (Cl_i) and decrease of potassium concentration (COOMBS et al. 1955 b; ECCLES et al. 1964 a, b), and hence would be indicative of Cl_i and K_i.

The recoveries of these ionic changes were followed by measuring the sizes of the motoneural potentials at various stages after the sodium injection. Their time courses were found to be exponential as a first approximation, and to parallel each other. For example, in Fig. 2 A, V_2 and V_4 are plotted on a logarithmic scale as function of time after an injection. The time constant for recovery curves such as these was determined as 91 ± 9 sec (S. D. of eleven motoneurones). The parallel recovery between V_2 and V_4 would indicate a compensatory change in Na_i and K_i. This would occur due to osmotic balance, and electroneutrality between the inside and the outside of the cell, which, in effect, keeps the total cation concentration in the cell constant (ITO and OSHIMA 1964 b). The change in $(Cl)_i$, which is also parallel with the other changes, can be explained by the operation of an inward chloride pump which slows the net outward chloride movement (ECCLES et al. 1964 a, b).

The exponential nature of the recovery curves of these sodium effects suggests that the motoneurones extrude sodium ions at a rate proportional to Na_i (ITO and OSHIMA 1964 b; cf. also COOMBS et al. 1955 b). The properties of the sodium efflux controlled by the intracellular sodium concentration have been studied extensively in frog muscles (LEVI and USSING 1948; KEYNES 1954; FRAZIER and KEYNES 1959; HODGKIN and HOROWICZ 1959), squid axons (HODGKIN and KEYNES 1955, 1956) and rat and mouse muscles (CREESE 1954,

1964; CREESE and EL-SHAFIE 1964; McLENNAN 1957, 1958). Table 1 lists the rate constants of sodium efflux estimated in these tissues with radioactive tracer methods. The comparative figure for motoneurones is given as the reciprocal of the recovery time constant, 91 sec (ITO and OSHIMA 1964 b). The value thus derived (40 hr^{-1}) is higher than the rate constants in the tissues listed above. When it is considered that the sodium efflux measured by the radioactive tracer method may involve a non-energized fraction (LEVI and USSING 1948; FRAZIER and KEYNES 1959; KEYNES and SWAN 1959 a), the difference would be greater than is indicated in Table 1 between the motoneurone and the other cells.

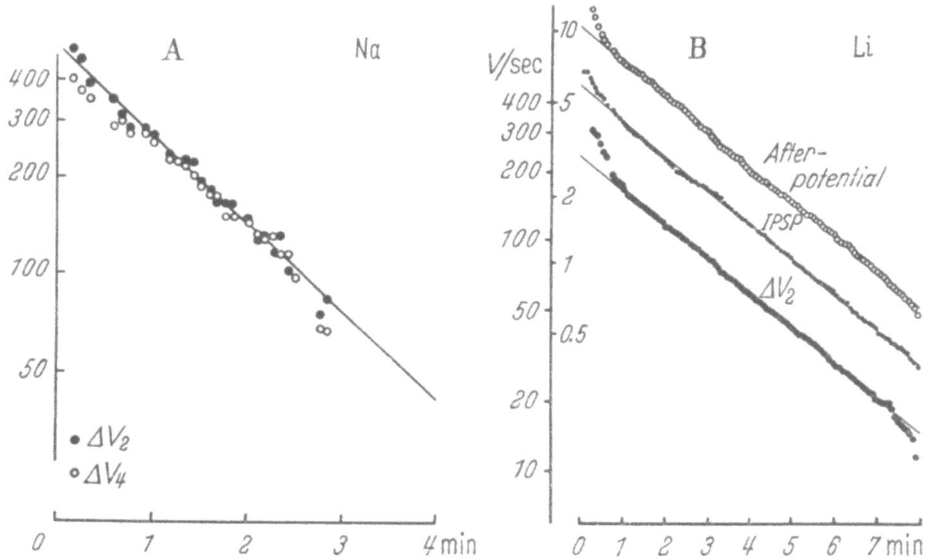

Fig. 2. A, recovery time courses of the spike of a posterior biceps-semitendinosus motoneurone following a sodium injection. Differences of V_2 and V_4 from their respective values at full recovery (ΔV_2 and ΔV_4) are plotted on a logarithmic scale against the time on a linear scale after the injection. B, similar plots of ΔV_2, and amplitudes of the after-potential and the IPSP after a lithium injection into an anterior biceps motoneurone. Li_2SO_4-filled microelectrode. V/sec scale applies to V_2 and V_4, and mV to the after-potential and the IPSP

The high rate constant of sodium extrusion is characteristic not only of motoneurones but may be also of other species of mammalian nerve cells. In cat spinal ganglion cells the electrophoretic sodium injection produced changes in the action potential similar to those in motoneurones (ITO and OSHIMA, unpublished). The recovery process thereafter could be approximated by an exponential curve, for which the time constant was comparable to that of motoneurones.

On the other hand, an extremely high rate constant has been obtained in electric organs of electric eel (Table 1). The recoveries of the heat production in the electric organ and of its electric response after repetitive electrical stimulation were approximately exponential curves with a time constant of 15 sec. The sodium efflux at rest could be estimated approximately as high as 1000 pmole cm^{-2} sec^{-1} (AUBERT, CHANCE and KEYNES 1964), a value which is in good agreement with the results obtained previously with the radioactive tracer method (SCHOFFENIELS 1959).

The rate constant of sodium extrusion is given as the product of two factors: the volume-to-surface ratio (V/A) and the constant k which is determined by the cell membrane. In cat motoneurones V/A was conveniently given as 4.6 μ from a simplified cell model (COOMBS et al. 1955 a; ITO and OSHIMA 1964 b). The fast rate of sodium extrusion in motoneurones or spinal ganglion cells, the latter having a smaller V/A due to lack of dendrites, appears to be caused partly by the small V/A value, but predominantly by the large value of k, relative to nerve and muscle fibres. However, the V/A value would be an important factor in characterizing the ionic movements in other central nerve cells which generally are much smaller than motoneurones or spinal ganglion cells.

The high vulnerability of motoneuronal potentials to asphyxia or anoxia (LLOYD 1953; KOLMODIN and SKOGLUND 1959; ECCLES, LØYNING and OSHIMA, unpublished) and to an intracellular injection of a metabolic inhibitor such as azide ions (ARAKI et al. 1961; ITO and OSHIMA 1964 c), can be explained by the high rate of sodium extrusion; an increase of Na_i at a rate of about 10 mM per min would follow a sudden blockade of the sodium pump, this value being calculated from the rate constant of 40 hr^{-1} (ITO and OSHIMA 1964 c). In the cerebral cortex, circulatory arrest induces rapid increase of electrical impedance of the brain material with swelling of apical dendrites and dominance of chloride in the cytoplasm of nerve cells (VAN HARREVELD and OCHS 1956; VAN HARREVELD and SCHADÉ 1959). These changes were explained by the movement of sodium chloride from the extracellular to the intracellular space, and can also be related to depression of the sodium pump activity.

In the comparative study with various alkaline cations (COOMBS et al. 1955 a, b; ARAKI et al. 1962, 1965), it is of particular interest that the motoneurone can extrude lithium ions at about half the rate for sodium ions. An electrophoretic injection of lithium ions changed motoneuronal potentials in exactly the same way as the sodium injection. Restoration of lithium concentration was thus followed by the same recovery processes of the altered potentials which were observed after sodium. Fig. 2 B illustrates the recovery curves of three types of motoneuronal potentials which had approximately linear falls on a logarithmic plotting. The time constant was calculated as 173 ± 29 sec (S. D. of 10 cells) (ARAKI et al. 1965). The order of sodium > lithium > quaternary ammonium ions in the effectiveness of their active extrusion agrees qualitatively with that obtained on frog muscles (KEYNES and SWAN 1959 b; RENKIN 1961). However, the fast rate of the lithium extrusion seems to be a unique feature of mammalian nerve cells in contrast with the data showing very low rates in other tissues (MAIZELS 1954; ZERAHN 1955; GREENGARD and STRAUB 1958; KEYNES and SWAN 1959 b).

This brief review has revealed that there is a considerable lack of direct information of the physicochemical processes which are involved in the active transport of cations in nerve cells. Although further investigation is required before one can visualize all the processes maintaining the ionic equilibria across the resting membrane, the following conclusion appears to be justified: the mammalian nerve cell membrane has a sodium pump activity similar to that of other tissues, and the rate of operation is comparatively high. The electrophoretic injection technique has yielded valuable results from cat motoneurones. Its application to other types of nerve cells may give opportunities to obtain more extensive information on their ion transport mechanisms.

Table 1. *Sodium extrusion in various tissues*

Tissue	Reference	rate constant (hr^{-1})	V/A (μ)	$k \times 10^{-7}$ ($cm\ sec^{-1}$)	Na_i^{s}* (mM)	Me^{s}* (pmol $cm^{-2}\ sec^{-1}$)	Temperature (°C)
Squid giant axon	HODGKIN and KEYNES (1956)	0.20	136	7.6	40	30	18—20
Frog sartorius muscle	KEYNES (1954)	0.81	20	4.5	12	5.4	17
Frog toe muscle	KEYNES (1954)	1.73	10.3	5.0	—	10.9	17
Frog semitendinosus muscle	HODGKIN and HOROWICZ (1959)	0.67	27	4.7	9.2	4.0	17
Rat diaphragm muscle	CREESE (1954)	3.8	4.3	4.5	60 in vitro (18 in vivo)	28	38
	CREESE (1964)	7.32	—	—	—	—	38
Mouse toe muscle	CREESE and EL-SHAFIE (1964)	4.32	—	—	—	—	—
Cat motoneurone	ITO and OSHIMA (1964 b)	40	4.6	51	15	77	37—38
Electric eel electric organ	AUBERT, CHANCE and KEYNES 1964	240	15	1000	10	1000	24

* Na_i^s and Me^s are the intracellular sodium concentration and the sodium efflux at the steady state respectively.

References

ARAKI, T., ITO, M. & OSCARSSON, O. (1961). *J. Physiol.* (Lond.) **159**, 410.

ARAKI, T., ITO, M., KOSTYUK, P. G., OSCARSSON, O. & OSHIMA, T. (1962). *Nature* (Lond.) **196**, 1319.

ARAKI, T., ITO, M., KOSTYUK, P. G., OSCARSSON, O. & OSHIMA, T. (1965). *Proc. Roy. Soc.* B, in press.

AUBERT, X., CHANCE, B. & KEYNES, R. D. (1964). *Proc. Roy. Soc.* B, **160**, 211.

BONTING, S. L., SIMON, K. A. & HAWKINS, N. M. (1961). *Arch. Biochem. Biophys.* **95**, 416.

BONTING, S. L., CARAVAGGIO, L. L. & HAWKINS, N. M. (1962). *Arch. Biochem. Biophys.* **98**, 413.

BROOKS, C. McC. & ECCLES, J. C. (1947). *J. Neurophysiol.* **10**, 349.

COOMBS, J. S., ECCLES, J. C. & FATT, P. (1955 a). *J. Physiol.* (Lond.) **130**, 291.

COOMBS, J. S., ECCLES, J. C. & FATT, P. (1955 b). *J. Physiol.* (Lond.) **130**, 326.

CREESE, R. (1954). *Proc. Roy. Soc.* B. **142**, 497.

CREESE, R. (1964). *Nature* (Lond.) **201**, 505.
CREESE, R. & EL-SHAFIE, A. L. (1964). *J. Physiol.* (Lond.) **171**, 33 P.
ECCLES, J. C., ECCLES, R. M. & ITO, M. (1964 a). *Proc. Roy. Soc.* B. **160**, 181.
ECCLES, J. C., ECCLES, R. M. & ITO, M. (1964 b). *Proc. Roy. Soc.* B. **160**, 197.
FRAZIER, H. S. & KEYNES, R. D. (1959). *J. Physiol.* (Lond.) **148**, 362.
GELFAN, S. & TARLOV, I. M. (1955). *J. Neurophysiol.* **18**, 170.
GREENGARD, P. & STRAUB, R. W. (1958). *J. Physiol.* (Lond.) **144**, 442.
HARREVELD, A. VAN (1946). *Amer. J. Physiol.* **147**, 669.
HARREVELD, A. VAN & OCHS, S. (1956). *Amer. J. Physiol.* **187**, 180.
HARREVELD, A. VAN & SCHADÉ, J. P. (1959). *J. Cell. Comp. Physiol.* **54**, 65.
HELLER, I. H. & ELLIOT, K. A. C. (1955). *Canad. J. Biochem. Physiol.* **33**, 395.
HODGKIN, A. L. (1951). *Biol. Rev.* **26**, 339.
HODGKIN, A. L. (1958). *Proc. Roy. Soc.* B. **148**, 1.
HODGKIN, A. L. & HOROWICZ, P. (1959). *J. Physiol.* (Lond.) **145**, 405.
HODGKIN, A. L. & KATZ, B. (1949). *J. Physiol.* (Lond.) **108**, 37.
HODGKIN, A. L. & KEYNES, R. D. (1955). *J. Physiol.* (Lond.) **128**, 28.
HODGKIN, A. L. & KEYNES, R. D. (1956). *J. Physiol.* (Lond.) **131**, 592.
HOLMES, E. G. (1930). *Biochem. J.* **24**, 914.
ITO, M. & OSHIMA, T. (1964 a). *Proc. Roy. Soc.* B **161**, 92.
ITO, M. & OSHIMA, T. (1964 b). *Proc. Roy. Soc.* B **161**, 109.
ITO, M. & OSHIMA, T. (1964 c). *Proc. Roy. Soc.* B **161**, 132.
KEYNES, R. D. (1954). *Proc. Roy. Soc.* B. **142**, 359.
KEYNES, R. D. & SWAN, R. C. (1959 a). *J. Physiol.* (Lond.) **147**, 591.
KEYNES, R. D. & SWAN, R. C. (1959 b). *J. Physiol.* (Lond.) **147**, 626.
KOLMODIN, G. M. & SKOGLUND, C. R. (1959). *Acta physiol. scand.* **45**, 1.
KOREY, S. R. & ORCHEN, M. (1959). *J. Neurochem.* **3**, 277.
LEVI, H. & USSING, H. H. (1948). *Acta physiol. scand.* **16**, 232.
LLOYD, D. P. C. (1953). *J. gen. Physiol.* **36**, 673.
MAIZELS, E. (1954). *Symp. Soc. exp. Biol.* **8**, 202.
MCLENNAN, H. (1957). *Biochem. biophys. Acta* **24**, 333.
MCLENNAN, H. (1958). *Pflüg. Arch. ges. Physiol.* **267**, 453.
RENKIN, E. M. (1961). *J. gen. Physiol.* **44**, 1159.
SCHOFFENIELS, E. (1959). *Ann. N. Y. Acad. Sci.* **81**, 285.
ZERAHN, K. (1955). *Acta physiol. scand.* **33**, 347.

COMPARISON OF TENDON AND BONE REPAIR

by J. J. PRITCHARD

Science has been defined as the search for invariants, or the attempt to "see the wood for the trees". In a campaign to sort out the essential features of a complex phenomenon, it is usually profitable to begin by analysing a simpler, but analogous phenomenon. Thus Sir JOHN ECCLES spent some years in the study of synaptic behaviour in the superior cervical ganglion before tackling the spinal cord. By the same token, the analysis of the events following transection of a tendon makes it easier to appreciate the nature and significance of the events following transection of a bone. Yet, strangely, the process of tendon repair is very poorly documented when compared with the great wealth of literature dealing with bone repair. I propose therefore to summarise the results of some recent investigations on tendon undertaken in the Department of Anatomy in Belfast (MILLER 1962; MILLER, PAISLEY and PRITCHARD 1963; PAISLEY 1963), and to discuss their implications for bone repair.

When the tendo Achillis of a rat, guinea-pig or rabbit is cleanly severed half-way along its length, the two ends are immediately pulled apart by contraction of the soleus-gastrocnemius-plantaris complex, and its anterior tibial antagonists, producing a gap 1—2 cm. wide, which persists. Each retracted tendon end consists of the tough, coarse, parallel, collagenous bundles of the *tendon proper;* the loose, vascular connective tissue network, or *endotenon,* lying between the tendon bundles; and the inner tendon sheath or *epitenon,* a cellular, vascular membrane closely bound to the outer surface of the tendon proper. The outer tendon sheath, or *paratenon* is really the innermost layer of the connective tissue wall of the tunnel through which the tendon glides; between it and the epitenon there is normally a gliding plane with few fibres and almost no cells. The paratenon therefore does not retract with the tendon, as the epitenon does, and it remains intact except where it has been damaged by the knife at the time of operation, though it does collapse down on the small amount of extravasated blood which occupies the former site of the tendon.

Within a few hours the tendon ends, and the damaged paratenon, show signs of inflammation, with hyperemia and oedema; and then the connective tissue cells begin to proliferate. The earliest, most constant, and most striking feature is the widespread multiplication of the epitenon cells. Paratenon cells also multiply rapidly where the paratenon has been damaged. Later the endotenon cells and the so called "tenocytes" of the tendon proper begin to multiply. This last statement has been disputed (SKOOG and PERSSON 1954), but from the distribution of mitotic figures I believe there is little room for doubt that all the cells take part, and not just the epitenon cells as has sometimes been affirmed. A few polymorphonuclear leucocytes, lymphocytes, macrophages and fibroblasts invade the blood clot in the paratenon tube between the tendon ends, but organisation of the clot is neither an early nor a conspicuous feature of tendon repair. As PAGET (1854) pointed out many years ago, repair goes on *around* rather than inside the clot, and virtually unaltered clot may be found inside what appears from the outside to be a perfectly repaired tendon.

Cellular proliferation is naturally most intense near the cut edges of the tendon, so that soon the ends become onion-shaped. Sometimes the bulbous ends are called "reparative blastemas", "stump caps" or "regeneration cones". In the process of forming the cones the coarse collagen bundles of the original tendon gradually disappear, partly it would seem because the bundles have been broken up into individual strands as a result of cellular infiltration and oedema, but almost certainly also because of dissolution and resorption of fibres. The histological appearances are not unlike those of the pelvic ligaments in some rodents at the time of parturition, where fibrous structures with relatively few cells are transformed into cellular masses with relatively few fibres.

At first the regeneration cones grow in width, but soon they begin to grow in length, and extend towards each other inside the empty paratenon sheath. Within a few days the cones, and the proliferated paratenon, have fused into a common mass termed the tendon "callus". Whether the cones fuse end to end, or whether a length of thickened paratenon intervenes, depends on a number of factors, including the degree of trauma initially to the tendon ends and paratenon, and the amount of blood clot. In any case the callus is a composite structure derived from paratenon, epitenon, endotenon, tendon proper, and even from connective tissue outside the paratenon, in cases where trauma has not been confined to the tendon and its sheaths. Generally, however, the greatest contribution is made by the epitenon. To begin with the callus is a highly vascular, cellular mass with few fibres. The cells are chiefly elongated fibroblasts and most of them are orientated approximately parallel to the original long axis of the tendon. Where the paratenon has been severed, however, there is less regularity and the cells tend to be in whorls. Collections of macrophages and foreign body giant cells are also present, chiefly about residual fibrin from the original blood clot.

As time goes on, fine spirally-running fibres appear between the cells. These increase and form thicker and straighter bundles, orientated, like the fibroblasts, parallel to the long axis of the tendon, though they never quite achieve the straightness and regularity of the fibre bundles of the original tendon.

After about four weeks the callus separates into a repaired segment of tendon, and a repaired paratenon sheath, with a new gliding plane in between. Functional recovery is now virtually complete, with a tendon of normal tensile strength able to slide freely inside its connective tissue tunnel. Anatomical recovery, however, is never perfect, though slow remodelling appears to go on for a very long time. Histological studies suggest that collagenous fibres continue to be resorbed, and new ones to be laid down, for some months. Surprisingly, the tendon and muscle never regain their original lengths, the tendon remaining longer, and the muscle bellies shorter, than normal. Whether or not this leads to appreciable inefficiency has not been ascertained.

A curious feature of tendo Achillis repair in the rat, and one of considerable theoretical importance, is that about six weeks after operation fibrocartilage and bone are almost always to be found replacing parts of the repaired tendon (BUCK 1953; MILLER 1962; PAISLEY 1963). These ectopic tissues are most abundant in young animals: they mature into ossicles with a cortex of lamellar bone and a medulla filled with either haemopoietic or fatty marrow. For some reason this does not happen in the rabbit and guinea-pig, or in other tendons so far studied in the rat. This suggests that the conditions, presumably mechanical in nature, which lead to heterotopic ossification, are rather narrowly defined.

The repair of severed toe-extensor tendons in these animals followed similar lines, except for the absence of heterotopic ossification. By contrast, other workers have found that the repair of flexor tendons, severed within their tough fibrous sheaths, is either imperfect, or takes place through the migration of paratenon cells into the gap (LIPSCOMB and WAKIM 1961).

From the welter of events which follow the transection of a long bone, it is relatively simple, and highly instructive, to abstract ones which parallel those just described as taking place in the course of tendon repair. The following account is based on the study of a wide range of bony fractures in rats, rabbits, guinea-pigs, lizards, frogs and salamanders (PRITCHARD and RUZICKA 1950; PRITCHARD 1961; PRITCHARD 1963; PRITCHARD 1964). Immediately on fracture the ends of the bone tend to be displaced by muscular, gravitational and other applied forces, though the actual amount of separation, angulation, overlap and distraction varies enormously from fracture to fracture. In general, however, there will be a bony gap, which is at least partly occupied by a collection of extravasated blood.

Unlike the tendon, the bone has no gliding plane between outer and inner sheaths, and the epitenon and paratenon are represented by the single periosteum. When the bone has been cleanly incised at operation the periosteum is cut cleanly with it, and adheres to the bone fragments. The fracture gap is therefore in such cases bounded on either side by the bone ends, but externally by extra-periostal connective tissues, and muscles. In the case of fractures produced by external violence, however, the periosteum is often stripped from the bone rather than torn across, and then the fracture gap is bounded externally by separated periosteum. It is reasonable to regard the bone marrow and the contents of the vascular canals of the cortex as the analogue of the endotenon, and to use the term *endosteum* to signify them (though admittedly the term is not always used in this sense). The analogue of the tendon proper is then the bone proper, and of the tenocytes, the osteocytes.

Within a few hours of fracture the periosteum and extraperiosteal tissues become grossly hyperemic and oedematous: the endosteum in its rigid container also becomes hyperemic, but only as space is made available through extrusion of marrow into the fracture gap, or resorption of marrow cells into the general circulation. Surprisingly little blood is extravasated in experimental fractures in small laboratory animals, and as in the tendon, repair tends to go on around the blood clot rather than at its expense, at least in the early stages. There is a great deal of variation here. Where the fracture is almost perfectly immobilized from the start, organization of the blood clot begins almost immediately; but where there is movement, organization is delayed until sufficient firm callus has been formed around the fracture site to immobilize it secondarily. If reasonable immobilization is never achieved, then the blood clot forms the fibrinous lining of the pseudo-synovial cavity of a false joint.

A day or so after injury the connective tissue cells of the periosteum, extra-periosteal tissues and endosteum in the vicinity of the fracture enlarge, and then multiply very rapidly. To begin with the periosteal changes are most marked. Some distance from the fracture conspicuous basophilic cells accumulate in the depths of the periosteum, adjacent to the bone, resurrecting the fibrous and osteogenic layers of the immature periosteum. Nearer the fracture, however, so great is the proliferation that the periosteum becomes many times its normal thickness, and the typical fibrous structure is lost; in its place there is a mass of plump cells, coursing between which a few widely separated, residual collagen fibres are to be seen. The bone ends, with their thickened periosteum, are now not unlike the "regeneration cones" which replace the ends of the divided tendon.

The extraperiosteal tissues around the fracture meanwhile, through cellular proliferation, have developed into a tissue resembling the proliferated periosteum, and fusing with the latter, help to form a single spindle-shaped mass of cellular callus around the fracture. Sooner or later this is joined by cells from the endosteum, though in many fractures, and for some considerable time, remains of the blood clot persist at the centre of the callus, and impede the advance of the endosteal contribution.

In tendon repair the resorption of the original tendon bundles, though very probable, is difficult to prove. In bone repair, however, resorption of the original bone near the fracture is a very prominent feature, the bone being replaced by cellular callus derived principally from the proliferated endosteum.

In both bone and tendon repair, orientated collagen fibres appear at an early stage between the callus cells: in tendon repair collagen fibres continue to accumulate, and this happens in the peripheral parts of fracture callus also: but next to the old bone, both inside and out, the young connective tissue is rapidly transformed into either bone or cartilage, in contrast to tendon callus which only develops bone and cartilage in rare circumstances, and then only at a late stage. Bone and cartilage continue to extend out from the original bone ends at the expense of the fibrous callus. In favourable circumstances a complete bridge of bone develops across the fracture, either directly, or following the prior transformation of the fibrous callus into fibro-cartilage. In less favourable circumstances fibrous union persists indefinitely.

Assuming that a bony bridge has been formed, to begin with this will be cancellous and voluminous. Gradually over the weeks and months it is remodelled into a cortex and medulla which harmonize with the remains of the old bone on each side. The residual fibrous callus externally forms a new periosteum.

Two major questions emerge from this analysis. Why do bone and cartilage always develop in fracture callus, but not as a rule in tendon callus: and why does ossification sometimes bridge the fracture gap quickly, sometimes slowly, and sometimes not at all?

Two main theories have been put forward in answer to the first question. One is that the cells of the endosteum and deeper layers of the periosteum are specialized for bone and cartilage formation, and warrant separate designation as *osteogenic cells* (HAM 1930) unlike the cells of the paratenon, epitenon and endotenon which are ordinary fibroblasts. The other is that some substance (osteogenin) exudes from the damaged bone and induces the cells of the callus to make bone and cartilage instead of fibrous tissue (LACROIX 1951). At present, histological evidence from in vivo and in vitro studies favours the "specific osteogenic cell" theory in so far as the early production of bone and cartilage is concerned. Later on, in many fractures, bone and cartilage appear to develop in fibrous tissue derived from non-specific fibroblasts, and it is probable that some chemical or physical induction is involved, though the inducing agent has not yet been discovered. A plausible theory is that active osteoblasts are able to bring about the infective transformation of neighbouring fibroblasts into osteoblasts along an "advancing ossification front" (URIST and JOHNSON 1943).

In answer to the second question it may be said that bony union is rapidly achieved when the anatomy of the fracture is such that osteogenic cells, and not fibroblasts, have early access to the fracture gap. This is the case with incomplete fractures, especially in cancellous bone; when the gap is exceedingly narrow; when the periosteum has not been completely torn across, especially in young animals; and when immobilization is perfect. In most experimental fractures, however, fibroblasts enter the gap, and a stage of fibrous union is reached. Transformation to bony union is then much slower, and may fail. At present we do not know the mechanism by which ossification spreads at the expense of fibrous tissue, though it is a common enough phenomenon in skeletal development; for example, where tendons attach to bones, and at sutures. From our experience with experimental fractures, however, we would conclude that ossification is favoured by a narrow gap, and by circumstances which impose strong and constant tension or pressure on the fibrous tissue. Fibrous union tends to persist in wide gaps; with imperfect immobilization; and with neutral or variable tensile and compressive forces.

Tendon repair is much easier to achieve because it is a fibrous union that is required in any case. However, this too is achieved under the direction of mechanical forces, and in the absence of tension on the callus the new tendon tissue has little strength and the fibres are not orientated properly (BUCK 1953).

Most physiological processes are controlled through the integrative actions of the nervous and endocrine systems. Repair processes, however, are ad hoc and autonomous. In the case of bone and tendon repair the complex and highly organized cellular activity seems to be controlled, as in the embryo, by instructions emanating from the cells themselves, and passed directly from one to the other, except in so far as the field of mechanical forces acts as an overall directing influence.

In bone and tendon repair we witness homeostatic regulation on the grand scale, explicable partly in terms of the capacity for self-integration possessed by a mass of young connective tissue cells, and partly in terms of the integrative action of the "mechanical field". Unconsciously applied mechanical forces, however, are commonly not sufficient of themselves to guide the repair process to the desired end. As a general rule the intelligent intervention of a medical team, and intelligent behaviour on the part of the patient, are also necessary if an acceptable degree of approximation to the *status quo ante* is to be achieved.

References

BUCK, R. C. (1953). *J. Path. Bact.* **66,** 1.
HAM, A. W. (1930). *J. Bone Jt Surg.* **12,** 827.
LACROIX, P. (1951). *The Organization of Bones.* London: Churchill.
LIPSCOMB, P. R. & WAKIM, K. G. (1961). *Proc. Staff Mayo Clin.* **36,** 271.
MILLER, R. L. (1962). B. Sc. (Hons.) *Thesis.* Univ., Belfast.
MILLER, R. L., PAISLEY, P. B. & PRITCHARD, J. J. (1963). *J. Anat.* (Lond.) **97,** 304.
PAGET, J. (1854). *Lectures on Surgical Pathology.* Philadelphia: Lindsay and Blakiston.
PAISLEY, P. B. (1963). B. Sc. (Hons.) *Thesis.* Univ., Belfast.
PRITCHARD, J. J. (1961). *Bone and Bones.* In *Recent Advances in Anatomy* (ed. GOLDBY, F. & HARRISON, R. J.) p. 204. London: Churchill.
PRITCHARD, J. J. (1963). *Bone Healing.* In *The Scientific Basis of Medicine Annual Reviews.* p. 286.
PRITCHARD, J. J. (1964). *Histology of Fracture Repair.* In *Modern Trends in Orthopaedics.* p. 69.
PRITCHARD, J. J. & RUZICKA, A. J. (1950). *J. Anat.* (Lond.) **84,** 236.
SKOOG, T. & PERSSON, B. H. (1954). *Plastic and Reconstructive Surg.* **13,** 384.
URIST, M. R. & JOHNSON, R. W. (1943). *J. Bone Jt Surg.* **25,** 375.

DENDRITIC SYNAPTIC PATTERNS: EXPERIMENTS WITH A MATHEMATICAL MODEL

by WILFRID RALL

Comment on real and simulated experiments

Progress in experimental biology occurs when a new technique or combination of techniques makes it possible for an investigator to seek and obtain answers to new questions; each of us can give numerous illustrations of this. A similar statement holds also for progress in theoretical biology. In particular, the combination of a well-chosen mathematical model and a modern computation facility make it possible to obtain answers to new questions; some of the resulting investigations can be characterized as computer simulation of experiments, or as experiments with a mathematical model.

At first glance, many neurophysiologists might be tempted to believe that the difference between a real biological system and a mathematical model of this system would be too great for anything of relevance to come from a simulated experiment. However, a more careful look should temper such views. On the one hand, any good experimental neurophysiologist can provide an extensive catalog of the unphysiological conditions of most experiments; the fact that we have come to accept the artificiality of such experiments does not do away with a fundamental uncertainty about the relation of the experimental preparation to the normal intact animal. On the other hand, the development of high speed digital computers makes it possible to include at least some biological complexity in our mathematical models; it is no longer necessary to make as many oversimplifications for the sake of computability.

Once a suitable mathematical model has been programmed for a high speed computer, it is easier to control and explore the effects of various combinations of input and various combinations of model parameters in simulated experiments, than it is to achieve corresponding control and exploration of a complicated experimental preparation. Consider, for example, a complicated neurophysiological experiment which is possible in principle but extremely difficult in practice; a simulation of this experiment can help one determine what results might be expected and thus help one decide whether the proposed expenditure of experimental effort is likely to be worthwhile.

There is, of course, no substitute for those experiments whose great value (*a posteriori*) arises from an unexpected result (i. e. one which differs significantly from our expectations). What, however, is the source of our expectations? In simple situations, we rely upon physical intuition; in more complicated situations, we can supplement physical intuition with soundly based mathematical models. Simulated experiments permit us to explore complicated systems and demonstrate properties that are not intuitively obvious. By providing quantitative expectations, they can contribute to the design of biological experiments with high information value.

The comparison of real and simulated experiments is essentially the well established discipline of subjecting the predictions of our scientific hypotheses to experimental test. In neurophysiology, as in other fields of science, the achievement of insight and greater theoretical understanding of actual biological systems depends upon such comparisons and such testing.

Soma-dendritic distributions of synaptic input

Consider the following question: what are the quantitative differences between an excitatory postsynaptic potential (EPSP) that results from brief synaptic excitatory input delivered to a neuron soma and the EPSP that results from an equal synaptic input delivered to the dendritic periphery of the same neuron? Suggestive experimental evidence has been provided by FADIGA and BROOKHART (1960), cf. also KATZ and MILEDI (1963). However, even where it may be possible to find two afferent pathways which synapse selectively (i. e. one set of synapses near the soma and the other set distributed over the dendritic periphery), it is very difficult to obtain quantitative control. How can the two inputs be made equal? Can they be made equal in time course? Can they be made equally free of complication from delayed synaptic input delivered by inter-neurons?

All of these uncertainties can be avoided in the case of a simulated experiment. The uncertainties of experimental control are replaced by uncertainties about the adequacy of the mathematical model. Taken together, real and simulated experiments complement each other; when well matched, each can help to reduce the uncertainty associated with the other.

A detailed presentation of mathematical models is not appropriate to the present volume. A mathematical treatment of excitatory and inhibitory conductance changes has been presented (RALL 1962) and applied to certain classes of dendritic trees. By means of a compartmental model (RALL 1964), it is now possible to compute the effects of any specified spatio-temporal pattern of membrane conductance changes distributed over any specified branching system.

Computed effects of four dendritic locations

The neuronal receptive surface is here represented as a chain of ten compartments; each compartment represents an equal amount of membrane surface area; see uppermost diagram in Fig. 1. The neuron soma is represented by compartment No. 1, while compartment No. 10 represents a lumping of the most peripheral portions of all the dendritic trees of this neuron model; compartments 2 through 9 represent divisions of electrotonic distance, beginning with the dendritic trunks and extending to the dendritic periphery.

The four diagrams (A, B, C and D in Fig. 1) show four different locations of otherwise identical square pulses of excitatory conductance. Each of the four curves in Fig. 1 represents a simulated EPSP as computed at the soma compartment for the indicated synaptic input location. These computed transients are not unlike the muscle end plate potentials of FATT and KATZ (1951, Figs. 5 and 8). The situations are not, of course, fully comparable.

Effects of two simulated spatio-temporal patterns

Here we consider the effect of combining the four input locations of Fig. 1 into the two spatio-temporal sequences, A → B → C → D and D → C → B → A. The results shown in Fig. 2 do not represent superpositions of the curves in Fig. 1;

they represent soma transients computed for the two specified sequences of membrane conductance change. The amplitude of the EPSP computed for sequence A → B → C → D is not significantly greater than that computed for case A in Fig. 1; however, the amplitude of the EPSP computed for sequence D → C → B → A is nearly twice as great. Both of these curves differ significantly from the dotted curve, which shows the computed effect of an equal amount of

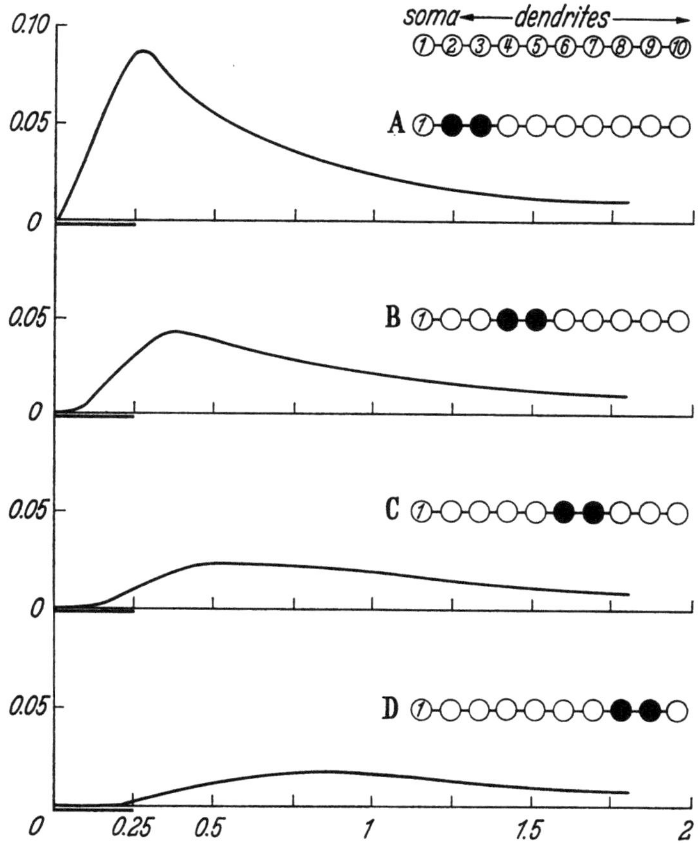

Fig. 1. Effects of four dendritic locations of a square excitatory conductance pulse upon computed EPSP at soma. In each case, the excitatory conductance in two compartments (shown black) is increased from a zero resting value to a value equal to the resting conductance of the membranes; this square conductance change has a duration (indicated by the black bar) equal to one fourth of the passive membrane time constant. Time scale is expressed relative to the passive membrane time constant; ordinate scale expresses membrane depolarization relative to the excitatory driving potential; see RALL (1964) for further details

excitatory membrane conductance that was distributed uniformly over the space and time occupied by the two spatio-temporal patterns. Three conclusions seem warranted.

(i) A temporal sequence of synaptic inputs does not result in a unique EPSP unless the soma-dendritic locations of these inputs are specified.

(ii) For maximal peak of the soma membrane depolarization, the peripheral dendritic synaptic input should be delivered earlier than the somatic or proximal dendritic synaptic input.

(iii) For rapid achievement and maintenance of a steady soma depolarization level, a brief somatic or proximal dendritic synaptic input followed by a prolonged sequence of peripheral dendritic synaptic input would be very effective.

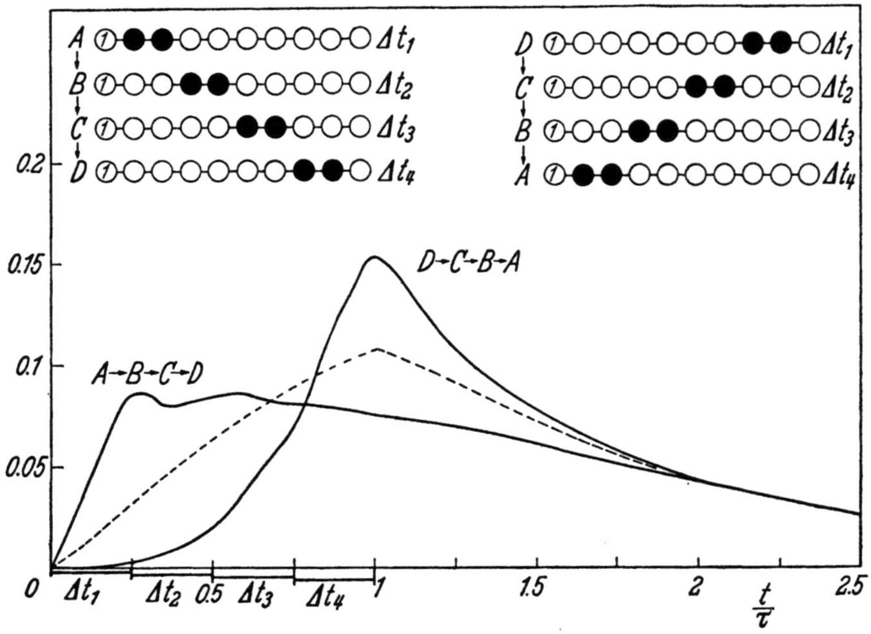

Fig. 2. Effect of spatio-temporal input pattern; see text. The component conductance changes are the same as in Fig. 1; the scales are in the same units as in Fig. 1

Concluding comment

Additional computations have been carried out to simulate inhibitory synaptic input, and to simulate various spatio-temporal combinations of excitation and inhibition (RALL 1962, 1964, and unpublished computations). These results have convinced several neural modelers that they should not neglect spatio-temporal dendritic patterns in their attempts to understand neuronal input-output relations, and in their attempts to simulate small systems of neurons. I hope that experimental neurophysiologists will also find these results to be of interest to them in the design of their experiments, and in the interpretation of their results.

Acknowledgements

I wish to express my appreciation to Dr. MONES BERMAN and Mrs. MARJORY WEISS for their advice and help in the use of their general computer program to obtain the results in Figs. 1 and 2; the computations were done on the IBM 7090 at the National Bureau of Standards, with the support of the National Institutes of Health. Reproduction of this article is permitted for any purpose of the

United States Government. These results were presented at the Symposium on Neural Theory and Modeling which was held at Ojai, California, December 4, 5, and 6, 1962.

References

FADIGA, E. & BROOKHART, J. M. (1960). *Am. J. Physiol.* **198**, 693.
FATT, P. & KATZ, B. (1951). *J. Physiol.* (Lond.) **115**, 320.
KATZ, B. & MILEDI, R. (1963). *J. Physiol.* (Lond.) **168**, 389.
RALL, W. (1962). *Ann. N. Y. Acad. Sci.* **96**, 1071.
RALL, W. (1964). In *Neural Theory and Modeling* (ed. by. REISS, R. F.) p. 73. Palo Alto: Stanford University Press.

THE EFFECT OF DRUGS ON THE REFLEX PATHS TO PRIMARY AFFERENT FIBRES

by ROBERT F. SCHMIDT

In this review it is intended to present a short account of the pharmacology of primary afferent depolarization and of presynaptic inhibition of the spinal cord. In contrast to the detailed knowledge of the functional organization of the presynaptic inhibitory reflex pathways (for reviews see ECCLES 1964 a, b), there are relatively few studies on the pharmacological aspects of presynaptic inhibition. This is to be deplored because these few studies have not only shown that in the spinal cord presynaptic inhibition has a pharmacology quite distinct from the various types of postsynaptic inhibition, but they have also provided evidence for a better understanding of the functional organization and synaptic mechanism of presynaptic inhibition.

Primary afferent depolarization (PAD) can most easily be recorded as a positive potential wave from the dorsal surface of the spinal cord (P-wave), or as the negative going dorsal root potential (DRP) from a dorsal rootlet (BARRON and MATTHEWS 1938). The recording of DRPs is a very convenient, though indiscriminative, procedure for the study of pharmacological actions on PAD. These potentials are sufficiently stable over the long periods of time that are required for the full development of a drug action and even for its decline. They have therefore been used extensively for pharmacological studies on PAD.

Anaesthetics. ECCLES and MALCOLM (1946) reported that pentobarbital (nembutal) greatly prolongs the decay of the DRP. This observation has meanwhile been confirmed both in the cat (Lloyd 1952; Eccles et al. 1963) and in the isolated frog and toad spinal cord (Schmidt 1963). The studies of Eccles, Schmidt and Willis (1963) and Schmidt (1963) have revealed that small doses of nembutal not only prolong but also increase the DRP, whereas higher doses lead to a further prolongation together with a reduced amplitude. Other barbiturates and chloralose have similar effects. The amplitude and time courses of presynaptic inhibition follow closely the variations of the DRP (Eccles et al. 1963, Figs. 1—3).

All other anaesthetics so far tested (chloral-hydrate, paraldehyde, ether, urethane) shorten the decaying phase of the DRP instead of prolonging it (Schmidt 1963). This shortening is the more obvious the longer the declining phase of the DRP was before the application of the anaesthetic. The most likely explanation for this phenomenon is that the anaesthetic curtails the after-discharge of the interneurones on the pathway responsible for the PAD. Regularly with smaller doses of anaesthetics the DRPs are also increased. This is probably caused by a hyperpolarization of the primary afferent fibres due to the removal of spontaneous background activity of the interneurones mediating the PAD. Such a removal would increase the efficiency of the primary afferent fibres producing the PAD, and would also increase the potential difference between the resting potential and the equilibrium potential of PAD, thus enhancing the synaptic drive during activation.

There is as yet no explanation for the prolongation of the DRP during barbiturate anaesthesia. Eccles et al. (1963) proposed that the prolongation was due either to the inhibition of an enzyme that destroys the transmitter substance producing the PAD, or that the anaesthetics act by impeding the diffusional spread of the transmitter substance from its site of action (cf. also Schmidt 1964). More recently Wall (1964) and Mendell and Wall (1964) have postulated that nembutal acts by prolonging the discharge of the interneurones on the pathway of PAD. They argue that the declining phase of the PAD is far too long to be a passive process, and that it must be sustained by a continuous interneuronal discharge. Eccles et al. (1963) had found that nembutal invariably decreased the rate of discharge of interneurones firing spontaneously or after orthodromic activation. This fits well with the observation that all anaesthetics, even the barbiturates, lead in the beginning of anaesthesia to a shortening of the PAD together with an increase of its amplitude. Wall (1964) and Mendell and Wall (1964) could not find a prolongation of interneuronal discharge during nembutal anaesthesia either. They therefore concluded that PAD is not mediated by interneurones like the D-cells proposed by Eccles, Kostyuk and Schmidt (1962), but by very small cells like those of the Substantia gelatinosa Rolandi which are too small to be detected by a searching microelectrode. It seems difficult to imagine that these cells should be the only ones in the spinal cord which prolong their discharge in the course of barbiturate and chloralose anaesthesia, but behave similarly to all other cells when exposed to any other type of anaesthetic.

Convulsants. It is well established that strychnine depresses postsynaptic inhibition in the spinal cord, and the convulsant activity of strychnine and similarly acting drugs seems to be fully explained by this depression (for review see CURTIS 1963; ECCLES 1964 a). There are several early reports in the literature (UMRATH 1933; BREMER 1941, 1944, 1953; DUN and FENG 1944; ECCLES and MALCOLM 1946) of the action of strychnine on the DRP of the frog spinal cord. All authors agree that the size of the DRP evoked by dorsal root stimulation is not appreciably altered, but that the declining phase is greatly prolonged. More recent re-investigations of the action of strychnine on the PAD both in the cat and in the isolated toad spinal cord have confirmed the prolongation of the declining phase of the DRP, and have shown that the amplitude of the PAD is increased in the early stages of strychnine poisoning whereas in the later stages there is a reduction of the PAD amplitude (cf. SCHMIDT 1963; Fig. 7). In the cat the increased and prolonged DRPs are paralleled by an increased and prolonged presynaptic inhibition of monosynaptic reflexes (ECCLES et al. 1963; Fig. 8).

The prolongation and increase of the DRP are likely to be caused by a removal of postsynaptic inhibition on the cells of the pathway of PAD, and a consequently enhanced and prolonged interneuronal discharge, whereas the decrease of the amplitude in the later stages of strychnine poisoning is probably due to the occlusion produced by the "spontaneous" activity of the preparation. Thus strychnine does not seem to have any direct action on the presynaptic inhibitory synapse.

Picrotoxin, and several other convulsants such as pentamethylentetrazol, do not affect post-synaptic inhibition, and there is as yet no fully satisfactory explanation of their convulsant action. A possible suggestion would be that these drugs depress presynaptic inhibition. In the isolated toad spinal cord picrotoxin depressed the DRPs in concentrations which enhanced the ventral root reflex discharges (SCHMIDT 1963, Fig. 8). ECCLES et al. (1963) found in the spinal cord of the cat that picrotoxin reduced the dorsal root potentials and diminished presynaptic inhibition. With higher concentrations the decreased DRP was accompanied by convulsant activity of the preparation. It was pointed out by these authors that this depression was probably not due to convulsant activity occluding the interneuronal pathways responsible for the PAD, since strychnine enhanced and prolonged the PAD rather than depressed it. Furthermore picrotoxin depresses presynaptic inhibition at a dosage below that causing convulsions. It has therefore been concluded that picrotoxin produces a genuine depression of the presynaptic inhibitory synapses, and that the convulsant activity of picrotoxin is at least in part due to this depression of presynaptic inhibition. Possibly this action is attributable to a competitive occupation of the receptor sites for the presynaptic inhibitory transmitter substance. Such a mechanism has already been proposed to account for the blocking action of picrotoxin at the inhibitory synapse of the crayfish neuromuscular junction (Robbins and van der Kloot 1958; Grundfest, Reuben and Rickles 1959; Grundfest and Reuben 1961). The striking difference between the inhibitory depressant actions of strychnine and picrotoxin suggests further that either different transmitter substances or quite different receptor sites, or both, are present at postsynaptic and presynaptic inhibitory synapses.

Cholinergic drugs. In isolated toad spinal cords Kiraly and Phillis (1961) have shown that the DRP which is generated by stimulation of ventral roots can be depressed by dihydro-β-erythroidine, and to a smaller extent by curare. This indicates that the reflex pathway contains a cholinergic synapse. Possibly this synapse corresponds to the synapse made by the motor axon collaterals at the Renshaw cells in the cat spinal cord.

The DRPs evoked both in the cat and in the toad by all other types of afferent inputs are not affected by a wide variety of drugs active at cholinergic synapses (Kiraly and Phillis 1961; Eccles et al. 1963). Thus dihydro-β-erythroidine, gallamine triethiodide (Flaxedil), atropine, nicotine, tetraethyl-pyrophosphate and eserine had no action on presynaptic inhibition when injected intravenously or applied topically in concentrations active on known cholinergic synapses. Therefore, it seems unlikely that these pathways contain a cholinergic synapse.

Amino acids. A wide variety of amino acids is known to increase or decrease the excitability of postsynaptic elements and several of the amino acids have been proposed as possible transmitter substances (for review see Curtis 1963). Among the most potent members of the depressant group are the neutral amino acids γ-amino-butyric acid (GABA) and 3-amino-1-propane-sulphonic acid. In crustacea GABA has an action which is identical with the inhibitory transmitter substance at several types of inhibitory synapses (for literature see Grundfest et al. 1959; Florey 1961; Curtis 1963); and this similarity also obtains for presynaptic inhibitory action (Dudel and Kuffler 1961). Picrotoxin acts as a blocking agent at these synapses, depressing both the inhibitory synaptic action and the action of GABA (Elliott and Florey 1956; Robbins and van der Kloot 1958; van der Kloot 1960; Florey 1961).

Since presynaptic inhibition in the cat is reversibly antagonized by picrotoxin, the effect of GABA and 3-amino-1-propane-sulphonic acid on PAD has been tested (Eccles et al. 1963). It was found that both substances reduce and shorten the DRPs and increase the dorsal root reflexes. Both amino acids also depressed the DRP in the isolated toad spinal cord (Schmidt 1963). Simultaneously they increased the excitability of the primary afferent fibres. However, the excitability increases were much smaller than would have been expected from the large DRP reduction. It has to be assumed that these amino acids reduce the DRP, not only by depolarizing the presynaptic fibres, but also by depressing the interneurones on the pathway to the presynaptic synapses and possibly by decreasing the space constant for electrotonic transmission along the primary afferent fibres. A depressant action of GABA and 3-amino-1-propane-sulphonic acid on the neurones of the motor-reflex pathways has already been reported by CURTIS, PHILLIS and WATKINS (1961).

Acidic amino acids such as glutamic acid act as excitants on postsynaptic elements. CURTIS et al. (1961) found that glutamic acid increased the ventral root reflex discharge of isolated toad spinal cords and depolarized motoneurones. PHILLIS (1960) reported that in the same preparation glutamic acid reduced the DRP evoked by dorsal root stimulation and this finding has been confirmed by SCHMIDT (1963). A comparison of the DRP depression and the excitability increase of primary afferent fibres showed that glutamic acid, unlike GABA and 3-amino-1-propane-sulphonic acid, probably directly depolarizes the primary afferent fibres. An alternative explanation would be that glutamic acid stimulates the interneurones liberating the presynaptic transmitter. Rather large concentrations of all three amino acids are required to obtain appreciable excitability increases and it seems unlikely therefore, as has been tentatively suggested by ECCLES et al. (1963), that any of these acids are closely related to the transmitter substance.

Except for the specific blocking action of dihydro-β-erythroidine on the DRP evoked by ventral root stimulation, all pharmacological effects reported so far have been found in all types of PAD independent of their origin. Meanwhile there are also reports of drug actions affecting only one or the other system of PAD. ANDÉN, LUNDBERG, ROSENGREN and VYKLICKÝ (1963) found that the DRPs caused by depolarization of the "flexor reflex afferents" (FRA) are markedly reduced after intravenous injection of L-3,4-dihydroxy-phenylalanine (DOPA), whereas DRPs evoked from Group I muscle afferents are not affected. The transmission from the FRA to motoneurones and ascending pathways is also depressed. When the short latency effects of FRA stimulation are suppressed by DOPA, sural nerve stimulation evokes a powerful long latency reflex in the tenuissimus nerve (flexor) together with a PAD of Group I fibres from flexors and extensors (ANDÉN, JUKES, LUNDBERG and VYKLICKÝ 1964). The various effects of DOPA can be inhibited by the adrenergic antagonists phenoxybenzamine and chlorpromazine.

Furthermore, both reserpine treatment and administration of the decarboxylase inhibitor meta-hydroxy-benzylhydrazine prevent the action of DOPA, whereas the monamine oxidase inhibitor nialamine potentiates the action of DOPA about tenfold. It was therefore suggested (ANDÉN, JUKES and LUNDBERG 1964) that DOPA has no direct central action but acts only through the formation of noradrenaline. DL-5-hydroxy-tryptophane (5 HTP), the precursor of 5-hydroxy-tryptamine (5 HT), has effects similar to those of DOPA (ANDÉN et al. 1964). Here tests with the monamine oxidase and decarboxylase inhibitors indicated that at least part, if not all, of the effects of 5 HTP are caused by 5 HT liberation.

Since there is histochemical and biochemical evidence of descending pathways with monoaminergic nerve terminals in the spinal cord (Carlsson, Falck, Fuxe and Hillarp 1964), Lundberg and his co-workers suggested from the foregoing evidence that the powerful descending inhibitory pathways to primary afferents may be monoaminergic, and that DOPA and 5 HTP act by inducing synthesis and overflow of noradrenaline or 5 HT from their synaptic terminals.

Conclusion. In the reflex paths to primary afferents there are numerous sites at which a pharmacological substance can act. By recording the DRP or testing the excitability of primary afferent fibres only the endproduct of a long chain of events is registered. Therefore, our ability to identify the exact point of action of a particular drug is rather limited. However, the experiments reviewed above have shown that, despite the rather crude methods, pharmacological studies have considerably increased our knowledge and our understanding of PAD and presynaptic inhibition.

References

Andén, N. E., Jukes, M. G. M. & Lundberg, A. (1964). *Nature* (Lond.) **202**, 1222.

Andén, N. E., Jukes, M. G. M., Lundberg, A. & Vyklický, L. (1964). *Nature* (Lond.) **202**, 1344.

Andén, N. E., Lundberg, A., Rosengren, E. & Vyklický, L. (1963). *Experientia* (Basel) **19**, 654.

Barron, D. H. & Matthews, B. H. C. (1938). *J. Physiol.* (Lond.) **92**, 276.

Bremer, F. (1941). *Arch. int. Physiol.* **51**, 211.

Bremer, F. (1944). *Arch. int. Pharmacodyn.* **69**, 249.

Bremer, F. (1953). In: *Ciba Found. Symp. The Spinal Cord.* p. 78. London: Churchill.

Carlsson, A., Falck, B., Fuxe, K. & Hillarp, N. (1964). *Acta physiol. scand.* **60**, 112.

Curtis, D. R. (1963). *Pharmacol. Rev.* **15**, 333.

Curtis, D. R., Phillis, J. W. & Watkins, J. C. (1961). *Brit. J. Pharmacol.* **16**, 262.

Dudel, J. & Kuffler, S. W. (1961). *J. Physiol.* (Lond.) **155**, 543.

Dun, F. T. & Feng, T. P. (1944). *J. Neurophysiol.* **7**, 327.

Eccles, J. C. (1964 a). *The Physiology of Synapses.* Berlin: Springer-Verlag.

Eccles, J. C. (1964 b). In: *Progress in Brain Research,* Vol. 12, *Physiology of Spinal Neurones* (ed. by Eccles, J. C. & Schadé, J. P.) p. 65. Amsterdam: Elsevier.

Eccles, J. C., Kostyuk, P. G. & Schmidt, R. F. (1962). *J. Physiol.* (Lond.) **161**, 237.

Eccles, J. C. & Malcolm, J. L. (1946). *J. Neurophysiol.* **9**, 139.

Eccles, J. C., Schmidt, R. F. & Willis, W. D. (1963). *J. Physiol.* (Lond.) **168**, 500.

Elliott, K. A. C. & Florey, E. (1956). *J. Neurochem.* **1**, 181.

Florey, E. (1961). *Ann. Rev. Physiol.* **23**, 501.

Grundfest, H. & Reuben, J. P. (1961). In: *Nervous Inhibition* (ed. by Florey, E.) p. 92. Oxford: Pergamon Press.

Grundfest, H., Reuben, J. P. & Rickles, N. H. (1959). *J. gen. Physiol.* **42**, 1301.

Kiraly, J. K. & Phillis, J. W. (1961). *Brit. J. Pharmacol.* **17**, 224.

Lloyd, D. P. C. (1952). *Cold Spr. Harb. Symp. quant. Biol.* **17**, 203.

Mendell, L. M. & Wall, P. D. (1964). *J. Physiol.* (Lond.) **172**, 274.

Phillis, J. W. (1960). Ph. D. Thesis, Australian National University, Canberra.

ROBBINS, J. & VAN DER KLOOT, W. G. (1958). *J. Physiol.* (Lond.) **143**, 541.
SCHMIDT, R. F. (1963). *Pflügers Arch. ges. Physiol.* **277**, 325.
SCHMIDT, R. F. (1964). In: *Progress in Brain Research*, Vol. 12, *Physiology of Spinal Neurones* (ed. by. ECCLES, J. C. & SCHADÉ, J. P.) p. 119. Amsterdam: Elsevier.
UMRATH, K. (1933). *Pflügers Arch. ges. Physiol.* **233**, 357.
VAN DER KLOOT, W. G. (1960). In: *Inhibition in the Nervous System and Gamma-Aminobutyric Acid.* (ed. by ROBERTS, E.) p. 409. Oxford: Pergamon Press.
WALL, P. D. (1964). In: *Progress in Brain Research*, Vol. 12, *Physiology of Spinal Neurones* (ed. by ECCLES, J. C. & SCHADÉ, J. P.) p. 92. Amsterdam: Elsevier.

THE ROLE OF SEGMENTAL REFLEX MECHANISMS IN THE REGULATION OF BREATHING

by T. A. SEARS

Comparatively little information is available concerning the role of proprioceptive discharges in the regulation of breathing, especially with regard to their segmental reflex actions. Some information concerning their importance has come from studies of the effects of cutting the cervical and thoracic dorsal roots. COOMBS (1918) found that section of thoracic dorsal roots greatly diminished intercostal movements in the cat, and even abolished them in the kitten (COOMBS and PIKE 1930). STELLA (1938) claimed that cutting all the thoracic and cervical roots is without effect on breathing, but his cats showed only diaphragmatic respiration before the dorsal roots were cut, so that his results cannot be held as being at variance with those of COOMBS on costal respiration as stated by some authors. In confirmation of STELLA (1938), SANT'AMBROGIO, WILSON and FRAZIER (1962) found no evidence from cervical dorsal root section for a significant proprioceptive afferent inflow from the cat diaphragm. On the other hand, a temporary paralysis of the hemi-diaphragm results from unilateral cervical dorsal root section in man (NATHAN and SEARS 1960), thus indicating a possible species variation. SEARS (1958; see also 1964 c) limited the section of thoracic dorsal roots usually to one segment and determined electromyographically its effects on the intercostal and abdominal muscles innervated from that segment. The recruitment both in frequency and number of expiratory motor units evoked by closing the trachea at the height of inspiration (the inspiratory-inhibitory, expiratory activating reflex of BREUER) was greatly diminished by section of the dorsal roots. A weakening of costal respiratory movements after limited dorsal root section has also been observed in man (NATHAN and SEARS 1960).

Reflex contractions in response to stretching of respiratory muscles (abdominal wall) were first described by Cleland and Tait (1927). Sustained, stretch-induced reflex contractions of the intercostal muscles, dependent on the integrity of the dorsal roots, were clearly demonstrated electromyographically by Ramos and Mendoza (1959) in spinal rabbits and cats. They also described a spinal reflex activation of the intercostal muscles by inflation and deflation of the chest. Spinal reflex activation of inspiratory muscles during deflation of the chest was described by von Euler and Fritts (1963). Sears (1958) reported an "inflation" reflex of exspiratory muscles during chest inflation in spinal cats under artificial respiration, and he later confirmed this in recordings of efferent discharges in nerve filaments innervating the internal intercostal muscles (Sears 1963, 1964 a).

The weakening of respiratory movements by sectioning the dorsal roots was studied quantitatively by von Euler and Fritts (1963). They measured the maximal intra-tracheal pressures developed by the respiratory efforts against a closed trachea, after inflating and deflating the lungs to known volumes. The changes in pressure (tension or "output") per unit change in volume (length or "input") before and after either cutting the dorsal roots or vagotomy, gave a measure of the closed loop gain (cf. Granit 1958) of the somatic proprioceptive and pulmonary control systems. The gain of the two systems was approximately equal, although for increasing lung volumes it was of opposite sign, the inspiratory pressure at a given lung volume being diminished by dorsal root section and increased by vagotomy.

Tracheal closure may be regarded as the extreme case of added resistance to airflow. From his analysis of the effects of sudden resistive loading of respiration in man, Fleisch (1928) described a series of proprioceptive reflexes arising from the respiratory musculature. However Nierkerk and Ter Braak (1935) obtained similar results which they attributed to the viscous-elastic properties of the respiratory muscles. The same view is also held by Riedstra and Dirken (1953) because they obtained similar results from simulated experiments on cadavers.

The effects of respiratory loading on breathing in man have been examined by Campbell, Dinnick and Howell (1961) and Campbell, Dickinson, Dinnick and Howell (1961), and also discussed in an important theoretical paper by Campbell and Howell (1962). Using both "elastic" and "threshold" loading there was a progressive increase in the force of inspiratory contraction during the first to fifth loaded breaths. From these studies it was concluded that the effects of loading are probably entirely due to chest wall reflexes (Campbell and Howell 1962), although pulmonary afferents in the vagus may play a "permissive" role (Campbell, Dickinson et al. 1961). A complicating factor is that the responses to "threshold" and large "elastic" loads in conscious, spontaneously breathing patients with traumatic tetraplegia (transections at C.6) are similar to those of normal subjects (private communication from Dr. J. Newsom Davis).

Collectively, the experiments cited above clearly demonstrate the importance of somatic proprioceptive reflexes from the respiratory musculature, but they provide no direct information as to the essential nature of the reflexes nor of the mechanism of their regulation. The remainder of this brief review is devoted to these problems.

Using intracellular recording ECCLES, SEARS and SHEALY (1962) found that excitatory postsynaptic potentials (EPSPs) are evoked monosynaptically in thoracic motoneurones by excitation of the low threshold afferent fibres that give rise to the Ia component (cf. ECCLES, ECCLES and LUNDBERG 1957 a) of the afferent volley recorded at the root entry zone (SEARS 1964 c). As the largest diameter afferent fibres of the intercostal nerves are present in the muscular branches (SEARS 1964 b), the origin of the Ia fibres from the intercostal spindles may be assumed with some certainty. Contrary to expectations, no reciprocal "direct" inhibition was found between inspiratory and expiratory motoneurones with regard to the Ia afferent fibres from their respective muscles (SEARS 1964 d). The relatively low amplitudes of the monosynaptic EPSPs (mean I.ImV) evoked by stimulation of the ipsi-segmental internal intercostal nerves probably accounts for the absence of reflex discharges with a monosynaptic latency in previous studies on the intercostal reflexes (DOWNMAN 1953, 1955). When advantage is taken of the convergence onto expiratory motoneurones of the Ia afferent fibres of the same and adjacent segments, by stimulating all three nerves, monosynaptic firing is obtained (unpublished results), although a pre-requisite may still be summation with the depolarizing phase of the "central respiratory drive potentials" (ECCLES et al. 1962; SEARS 1964 e).

At high frequencies of repetition (greater than 60 c/sec) the temporal summation of monosynaptic EPSPs causes a relatively smooth depolarization of the motoneurones sufficient to cause their sustained, repetitive firing. Discharge frequencies as high as 40 c/sec have been so evoked in previously inactive motoneurones (SEARS 1964 a). This finding demonstrates the effectiveness of this pathway (cf. EULER and FRITTS 1963), which presumably subserves the reflex responses of the intercostal muscles to stretch.

Recent physiological studies on the efferent and afferent innervation of the intercostal muscle spindles have revealed an important role for this receptor and its reflexes in breathing. By recording efferent discharges in intercostal nerve filaments SEARS (1962) found that the fusimotor neurones innervating intercostal muscle spindles fire with a respiratory rhythm during normal breathing. Fusimotor neurones innervating inspiratory muscle spindles fire maximally during inspiration, while those innervating expiratory muscle spindles fire maximally in expiration. This reciprocal activation during breathing is supra-spinal in in origin and is altered in a similar way to the response of alpha respiratory motoneurones by such factors as hypoxia, hypercapnia, hyperventilation and the Hering Breuer reflexes (SEARS 1963, 1964 c). Independently, CRITCHLOW and von Euler (1962, 1963) showed that the inspiratory muscle spindles (primary endings) discharged maximally during inspiration and so indicated that a rhythmic, fusimotor activation must exist to offset the unloading of the spindle that would otherwise occur during shortening of the extrafusal muscle fibres. This activation was partly due to a spinal reflex drive of the fusimotor neurones from the rhythmic movements of the chest wall. However, the close temporal relation between the increased afferent discharge and the volume (chest) record, indicated a close linkage between the fusimotor and alpha motor systems, most probably at a supraspinal level. Subsequently Eklund, von Euler and Rutkowski (1963, 1964) confirmed their previous conclusions also by recording efferent activity in the intercostal nerve filaments.

From these studies a concept of the role of the spindle loop in breathing has emerged which both confirms the suggestion by Nathan and Sears (1960) that the "indirect" route of muscle activation (cf. Eldred, Granit and Merton 1953) is involved in breathing, and fulfills certain of the requirements suggested by Campbell and Howell (1962) in their evaluation of the significance of such a regulatory mechanism.

The rhythmic activation of the alpha and fusimotor neurones during breathing represents the demand for a certain change in length of the respiratory muscles, i. e., to a demand for a tidal volume appropriate to the prevailing chemical drive. This demand applies equally to the expiratory phase, since in experiments with recordings from the lower, dorsal part of the chest, periodic, expiratory fusimotor neurone discharge has been an invariable finding, and also because the expiratory alpha motoneurones are acted upon by sub-liminal excitatory synaptic drives during quiet breathing (Eccles et al. 1962; Sears 1964 e). Under these circumstances the spindle measures any misalignment between the rate of intrafusal and extrafusal muscle shortening (cf. Eldred et al. 1953). When the latter lags on the former, for example, the increased afferent discharge would monosynaptically excite the relevant alpha motoneurones so increasing the force of contraction until the misalignment is reduced and the demanded muscle length attained. Conversely, monosynaptic excitation may be withdrawn from the motoneurones. This control mechanism would of course also respond in a similar way to an increased chemical drive on respiration. In the cat this regulation appears to be confined principally to the thorax (cf. Stella 1938; Sant'Ambrogio et al. 1962). Gill and Kuno (1963) found no evidence for monosynaptic excitation of cat phrenic motoneurones by phrenic afferents, although the latter carry proprioceptive discharges (Yasargil 1962).

The operation of this control mechanism is at present understood only in the most simple terms. But, as Eklund et al. (1964) so pertinently point out, "A naturally occurring regulated process, such as respiration, offers the promise of revealing principles of motor control which are inaccessible to an experimenter restricted to imitation of motor performance by electrical stimuli".

References

CAMPBELL, E. J. M., DICKINSON, C. J., DINNICK, O. P. & HOWELL, J. B. L. (1961). *Clin. Sci.* **20**, 223.
CAMPBELL, E. J. M., DINNICK, O. P. & HOWELL, J. B. L. (1961). *J. Physiol.* (Lond.) **156**, 260.
CAMPBELL, E. J. M. & HOWELL, J. B. L. (1962). In: *Ciba Foundation Symposium on Pulmonary Structure and Function* (ed. by REUCK, A. V. S. & O'CONNOR, M.) p. 29. London: Churchill.
CLELAND, J. G. P. & TAIT, J. (1927). *Quart. J. exp. Physiol.* **17**, 179.
COOMBS, H. C. (1918). *Amer. J. Physiol.* **46**, 459
COOMBS, H. C. & PIKE, F. (1930). *Amer. J. Physiol.* **95**, 681.
CRITCHLOW, V. & VON EULER, C. (1962). *Experientia* (Basel) **18**, 426.
CRITCHLOW, V. & VON EULER, C. (1963). *J. Physiol.* (Lond.) **168**, 820.
DOWNMAN, C. B. B. (1953). In: *Ciba Foundation Symposium on the Spinal Cord.* London: J. A. Churchill.
DOWNMAN, C. B. B. (1955). *J. Neurophysiol.* **18**, 217.
ECCLES, J. C., ECCLES, R. M. & LUNDBERG, A. (1957). *J. Physiol.* (Lond.) **136**, 527.
ECCLES, R. M., SEARS, T. A. & SHEALY, C. N. (1962). *Nature* (Lond.) **193**, 844.
EKLUND, G., VON EULER, C. & RUTKOWSKI, S. (1962). *Acta physiol. scand.* **57**, 481.
EKLUND, G., VON EULER, C. & RUTKOWSKI, S. (1964). *J. Physiol.* (Lond.) **171**, 139.
ELDRED, E., GRANIT, R. & MERTON, P. (1953). *J. Physiol.* (Lond.) **122**, 498.
FLEISCH, A. (1928). *Pflüg. Arch. ges. Physiol.* **219**, 706.
GILL, K. P. & KUNO, M. (1963). *J. Physiol.* (Lond.) **168**, 274.
GRANIT, R. (1958). *J. Physiol.* (Lond.) **143**, 387.
NATHAN, P. W. & SEARS, T. A. (1960). *J. Neurol.* **23**, 10.
NIEKERK, VAN, J. & BRAAK, TER, J. W. G. (1935). *Pflüg. Arch. ges. Physiol.* **236**, 44.
RAMOS, J. G. & MENDOZA, E. L. (1959). *Acta physiol. latinoamer.* **9**, 257.
RIEDSTRA, J. W. & DIRKEN, M. N. J. (1953). *Acta physiol. pharm. neerl.* **3**, 19.
SANT'AMBROGIO, G., WILSON, M. F. & FRAZIER, D. T. (1962). *J. appl. Physiol.* **17**, 829.
SEARS, T. A. (1958). *J. Physiol.* (Lond.) **142**, 35 P.
SEARS, T. A. (1962). *Aust. J. Science,* **25**, 102.
SEARS, T. A. (1963). *Nature* (Lond.) **197**, 1013.
SEARS, T. A. (1964 a). In: *Physiology of Spinal Neurones* (ed. by ECCLES, J. C. & SCHADÉ, J. P.) p. 259. Amsterdam: Elsevier.
SEARS, T. A. (1964 b). *J. Physiol.* (Lond.) **172**, 150.
SEARS, T. A. (1964 c). *J. Physiol.* (Lond.) **174**, 295.
SEARS, T. A. (1964 d). *J. Physiol.* (Lond.) **175**, 386.
SEARS, T. A. (1964 e). *J. Physiol.* (Lond.) **175**, 404.
STELLA, G. (1938). *J. Physiol.* (Lond.) **93**, 10.
VON EULER, C. & FRITTS, J. W. (1963). *Acta physiol. scand.* **57**, 284.
YASARGIL, G. M. (1962). *Helv. Physiol. Acta* **20**, 39.

IMPLICATIONS OF CENTRAL PROJECTION OF C FIBERS IN SENSORY PERCEPTION

by C. NORMAN SHEALY

Since the first description of the compound action potential of peripheral nerve by GASSER and ERLANGER (1927), there has been considerable debate concerning the function of the various components, later fractionated into A (alpha-beta-gamma-delta), and C fibers. Extensive investigations of the central afferent pathways have also been carried out. The question of function has not been settled, although there is evidence that impulses generated by painful stimuli are carried by the smaller fibers (gamma-delta and C), even though those groups of fibers are also concerned with other sensations; and that the primary function of the larger (beta) fibers is to conduct impulses associated with touch and proprioception (Douglas and Ritchie 1962). Central projections of the beta fibers have been traced over the ipsilateral dorsal columns with crossing to the contralateral medial lemniscus, contralateral thalamic nucleus ventralis and to the somatic sensory cortex. Gamma and delta responses have also been followed through dorsal roots to the contralateral anterior quadrant of the spinal cord, and through the medulla and ventral tegmentum to the posterolateral ventral nucleus of thalamus (Collins and Randt 1961).

Central projections of C fibers have been much more elusive. Beyond the dorsal roots, evoked potentials to C fibers stimulation have been reported in the caudal medulla and ventral tegmentum with no somatopic localization (Collins and Randt 1961). Clinical experiences with pain relief from anterolateral cordotomy suggest that most fibers concerned with pain cross in the anterior spinal commissure to the opposite anterolateral spinal cord and then ascend to the brain. In at least one previous electrophysiological investigation, attempts to find C responses in the spinal cord were unsuccessful (Collins, personal communication). Clarification of these pathways seems essential, however, in order to lead to a better understanding of the role C fibers in sensation.

Methods

Adult mongrel cats were anaesthetized with cyclopropane, and after tracheotomy anaesthesia was maintained with nitrous oxide and oxygen using a Bird respirator. A laminectomy was performed from the tenth thoracic to the seventh lumbar vertebrae and the animal suspended on a stereotaxic instrument. The left posterior tibial nerve was exposed for stimulation and recording; the site of recording was 8 to 10 cm proximal to the site of stimulation. Platinum electrodes electrically shielded to the tips were used. Evoked potentials were recorded by monopolar steel electrodes, which had been electrolytically ground to tips of 10 μ, insulated with Lucite 45 to within 30 μ of the tips and encased in glass shields. Both the nerve action potentials and cord evoked responses were observed on a 555 or 565 Tektronix oscilloscope and were recorded with a Grass Kymograph camera. An American Electronics Laboratory Stimulator was used. In some experiments recording electrodes were also introduced into the caudal medulla or tegmentum. At the end of the experiments significant recording loci were marked electrolytically and the spinal cord or brain was sectioned serially.

Results

A response corresponding to the delta wave of peripheral nerve can be obtained routinely from the contralateral anterior quadrant of the cord. Responses to C fiber stimulation are much more difficult to elicit. In the posterior tibial nerve the C wave in these experiments was about 10 to 25 microvolts in amplitude. The conduction velocity in the nerve was 0.8 to 1.0 meter/sec (animal's rectal temperature between 34° and 37°).

Many attempts to record C responses in the cord were unsuccessful. However, three patterns were found. In the spinal commissure at L-6 (the posterior tibial nerve of cat contains L-6, L-7 and S-1 fibers) repetitive firing was elicited by a single stimulus above C threshold (18—30 volts, 0.5 msec duration) (Fig. 1) This response was very similar to that seen in the tegmentum after a tetanizing stimulus and called by Collins a "flare" (Collins and Randt 1960). Undoubtedly the first portion of the response stems from delta fibers, although the late portion is from C fibers. This flare did not appear at stimulus intensities below the threshold for C fibers and the last two-thirds remained when the larger fibers were blocked by an inductorium coil. On the other hand, this response to single stimuli has not been found outside the commissure.

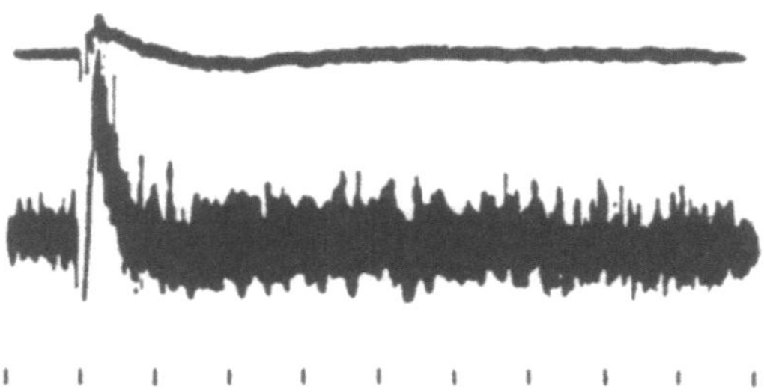

Fig. 1. Top line: Peripheral nerve response. Small "C" response occurs at 130 msec. Bottom line: Spinal cord potential recorded in spinal commissure at L-6. Note prolonged "flare". Time calibration: 50 msec

In the lower thoracic and upper lumbar anterior quadrant just lateral to anterior horn a small response (10 microvolts) could be detected when the background noise was low (Figs. 2 a and 2 b). This occured after the typical delta response,

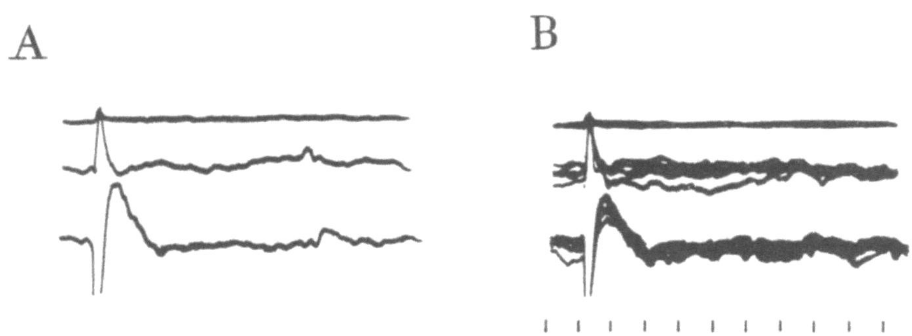

Fig. 2a. "C" response from anterior spinal cord. Top line: Peripheral nerve. Middle line: Cord surface. Bottom line: Anterior quadrant of cord. Amplifier filters cut to high gain of 50, low of 8 cycles/sec. Cord response occurs at 325 msec

Fig. 2b. Ten (10) stimuli superimposed to help overcome background noise. Note that the response at 350 msec on bottom line (anterior cord) remains. There is similarly a suggestion of same response on the surface of the cord (middle line). Time calibration: 50 msec

the threshold was that of peripheral C fibers and the conduction velocity was 0.7 meter/sec. Conditions have not yet been satisfactory to record this response after larger fibers were blocked, but the characteristics listed suggest that it is related to C fibers.

The third response was a typical flare to tetanic stimulation and was similar to that found in the tegmentum (Fig. 3). It was recorded at the same site as the second response described above and was elicited by 15—20 stimuli at a frequency of 50/sec and strength at or above the threshold for C fibers. The flare lasted for 500 to 1000 msec. and the latter two thirds of the response remained after inductorium blocking of the larger fibers.

Fig. 3. Anterior cord "flare", lasting 700 msec after 300 msec train of stimuli

A similar response could be elicited in the medulla, at the medial portion of the subnucleus ventrales, nucleus medulla oblongata centralis and in the ventral tegmentum at the dorsomedial margin of the red nucleus above the brachium conjunction. At these sites no response to a single C stimulus has been found, although there was a characteristic response for delta fiber activation. These findings were similar to those of COLLINS and RANDT (1956 and 1960). No response to C fiber stimulation has been found in the ventrolateral portion of the thalamus, either to single stimuli or to trains of stimuli. Further studies are in progress.

Discussion

Many studies have been carried out of the C fiber characteristics in peripheral nerve. Effects of temperature, various other physiological stimuli and of many drugs have been so studied. It is well known, for instance, that un-myelinated (C) fibers outnumber myelinated fibers in dorsal roots by 3 or 4 to 1 (DOUGLAS and RITCHIE 1962) and this suggests that these fibers are important in sensory perception. However, previous studies of central projections of C fibers have been sparse and incomplete, almost certainly because of problems inherent in recording central responses. Several papers are of note. COLLINS and RANDT (1956) recorded the delta response in the contralateral anterior quadrant of the cord, with a 30—40 second time course and 50—75 microvolt amplitude. These investigators were unable to find responses associated with stimulation of peripheral C fibers, although strenuous attempts were made to do so (COLLINS — personal communication). Later, however, COLLINS and RANDT (1958) were able to record a response from the medulla in the region of the nucleus oblongata centralis, subnucleus ventralis. This response could be obtained only by repetitive stimulation and consisted of a "flare" of firing which lasted for about 400 msec. In 1960, an almost identical response from the tegmentum was reported (COLLINS and RANDT 1960). This response was related to C fiber stimulation since it persisted after the larger fibers were blocked in the peripheral nerve. COLLINS and RANDT (1961) have also reported evoked potentials from the thalamus, but no response to C fiber stimulation was noted. GAZE and GORDON (1954) recording in the thalamus were able to obtain only 2 photographs of responses apparently due to C fiber activation (one in monkey; another in cat). Both of these show multiple spikes occurring considerably later than the beta-gamma-delta responses.

No other central responses to C fiber stimulation have been reported. Therefore, the current findings are of considerable interest. First, the "flare" to a single stimulus which was found only in the lumbar commissure may indicate that many impulses are never propagated beyond the local area, or that projection pathways are much more diffuse beyond the commissure. The occasional finding of a response in the spinal cord at an appropriate time further suggests that C fiber impulses evoked by a single stimulus may project more centrally, but it is possible that they are attenuated at higher levels, or that multiple stimuli are necessary for central activation. This deserves much more investigation. On the other hand, repetitive firing to a tetanic stimulus is shown now to be present in the lumbar and lower thoracic cord as well as in the medulla and tegmentum.

It is of considerable interest that repetitive discharges similar to the flare had been seen in peripheral nerve after injection of acetylcholine (DOUGLAS and RITCHIE 1960). Thus repetitive firing set up by single or repetitive electrical stimuli may have definite physiological significance. This pattern must be important in spontaneous, continuous, suffering pain as opposed to the brief sharp pain produced by pin-testing. If no central response to single stimulation of C fibers occurs, these two types of pain can be readily explained. Furthermore, other sensations known to be mediated by C fibers (cooling, itch) require the same prolonged firing for their sustained appreciation. WEDDELL (1955) suggested that sensation is the result of a complex interpretation of spatially and temporally dispersed patterns of impulses. He and many others strongly oppose the theory of specificity of sensory fibers. On the other hand pain has not been reported from activation of the larger fibers. Pain first appears at delta threshold. Furthermore, when these larger fibers are blocked by inductorium or cold, repetitive stimulation of the remaining C fibers elicits severe pain (COLLINS, personal communication). While it is quite possible that a few large fibers may be spared or that a few C fibers may be blocked under such conditions, at least the major portions of such fibers are separated. Also, destruction of the posterior columns of the spinal cords, into which beta fibers project, does not abolish the appreciation of either pain or temperature. It has also been observed that suffering pain can be relieved by either spinal commissurotomy or incomplete anterior spinal cordotomy, with preservation of pin-prick appreciation (BALLANTINE, JR. and

Bertrand, personal communications). Thus it can be concluded that both the type of fiber and pattern of activation participate in sensory appreciation. Available evidence suggests that brief sharp pain is mediated by delta fibers and that prolonged suffering pain results from repetitive C fiber stimulation.

Summary

Electrical potentials having characteristics related to C fiber stimulation have been recorded in the spinal cord of the cat for the first time. These consist of both repetitive firing and single responses in the contralateral anterior quadrant of the spinal cord. Similar repetitive responses have been found in the medulla and tegmentum. No response to a single stimulus applied to C fibers has been found in the brain. These findings are discussed in relation to sensory perception.

Acknowledgement

This research was supported by Grant B-779 from The National Institute of Neurological Diseases and Blindness, National Institutes of Health, Bethesda, Maryland.

References

Collins, W. F. & Randt, C. T. (1956). *J. Neurophysiol.* **19**, 438.
Collins, W. F. & Randt, C. T. (1958). *J. Neurophysiol.* **21**, 345.
Collins, W. F. & Randt, C. T. (1960). *J. Neurophysiol.* **23**, 47.
Collins, W. F. & Randt, C. T. (1961). *Arch. Neurol.* **5**, 100.
Douglas, W. W. & Ritchie, J. M. (1960). *J. Physiol.* (Lond.) **150**, 501.
Douglas, W. W. & Ritchie, J. M. (1962). *Physiol. Rev.* **42**, 297.
Gasser, H. S. & Erlanger, J. (1927). *Amer. J. Physiol.* **80**, 522.
Gaze, R. M. & Gordon, G. (1954). *Quart. J. Exp. Physiol.* **39**, 279.
Weddell, G. (1955). *Ann. Rev. Psychol.* **6**, 119.

SOME ASPECTS OF CEREBELLAR PHARMACOLOGY

by P. E. Voorhoeve

Recently a new impetus has been given to cerebellar research with the analysis of an inhibitory mechanism in the feline cerebellar cortex (Andersen, Eccles and Voorhoeve 1963 a, 1964). Concomitantly, and in accordance with Szenta-

GOTHAI's (1964) histological approach, these authors showed that basket cells in the cerebellum are inhibitory to Purkinje cells (P-cells). This was based on the intracellular recording of excitatory postsynaptic potentials (EPSPs) and inhibitory postsynaptic potentials (IPSPs) from P-cells upon stimulation of parallel fibers, on the analysis of focal potentials, and on the cessation of the spontaneous or the induced P-cell firing which was produced by parallel fibre stimulation.

SZENTAGOTHAI discussed the extent and distribution of the basket cell axons. He suggested that the simultaneous excitation of basket cells and P-cells by impulses in parallel fibers, which would be produced by the activation of a small cluster of granule cells by way of incoming mossy fibres, would result in an activated strip of P-cells with a zone of inhibited P-cells on either side. This is exactly what has been found electrophysiologically, and provides a beautiful example of lateral inhibition in the central nervous system.

Based on these findings, some experiments on drug induced activation and on the inhibition of P-cells have now been performed.

I. Cholinoceptive activation of Purkinje cells

Acetylcholine (ACh) had been administered earlier by other workers with varying results. Topical application caused depression or activation, depending on previous treatment with eserine (CREPAX, NIGRI and PARMEGGIANI 1956). Enhancement or suppression of spontaneous unitary activity, depending on concentration, has also been recorded (COCEANI, FADIGA and GESSI 1962). Intra-arterial injection of small amounts of ACh augmented the electrical activity as recorded from the surface of the cerebellum (CROSSLAND and MITCHELL 1956).

Cell activation induced by either of these methods need not necessarily be due to direct synaptic activation by the drug in question, and is open to many other interpretations. On the other hand, the electrophoretic administration of a drug in the immediate vicinity of a single cell gives more precise information with respect to the possible transmitter function of this substance. The requirements which such a substance has to fulfil before it can be accepted as a transmitter have been rigidly stated by PATON (1958) and CURTIS (1961).

On the basis of the findings of ANDERSEN et al. (1963 a), which aided the identification of P-cells, ACh was administered electrophoretically in the cerebellar cortex and was shown to activate P-cells (CRAWFORD, CURTIS, VOORHOEVE and WILSON 1963 a). The excitation of the P-cells by ACh differed in many respects from that of Renshaw cells in the spinal cord for which there is good evidence of cholinergic transmission from axon collaterals. In the cerebellum the onset of activation was usually slow and often followed a period in which the spontaneous activity was reduced. When the electrophoretic ejection ceased, the firing induced by ACh persisted for as long as one minute. Carbamylcholine and dl-muscarine were more potent than ACh whereas acetyl-β-methylcholine and nicotine were approximately equal in potency to ACh. The excitation of some cells by ACh was blocked by dihydro-β-erythroidine; in other cases no block was observed. A specific blocking action of atropine could not be demonstrated when this substance was administered electrophoretically but was readily observed after intravenous injection. Gallamine triethiodide excited P-cells.

These results conform in general with those obtained at other sites in the brain where cells have been found to be excited by ACh. It seems that the ACh effect has a more muscarinic character at these levels, whereas the Renshaw cell activation is predominantly nicotinic. The recent finding of both muscarinic and nicotinic receptors upon Renshaw cells (CURTIS and RYALL 1964) suggests, in the light of the above pharmacological findings, that P-cells have more muscarinic than nicotinic receptors.

The finding of cholinoceptive activation of P-cells is in contrast with a recent report by MCCANCE and PHILLIS (1964). These authors located ACh sensitive cells at greater depths, and supposed some of them to be within the granular layer. The possibility that granule cells are cholinoceptive, however, seems to be excluded by more recent results of CRAWFORD and CURTIS (personal communication). With one five-barrel micropipette a P-cell was located which could be excited with either DL-homocysteic acid (DLH) or carbamylcholine. With another similar micropipette DLH and carbamylcholine were ejected into the granule layer at distances up to 500 μ from the P-cell, both in and out of line of the parallel fibre connection between these cells. When DLH was passed into the granular layer both excitation and depression of the distant P-cell were observed. Such alterations in activity were of short latency and presumably were due to excitation of granule cells and basket cells by the excitant amino acid. Similar results were not observed with carbamylcholine, and it is unlikely therefore that granule cells are cholinoceptive.

The content of cholineacetylase and ACh within the cerebellum is low, whereas that of acetylcholinesterase (AChE) is relatively high. The latter enzyme appears to be associated with cells in the granular layer. According to KOELLE (1963) only scattered large cells in this layer exhibit an intense AChE staining, and numerous small granular cells show hardly any or no staining at all (cf. AUSTIN, PHILLIS and STEELE 1964; SNELL 1961). If, as has been supposed earlier (GEREBTZOFF 1959), the granule cells are cholinergic, the excitation of both P-cells and basket cells by impulses in parallel fibres should be by acetylcholine. This would be in accordance with DALE's postulate of the release of one identical transmitter from all terminals of the one neurone. One would therefore expect that a suitable ACh antagonist would depress the synaptic potentials evoked by impulses in the parallel fibres to both types of cells. Intravenously administered atropine, however, in doses as high as 1 mg/kg, did not notably alter the size of the focal potentials recorded in the molecular layer which represent the extracellularly recorded EPSPs and IPSPs of P-cells. Furthermore, neither dihydro-β-erythroidine nor atropine appeared to depress the excitation of P-cells produced by stimulation of the inferior olive, lateral cuneate nucleus or lateral reticular nucleus.

These results show that ACh excites P-cells and although this action may be specific, not all criteria for establishing it as a true transmitter have been fulfilled. Intracellular recording from P-cells during electrophoretic drug application has not been attempted and thus conductance changes produced by ACh have not been measured. Extracellular records have not been obtained from cells which could be identified as basket cells (ANDERSEN et al. 1964), and which should also be activated by ACh if the parallel fibres were cholinergic. If the parallel fibre system was indeed cholinergic a higher content of ACh and cholineacetylase might be expected in the molecular layer because the parallel fibre synapses are so numerous. It is possible that other cells in the granular layer which send axons into the molecular layer (Lugaro cells — Fox 1959) may be cholinergic. In view of the apparent non-cholinoceptive nature of granule cells, and the recent location of acetylcholinesterase at the mossy fibre synapses upon these cells (MEAD and VAN DER LOOS 1964; SHUTE and LEWIS 1965) it might even be proposed that neither the presence of AChE in the cerebellar cortex nor the excitation of P-cells by cholinomimetics is related to a role of acetylcholine as a cerebellar transmitter.

II. Convulsant drugs

Strychnine has always been a popular drug for topical administration in the central nervous system. On the cerebellum, however, it does not evoke synchronous discharges as it does on the cerebral cortex, but produces high-frequency bursts of individual cells as has been shown by many authors (cf. DOW and MORUZZI 1958). Intravenous injection, on the other hand, produces synchrony in the bursts, but it has been shown that this is induced by afferent volleys from bulbo-spinal convulsions (BREMER and BONNET 1953; BREMER and GERNANDT 1954). When administered electrophoretically to P-cells strychnine had an excitatory action and in high concentrations, a depressant effect not unlike that produced by local anaesthetics.

Strychnine, either locally of systemically applied, enhances the surface negative component of evoked potentials (BREMER 1958). It is difficult to interpret these results because many systems may be involved. For example, it was supposed that the surface negative component was produced by the discharge of P-cells, but this in questionable.

No detailed reports existed on the effect of strychnine on intrinsic inhibitory mechanisms in the cerebellum. In a long study on the differences between cerebral and cerebellar cortical reactions to strychnine, PURPURA and GRUNDFEST (1957) concluded that the cerebellar cortex was "poorly endowed with inhibitory synapses". This conclusion was based mainly on the interpretation of the alterations of surface negative and positive waves by drugs. These waves were interpreted as representing EPSPs and IPSPs respectively, which need not necessarily be so. A positive or negative potential can only be correctly interpreted by determining where it is generated, that is, where its reversal point is located on tracking with a recording electrode. Using this technique the IPSP from P-cells is recorded as a surface negative wave (ANDERSEN et al. 1964).

Strychnine is a specific competitor for certain spinal inhibitory transmitters (ECCLES 1964). Recently, however, a number of inhibitory synapses have been found at different locations in the brain which are resistant to the action of strychnine (ANDERSEN et al. 1963 b; CRAWFORD et al. 1963 b). In the cerebellum this was tested by the lack of effect of relatively high doses of strychnine on a known inhibitory system. There was no decrease in the size of the extracellularly recorded IPSP, and no change in the inhibitory effect on the second of two potentials evoked by local stimulation of the parallel fibres. Moreover, the duration of the period of inhibition of the spontaneous or drug-induced P-cell firing which was produced by parallel fibre stimulation did not diminish after the administration of strychnine.

Picrotoxin has been reported to increase the surface negative phase of evoked potentials (CURTIS 1940) but the substance had no effect on the duration of the inhibitory pause of discharging P-cells by parallel fibre stimulation. Similarly, pentamethylene-tetrazol and dihydro-β-erythroidine do not influence this inhibition.

As yet only a few aspects of the pharmacology of the cerebellar cortex have been investigated and most problems are still unsolved. It is hoped that with further elucidation of intracerebellar relations (cf. ECCLES, LLINAS and SASAKI 1964), and with the use of a wider spectrum of drugs, the chemical and pharmacological aspects of synaptic transmission in this organ will become clearer.

References

ANDERSEN, P., ECCLES, J. C., LØYNING, Y. & VOORHOEVE, P. E. (1963 b). **200**, 843.

ANDERSEN, P., ECCLES, J. C. & VOORHOEVE, P. E. (1963 a). *Nature* (Lond.) **199**, 655.

ANDERSEN, P., ECCLES, J. C. & VOORHOEVE, P. E. (1964). *J. Neurophysiol.* **27**, 1138.

AUSTIN, L., PHILLIS, J. W. & STEELE, R. P. (1964). *Experientia* **15**, 218.

BREMER, F. (1958). *Physiol. Rev.* **38**, 357.

BREMER, F. & BONNET, V. (1953). *Folia psychiat. neerl.* **56**, 438.

BREMER, F. & GERNANDT, E. E. (1954). *Acta physiol. scand.* **30**, 120.

COCEANI, F., FADIGA, E. & GESSI, T. (1962). *Abstr. XXII Int. Physiol. Congr.* 1148.

CRAWFORD, J. M., CURTIS, D. R., VOORHOEVE, P. E. & WILSON, V. J. (1963 a). *Nature* (Lond.) **200**, 579.

CRAWFORD, J. M., CURTIS, D. R., VOORHOEVE, P. E. & WILSON, V. J. (1963 b). *Nature* (Lond.) 200, 845.
CREPAX, P., NIGRO, A. & PARMEGGIANI, L. (1956). *Abstr. XX Int. Physiol. Congr.* 200.
CROSSLAND, J. & MITCHELL, J. F. (1956). *J. Physiol.* **132**, 391.
CURTIS, D. R. (1961). In: *Nervous Inhibitions* (ed. by. FLOREY, E.), p. 342. New York: Pergamon Press.
CURTIS, D. R. & RYALL, R. W. (1964). *Nature.* In Press.
CURTIS, H. J. (1940). *Proc. soc. exp. Biol.* (N. Y.) **44**, 664.
DOW, R. S. & MORUZZI, G. (1958). *The physiology and pathology of the cerebellum.* Minneapolis: The University of Minnesota Press.
ECCLES, J. C. (1964). *The physiology of synapses.* Berlin, Göttingen, Heidelberg: Springer-Verlag.
ECCLES, J. C., LLINAS, R. & SASAKI, K. (1964). *Nature* (Lond.) 203, 245.
FOX, C. A. (1959). *J. comp. Neurol.* **112**, 39.
GEREBTZOFF, M. A. (1959). In: *Cholinesterases,* p. 79. London: Pergamon Press.
KOELLE, G. B. (1963). In: *Handbuch der Experimentellen Pharmakologie,* vol. 15 (ed. by KOELLE, G. B.), p. 203. Berlin: Springer-Verlag.
MCCANCE, I. & PHILLIS, J. W. (1964). *Experientia* **15**, 217.
MEAD, C. O. & VAN DER LOOS, H. (1964). *Anat. Rec.* **148**, 311.
PATON, W. D. M. (1958). *Ann. Rev. Physiol.* **20**, 431.
PURPURA, D. P. & GRUNDFEST, H. (1957). *J. Neurophysiol.* **20**, 494.
SHUTE, C. C. D. & LEWIS, P. R. (1965). *Nature* (Lond.) **205**, 242.
SNELL, J. (1961). *Bibl. Anat.* **2**, 50.
SZENTAGOTHAI, J. (1964). In: *Progress in Brain Research* (ed. by. SINGER, M. & SCHADÉ, J. P.), vol. 14. Amsterdam: Elsevier Publishing Company.

SPECIFICITY IN REGENERATION OF OPTIC AND OLFACTORY PATHWAYS IN TELEOST FISH

by R. A. WESTERMAN

The remarkable regenerative capacity of various parts of the teleost central nervous system has been clearly demonstrated in numerous investigations (KOPPANYI 1955; KIRSCHE and KIRSCHE 1961) and several of these have strikingly illustrated the apparent high degree of specificity which may be achieved by regenerating fibres in making central reconnexions (ATTARDI and SPERRY 1963; WESTERMAN and ARORA 1965). Even some innate (SEGAAR and NIEWENHUYS 1963) and learned (ARORA and SPERRY 1963) behaviour has been shown to return after such central regeneration.

Although there is abundant experimental evidence for this regenerative capacity from behavioural, histological and, more recently, electrophysiological studies (GAZE and JACOBSON 1963; WESTERMAN and ARORA 1965), there have been in the case of optic nerve regeneration in amphibia apparently conflicting explanations of the phenomena (SPERRY 1963; GRAFSTEIN and BURGEN 1964).

Regeneration with recovery of function is suggested by Sperry (1963) to depend on chemotactic and mechanical guidance of regenerating optic axons to the tectum where selective central synapsis occurs, regulated by specific chemoaffinities between corresponding neurones of retinal and tectal fields. An orderly cytochemical mapping is proposed in terms of "gradients during embryonic differentiation that spread across and through each other with axes roughly perpendicular", which being superimposed successively on retinal and tectal fields would identify each cell with some matching code between corresponding units of retinal and tectal maps (Sperry 1963).

Recent elegant experiments of Gaze, Jacobson and Szekely (1963) provide strong evidence to support the postulated mechanisms of retino-tectal locus specification in terms of two crossed gradients. Their study points to the apparently discrepant findings of Grafstein and Burgen (1964) as being instructive about "gradient" organization, but not demanding any change in the hypothesis above.

Regeneration of optic tracts in goldfish

The experiments (Westerman and Arora 1965) described below attempted to extend Jacobson's (1961) observation that in incomplete regeneration of the optic nerve in amphibia, a wide lateral spread of the terminal arborizations of regenerating optic fibres might occur, with later shrinkage of their field down to the correct retino-topic locus. In order to test this and also to determine whether any connexions of functionally abnormal types are made during early regeneration a series of twenty-one goldfish were operated on as follows. Unilateral subtotal retinal ablation was performed leaving either an inferior temporal or nasal quadrantic remnant of the retina intact, together with ipsilateral optic nerve transection (Fig. 1). This was expected to provide a restricted tectal zone for potential synaptic reconnexions, with the greatest possibility of showing ectopic lateral spread of synapsis during regeneration. An essential preliminary electrophysiological mapping and analysis of the optic tectum in normal goldfish provided data similar to the findings of Jacobson and Gaze (1964) indicating point-to-focal-area retino-tectal projection in goldfish. Most unit responses were obtained extracellularly with coarse micropipettes from the deepest tectal layers.

Fifty per cent of the total 119 normal units were activated from elongated retinal receptive fields, some subtending angles as large as 30° (bottom right, Fig. 1). There was a preponderance of units in this group with vertically oriented receptive fields (29 units); of the remainder, 19 were horizontally and 11 obliquely oriented. Although the total sample was small, the results of shape discrimination testing in goldfish (Mackintosh and Sutherland 1963) indicate that it may have been reasonably representative.

Electrophysiological techniques used included (a) micropipette recording from deeper tectal layers, (b) coarse micropipette recording of flash-evoked field

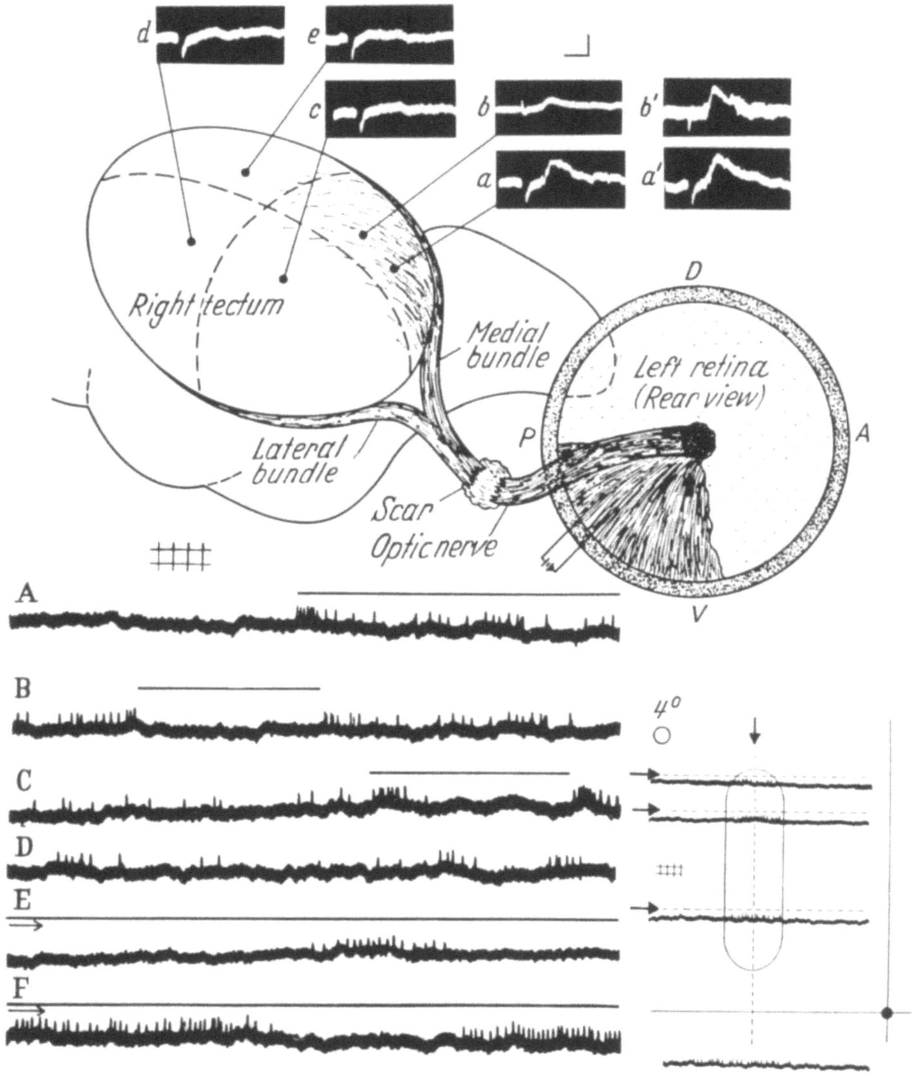

Fig. 1. Upper portion of figure shows schematically the right tectum and left eye of a goldfish viewed from the right. Records a—e of responses from corresponding tectal regions were obtained from one goldfish 45 days post operatively by direct electrical stimulation of that part of retinal remnant shown in the diagram. This was 15 days after the first behavioural indication of any visual recovery in this fish. Note absence of responses from areas, c, d, e. Records a′, b′ were obtained from a normal fish by stimulating and recording from approximately the same regions as a, b above. Calibrations are 50 msec and 200 μV

Typical extracellular records A—F obtained 45 days postoperatively from units in the periventricular cell layer of the same goldfish described above. The horizontal lines above records show the duration of stimulus-presentation on the screen facing the fish. The following types of responses are illustrated: A — "on" unit responding to diffuse illumination; B — unit inhibited by diffuse light; C — "on-off" unit response to diffuse light; D — some rhythmical activity with steady illumination; E — a unit discharging during illumination of its receptive field by a 4° moving light spot; F — a unit which ceases firing during illumination of its receptive field by a moving 4° light spot. Inset (right) illustrates a unit giving responses of the type shown in record E, and having a large elongated receptive field subtending approximately 30° at the eye. Unit discharges occur only during the passage of the light spot across the receptive field potentials and the depth at which they reversed during penetration, (c) direct electrical stimulation of retinal quadrants with surface recording from tectum.

Goldfish with subtotal retinal ablations and ipsilateral optic nerve sections were examined both electrophysiologically and histologically at post-operative intervals varying from two to seven weeks. The earliest recovery of vision, as tested with both baited lure and optokinetic responses, appeared 29 days after operation in these fish, and, with longer intervals than this, active visual units could be recorded in appropriate areas of the tectum. Surface responses were smaller in amplitude (Fig. 1) and unit discharges infrequent and difficult to record. However, most of the unit-types found in the normal tectum were represented, even if scantily, and no abnormal types of responses were detected from such units with regenerated optic nerve connexions. Several fish in which a few units were apparently inappropriately located subsequently proved to have a larger retinal remnant than estimated at the time of ablation. A good correlation between the occurrence and size of evoked surface response and the histological demonstration of regenerating fibres was obtained in several fish in which few or no units were recorded just at the time of visual recovery. Regenerating fibres were not demonstrable in inappropriate tectal locations, either microscopically or by electrical recording, and any early non-selective phase of regeneration, if present, was not detected using these techniques.

As yet there is no final proof that neurone specificities determine anatomical relationships during neurogenesis or regeneration, but particularly in the case of optic pathways, existing evidence rather supports this topographic "connectionist" view. If central connexions are laid down early in regeneration in random profusion, their subsequent response-specificity certainly suggests that only a selective minority of the new synapses ever becomes functional. This interpretation cannot be proven until more subtle methods are found for observing the establishment and activity of the finest terminal synapses.

Regeneration of olfactory tracts in carp

The limits of the regenerative capacity of teleost central nervous pathways to make specific and appropriate connexions have not yet been defined, but recent morphogenetic studies (Corner 1963) suggest that in amphibians no new neurotomes are ever formed. That is, after removal of a segment, no regeneration occurs in the cranio-caudal axis subsequent to the time the neural plate segments are formed. On the other hand, all types of laterally oriented regeneration are possible at later stages (Corner 1963; Teichmann 1964) which agrees well with the experimental findings of Kirsche and Kirsche (1961) and Segaar and Niewenhuys (1963). So far no functional regeneration of the Mauthner cells in teleosts has been demonstrated (Koppanyi 1955), nor has regeneration of the non-myelinated fila olfactoria connecting the olfactory receptor mucosa to the olfactory bulbs (NAGAHARA 1940; KOPPANYI 1955).

Some current investigations of regeneration of olfactory pathways in carp will be briefly described which have been aimed at extending knowledge of the limits of teleost regenerative capacity, and which also sought further evidence concerning the degree and manner in which regenerating afferent fibres re-establish appropriate central connexions in the brain. Carp were chosen for the preliminary portions of this study because of the well-defined anatomy of their olfactory pathways (SHELDON 1912) and particular accessibility for surgical manipulations (WESTERMAN and VON BAUMGARTEN (1964, 1965).

A refined method of testing olfaction in fish has been developed (inset Fig. 2).

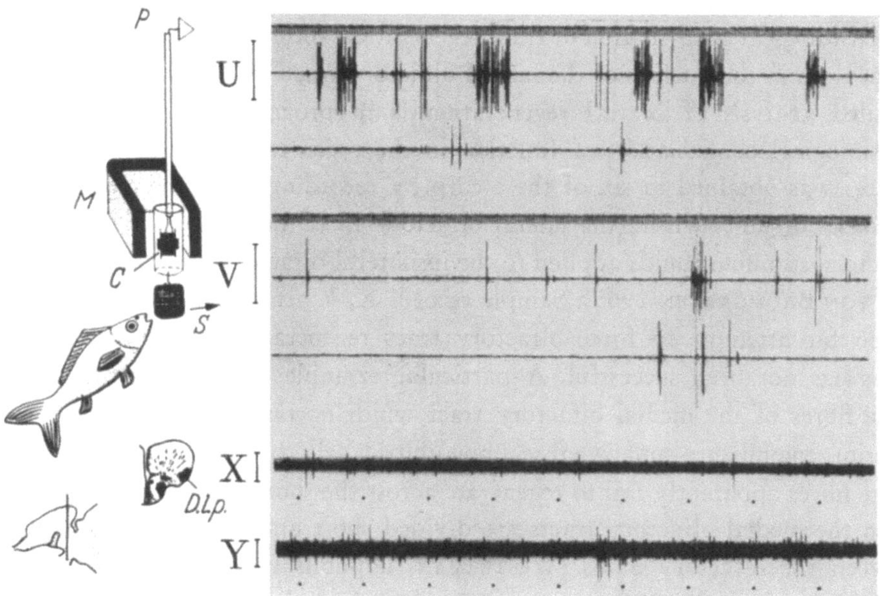

Fig. 2. As illustrated diagramatically at upper left, a fish touching and striking at the gauze sack (S) moves the waterproof coil (C) through the magnetic field of the magnet (M). Penwriter (P) amplifies and records induced potentials so produced from two similar coils and sacks hung in the aquarium
At right, typical records U, V, each register: top — a second time-mark, centre — a sack containing crushed worms, and lower — a sack concealing stones. Record U is that of four normal fish. Record V, from four fish whose olfactory tracts were severed bilaterally 49 days previously, showing fewer responses of shorter duration typical of early regeneration
Inset shows a cross-section through the posterior part of the forebrain to indicate the approximate recording site (Dlp) for records X, Y — in the ventrolateral caudal part of the area dorsalis telencephali. Trace X shows the unit activity prior to, and Y after application of 10^{-4} morpholin solution to the ipsilateral olfactory mucosa
Time mark sec. and amplitude calibration 500 μV for records U, V, X, Y

and this has provided a sensitive index of the recovery of olfaction with regeneration of central olfactory connexions in fourteen carp after complete bilateral tract transection (WESTERMAN and VON BAUMGARTEN 1964, 1965). No olfactory responses were observed earlier than forty days postoperatively, and even then some impairment of olfaction was apparent for a further two or three weeks. The ability of such fish with regenerated tracts to learn and display conditioned positive responses to usually repellant odours indicates considerable sophistication of the recovered olfactory function. This was further demonstrated by the ability of these fish to reach a high level of correct performance in a two-odour discrimination situation using a specially constructed discrimination aquarium (WESTERMAN and VON BAUMGARTEN 1965), and these results suggest a high degree of apparent specificity in central reconnexions. Although carp retain much olfaction if the ipsilateral (lateral olfactory tract) connexions only are intact, the crossed (medial olfactory tract) connexions of bulb to forebrain are essential for a high degree of correct performance in olfactory discrimination tasks (WESTERMAN and VON BAUMGARTEN 1965). This specificity approaches that in recovery of colour discrimination in fish (ARORA and SPERRY 1963), which requires tecto-bulbar connexions in addition to retinotectal ones.

Histological confirmation of tract regeneration and re-establishment of central connexions in the nucleus of the lateral olfactory tract was obtained, and a more detailed analysis of central regeneration is in progress. Preliminary electrophysiological confirmation of function in the regenerated olfactory tracts and centres was obtained in six of these carp by recording, with micropipettes, the responses of units within the lateral olfactory tract nucleus of the forebrain to olfactory stimulant fluids applied to the ipsilateral olfactory rosette (WESTERMAN and VON BAUMGARTEN 1965). Sample records X, Y are shown in Fig. 2.

So far attempts to force olfactory tract regeneration along inappropriate paths has not been successful. A particular example of this is constituted by those fibres of the medial olfactory tract which normally cross directly in the anterior commissure and synapse upon mitral cells of the contralateral bulb. These fibres apparently fail to regenerate across the short artificial path provided when the medial olfactory tracts are divided short and the ends approximated between the olfactory bulbs (WESTERMAN, unpublished). This argues strongly that there may be other factors operating here in addition to the chemo-affinity postulated between the regenerating tract fibres and their prospective synaptic target cells in the opposite bulb. It is hoped that the further investigation of this system will yield answers to many of the questions it has now posed.

The above results emphasize that a further understanding of the factors controlling growth and re-establishment of appropriate functional connexions during regeneration will only be achieved by the development and application of even more subtle and refined techniques, and by the exploration of new areas within the field.

References

ARORA, H. L. & SPERRY, R. W. (1963). *Develop. Biol.* **7**, 234.
ATTARDI, D. G. & SPERRY, R. W. (1963). *Exp. Neurol.* **7**, 46.
CORNER, M. A. (1963). *J. exp. Zool.* **155**, 301.
GAZE, R. M. & JACOBSON, M. (1963). *Proc. Roy. Soc.* B. **157**, 420.

GAZE, R. M., JACOBSON, M. & SZEKELY, G. (1963). *J. Physiol.* (Lond.) **165**, 484.
GRAFSTEIN, B. & BURGEN, A. S. V. (1964). In: *Topics in Basic Neurology* (ed. by BARGMANN, W. & SCHADÉ, J. P.), p. 126. Amsterdam: Elsevier.
JACOBSON, M. (1961). *Proc. Roy. Phys. Soc. Edinburgh* **28**, 131.
JACOBSON, M. & GAZE, R. M. (1964). *Quart. J. exp. Physiol.* **49**, 199.
KIRSCHE, W. & KIRSCHE, K. (1961). *Z. mikr.-anat. Forsch.* **67**, 40.
KOPPANYI, T. (1955). In: *Regeneration in the Central Nervous System* (ed. by WINDLE, W. F.), p. 1. Springfield: Charles C. Thomas.
MACKINTOSH, J. & SUTHERLAND, N. S. (1963). *Anim. Behav.* **11**, 135.
NAGAHARA, Y. (1940). *Jap. J. med. Sci. V. Path.* **5**, 165.
SEGAAR, J. & NIEWENHUYS, R. (1963). *Anim. Behav.* **11**, 331.
SHELDON, R. E. (1912). *J. comp. Neurol.* **22**, 177.
SPERRY, R. W. (1963). *Proc. nat. Acad. Sci.* (Wash.) **50**, 703.
TEICHMANN, H. (1964). *Wilhelm Roux' Arch. Entwickl.-Mech. Org.* **155**, 129.
WESTERMAN, R. A. & ARORA, H. L. (1965). *Exp. Neurol.* In the press.
WESTERMAN, R. A. & VON BAUMGARTEN, R. (1964). *Experientia* (Basel) **20**, 519.
WESTERMAN, R. A. & VON BAUMGARTEN, R. (1965). *Pflüg. Arch. ges. Physiol.* In the press.

THE ORIGIN OF LEARNING: A HYPOTHESIS

by J. Z. YOUNG

Solution of the problem of the nature of learning may be helped by considering a system that has acquired its learning powers relatively recently, by evolution from a purely "reflex" system. Among cephalopods the cuttlefishes and squids (decapods) have a complicated visual memory system, but the tactile centres are simple. The octopods have added lobes that give the capacity to learn to respond according to the texture of objects touched (WELLS 1959). By comparing the two types of brain we can obtain some idea how the "equipment" needed for learning was acquired and this may help considerably in discovering the principles by which it works.

I. Learning systems were evolved by evolutionary modification of reflex centres

Squids and cuttlefishes only use their arms for seizing prey. The centres that control the arms allow the performance of the simple reflex operations of drawing in the prey and holding it while it is poisoned and then eaten. These acts require the exercise of only a limited choice between a few alternatives. However in any nervous system, except perhaps the very simplest, there must be present the equipment for inhibition, if only to allow rhythmical or alternative actions. The drawing in and eating must stop if signals of pain arise and there is reason to think that small cells are concerned with such inhibition, as for example, in the spinal cord (see WALL 1962; ECCLES 1964).

In octopods the highly extensible arms are used as distance receptors. The four lobes that provide the chemotactile memory system have been developed from the single centre present in decapods. These lobes give the ability to produce distinct reactions to various combinations of chemotactile influences, according to past experience. The four lobes of the "new" tactile learning system are arranged on a plan of paired centres, with upper and lower circuits in parallel, exactly as in the visual memory system (Fig. 1) (YOUNG 1963). We can understand the significance of these lobes and circuits if we consider what is required to convert a "reflex" system into one where there is a choice of outputs, determined by experience.

II. The paired centres develop embryologically from single centres

In an unhatched octopus the visual and tactile centres each consist essentially of a single lobe. The four centres of each set differentiate within this. The larger cells of the lower pairs provide the output to motor centres. The upper lobes

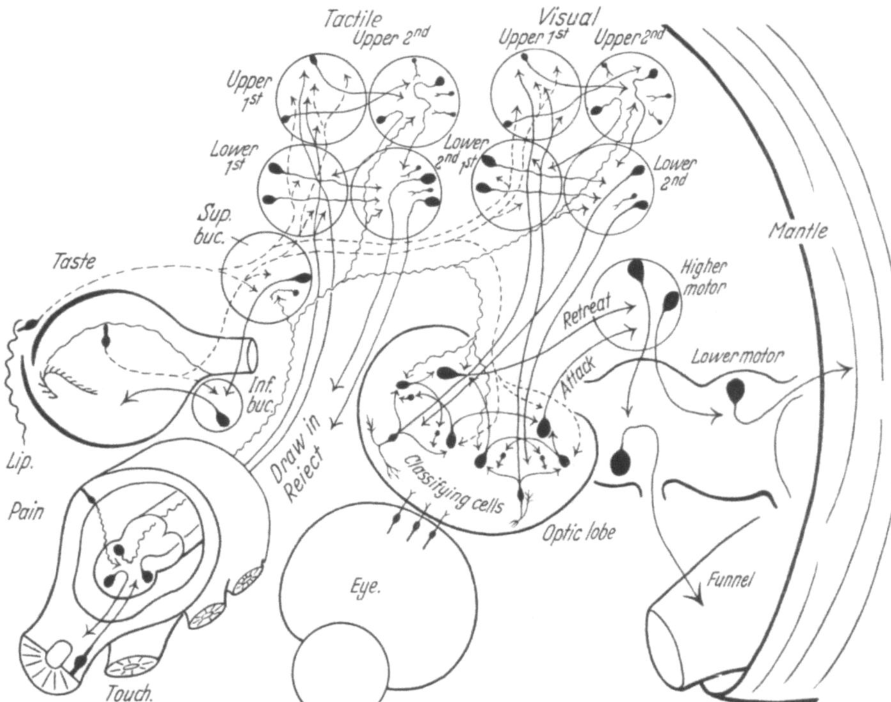

Fig. 1. The sets of centres concerned with tactile and visual learning in *Octopus*. All the pathways shown have been proved to exist except for the details of those at the centre of the optic lobe. Impulses from the suckers interact in the tactile lower and upper 1st. lobes with impulses from the lips and buccal mass, which presumably signal taste. The output of the lower 1st. passes to the lower 2nd lobe, which probably contains the actual memory record (Wells 1961). The output of the 1st upper lobe is to the 2nd upper, in which it meets further afferents from the arms, presumed to signal pain. The hypothesis is that the two first lobes tend to increase the tendency to take an object, *unless* pain signals intervene through the second lobes, causing rejection.
The same four lobes occur in the visual system but the optic lobes provide a complicated classifying and memory system. The classifying cells *(class.)* record some particular feature of the input (say horizontal contouring). Each has outputs that can lead to either "attack" or "retreat". The hypothesis is that learning occurs according to whether attack at a given figure yields food or pain, by the action of the small cells inhibiting the unwanted pathway

provide self-re-exciting circuits. Their cells are small, often minute and with axons that remain within the four lobes, often within one lobe. These masses of small cells are directly continuous with the layers of small cells that accompany the large ones in the lower parts of the system. The upper parts have thus become differentiated to provide cells and circuits that allow the possibility of different reactions to various combinations of inputs and the "printing" of a record of the results that followed each combination.

III. Choice and learning involve response to combinations of inputs

The first lobes of both lower and upper circuits of the tactile system contain interweaving bundles of fibres from the arms, running across the dendritic spines of the output cells of the lobes. Each of these cells is thus able to respond to activity in a particular combination of input channels. The axons of these cells pass solely to the second lobes of each circuit. They are thus a specialised addition to the reflex system, providing signals of a range of combinations. We can call them classifying cells (a name that has also been used for the cells of the optic lobes that respond to specific orientations of visual stimulus patterns, see below). Evidently a complex hierarchy of classification is involved. The first cells respond to particular features of the input, but at later stages there is equipment for extracting the integral of varied combinations of the primary classifiers.

IV. Learning requires signals of the results of action

Receptors for taste and for pain (trauma) are present even in nervous systems with a minimum of choice and learning. They operate the consummatory mechanisms, such as eating and avoidance. In learning, such signals must be available to teach the memory that particular combinations of afferents were followed by results that were good or bad for the organism. In an octopus, pathways for such signals have been identified anatomically though not yet confirmed with electrodes, and they lead to the interweaving fibre systems of the paired circuits. There must also be provision for the spreading of such signals so that they can reach all the afferent systems that may be involved. In mammals this spreading of signals may be one of the functions of the hypothalamic centres, which are also often arranged in pairs, with opposite effects, and are concerned both with consummatory actions and with memory.

V. An exploratory, learning homeostat must keep a balance between enquiry and caution

The paired circuits of an octopus serve to maintain this balance. After food reward the tendency to attack is increased, after pain decreased. After surgical interruption of the circuits the balance is disturbed. There is some evidence that the first lobe of each circuit increases the tendency to attack, passing its output through the second lobe, which further increases it *unless* pain signals arrive (Young 1964). The system probably also receives impulses indicating the "needs" of the animal (e. g. hunger). In mammals the conditions associated with "drive" and "motivation" are dependent on centres in the hypothalamus (see e. g. Olds and Olds 1961).

VI. The paired circuits provide a short memory during the interval between stimulation and consummation

After a figure has appeared in the visual field there must be some means for ensuring that a record of it is retained until the signals indicating the results of attacking it arrive. Similarly there is a delay before objects touched are tasted in the mouth. The paired centres may have arisen to bridge this time gap. They provide self-re-exciting circuits within the brain, around which impulses can circulate. In computer terms we might call this "maintaining the addresses" of the classifying cells that were stimulated.

The concept of self-re-exciting chains as a basis for memory was formulated for the visual learning system of the cuttlefish many years ago (Young 1938). Although it is probable that long term memories do not function in this way there are indications that a short term record involves a dynamic storage. The circuits of the paired centres probably serve other functions during what may be called "reading-in" to the memory. Thus they allow generalisation and transfer; without them, actions learned with one eye or one arm cannot be performed by the others (Muntz 1961; Wells 1959).

VII. Recording in the memory by the closure of one of two alternative pathways

It is characteristic of a learning system that a given input has more than one possible output. In the visual system of the octopus it is suggested that the classifying cells can produce either "attack" or "retreat". The hypothesis is that the record is "written" in the memory by closing one of these channels (Young 1963). There is no precise information as to which cells are involved, in either the visual or the tactile systems. In the visual system they must be in the optic lobes, for the visual memory can operate after removal of the paired centres (though imperfectly). However the latter normally play a large part in both the "reading-in" and "reading-out". The tactile memory cells are probably in the second lower tactile centre (Wells 1959, 1961). In both systems the regions presumed responsible for storage have large cells, accompanied by smaller one whose axons do not leave the lobes.

VIII. Learning evolved by increasing the duration of inhibition

The clue as to the mechanism that changes the probability of the use of one of these two outputs may be that the learning system has developed from the mechanism for inhibition in a reflex system. Small cells occur with the large ones in the regions that learn. The hypothesis is that in memory systems the small cells that previously provided temporary inhibition have become specialised to produce a lasting change of probabilities. It may be that they are "triggered" to change their metabolism when a particular set of inputs occurs. This change could be disinhibition of an enzyme system, allowing the cell to produce a flood of inhibitory transmitter, blocking or making less likely, the use of a particular pathway. There are several ways in which the nucleotide mechanism could be involved in such a change.

IX. Presynaptic inhibitory systems

Thus in the octopus, as in the spinal cord of the cat, systems allowing combination between various inputs have been superimposed upon the reflex pathways. In both situations there are sets of small cells associated with inhibition. In both also there are unusual synapses with vesicles on both sides suggesting pre-synaptic inhibition (Gray and Young 1964). After removal of this region an octopus is definitely "uninhibited" and cannot learn *not* to attack an object from which shocks have been received. However it is still not clear what the connection may be between this inhibitory mechanism and the change that is postulated to close one of two alternative pathways during learning.

References

Eccles, J. C. (1964). *The Physiology of Synapses.* Berlin: Springer-Verlag.

Gray, E. G. & Young, J. Z. (1964). *J. cell. Biol.* **21**, 87.

Muntz, W. R. A. (1961). *J. comp. physiol. Psychol.* **54**, 186.

Olds, J. & Olds, M. E. (1961). In: *Brain Mechanisms and Learning* (ed. by Delafresnaye, J.), p. 135. Oxford: Blackwell Scientific Publications.

Wall, P. D. (1962). *J. Physiol.* (Lond.) **164**, 508.

Wells, M. J. (1959). *J. exp. Biol.* **36**, 590.

Wells, M. J. (1961). *J. exp. Biol.* **38**, 811.

Young, J. Z. (1938). In: *Evolution; Essays presented to E. S. Goodrich* (ed. DeBeer, G. R.), p. 179. Oxford: Clarendon Press.

Young, J. Z. (1963). *Nature* (Lond.) **198**, 626.

Young, J. Z. (1964). *Proc. roy. Soc.* B. **159**, 565.

SUBJECT INDEX

MIX
Papier aus verantwortungsvollen Quellen
Paper from responsible sources
FSC® C105338

If you have any concerns about our products,
you can contact us on
ProductSafety@springernature.com

In case Publisher is established outside the EU,
the EU authorized representative is:
Springer Nature Customer Service Center GmbH
Europaplatz 3, 69115 Heidelberg, Germany

Printed by Libri Plureos GmbH
in Hamburg, Germany